COMPUTATIONAL MODELING OF SHALLOW GEOTHERMAL SYSTEMS

Multiphysics Modeling

Series Editors

Jochen Bundschuh
University of Southern Queensland (USQ), Toowoomba, Australia
Royal Institute of Technology (KTH), Stockholm, Sweden

Mario César Suárez Arriaga
Department of Applied Mathematics and Earth Sciences,
Faculty of Physics and Mathematical Sciences, Michoacán University UMSNH,
Morelia, Michoacán, Mexico

ISSN: 1877-0274

Volume 4

Computational Modeling of Shallow Geothermal Systems

Rafid Al-Khoury
Delft University of Technology, Delft, The Netherlands

CRC Press
Taylor & Francis Group
Boca Raton   London   New York   Leiden

CRC Press is an imprint of the
Taylor & Francis Group, an **informa** business

A BALKEMA BOOK

CRC Press
Taylor & Francis Group
6000 Broken Sound Parkway NW, Suite 300
Boca Raton, FL 33487-2742

First issued in paperback 2017

© 2012 by Taylor & Francis Group, LLC
CRC Press is an imprint of Taylor & Francis Group, an Informa business

No claim to original U.S. Government works

Version Date: 20111207

ISBN 13: 978-0-415-59627-5 (hbk)
ISBN 13: 978-1-138-07342-5 (pbk)

Visit the Taylor & Francis Web site at
http://www.taylorandfrancis.com

and the CRC Press Web site at
http://www.crcpress.com

About the book series

Numerical modeling is the process of obtaining approximate solutions to problems of scientific and/or engineering interest. The book series addresses novel mathematical and numerical techniques with an interdisciplinary emphasis that cuts across all fields of science, engineering and technology. It focuses on breakthrough research in a richly varied range of applications in physical, chemical, biological, geoscientific, medical and other fields in response to the explosively growing interest in numerical modeling in general and its expansion to ever more sophisticated physics. The goal of this series is to bridge the knowledge gap among engineers, scientists, and software developers trained in a variety of disciplines and to improve knowledge transfer among these groups involved in research, development and/or education.

This book series offers a unique collection of worked problems in different fields of engineering and applied mathematics and science, with a welcome emphasis on coupling techniques. The book series satisfies the need for up-to-date information on numerical modeling. Faster computers and newly developed or improved numerical methods such as boundary element and meshless methods or genetic codes have made numerical modeling the most efficient state-of-the-art tool for integrating scientific and technological knowledge in the description of phenomena and processes in engineered and natural systems. In general, these challenging problems are fundamentally coupled processes that involve dynamically evolving fluid flow, mass transport, heat transfer, deformation of solids, and chemical and biological reactions.

This series provides an understanding of complicated coupled phenomena and processes, its forecasting, and approaches in problem solving for a diverse group of applications, including natural resources exploration and exploitation (e.g. water resources and geothermal and petroleum reservoirs), natural disaster risk reduction (earthquakes, volcanic eruptions, tsunamis), evaluation and mitigation of human induced phenomena (climate change), and optimization of engineering systems (e.g. construction design, manufacturing processes).

Jochen Bundschuh
Mario César Suárez Arriaga
(Series Editors)

Editorial board of the book series

To Sawsan, Zain and Maryam

Contents

Preface

Geothermal heat is a viable source of energy and its environmental impact in terms of CO_2 emissions is significantly lower than conventional fossil fuels. Shallow geothermal systems are increasingly utilized for heating and cooling buildings and greenhouses. However, their utilization is inconsistent with the enormous amount of energy available underneath the surface of the earth. Projects of this nature are not getting the public support they deserve because of the uncertainties associated with them, and this can primarily be attributed to the lack of appropriate computational tools necessary to carry out effective designs and analyses. For this energy field to have a better competitive position in the renewable energy market, it is vital that engineers acquire computational tools, which are accurate, versatile and efficient. This book aims at attaining such tools.

This book is the first of its kind, focusing mainly on modeling ground source heat pumps (geothermal heat pumps). It effectively treats two main related aspects: mathematical modeling and computational procedures. Both aspects are considerably challenging because of the geometry and physical processes involved. Disproportionate geometry consisting of multiple components and highly slender heat exchangers embedded in a vast soil mass is difficult to simulate using both analytical and numerical methods. This difficulty exacerbates by the presence of thermal convection. Due to this difficulty, current computational models are either overly simplified or computationally very demanding, limiting engineering capabilities for producing efficient designs and innovative technologies. Therefore, a better approach involving models, which are comprehensive and lend themselves to computational efficiency, is essential. A good combination of mathematical modeling and computational procedures can greatly reduce the computational efforts. This book thoroughly treats this issue and introduces a straightforward methodology for developing innovative computational models, which are both rigorous and computationally efficient.

The mathematical models and their computational procedures introduced in this book are a revised and extended version of the author's papers, which have been published in prominent scientific journals: *International Journal for Numerical Methods in Engineering*, *International Journal of Numerical Methods for Heat and Fluid Flow*, and *Computers and Geosciences*. In addition, the book presents an elaborate treatment of the basic mathematical procedures: Fourier transforms, Laplace transforms, the spectral element method and the finite element method. It makes no attempt to present the fundamental theories in their most general form, rather it is primarily a presentation of their applications in solving heat flow problems. Chapters involving the basic mathematical procedures and their worked examples are concocted from highly acknowledged text books and monographs, including those of Kreyszig; Spiegel; Carslaw and Jaeger; Heinrich and Pepper; and Doyle.

The book addresses researchers and developers in the fields of computational mechanics, geoscience and geology working on geothermal energy systems. In covering the topics, explicit methodologies are introduced to enable the reader to acquire the skill to develop analytical, semi-analytical and numerical tools. The models presented here are generic and can be extended to other related fields, including deep geothermal energy systems. Moreover, the inclusion of the basic concepts and examples will assist engineering professionals to understand the involved mathematics and be aware of the latest numerical and analytical tools.

The book is structured in three parts: Part I: preliminaries (chapters 1–5); Part II: analytical and semi-analytical modeling (chapters 6–10); and Part III: numerical modeling (chapters 11–12).

In Part I, basic definitions and concepts of heat transfer are given. It includes two chapters for modeling heat flow in porous media and borehole heat exchangers. In Part II, focus is placed on modeling ground source heat pumps using analytical and semi-analytical methods. This part includes an introduction of the most well-known computational models for geothermal heat pump systems. It concludes by introducing alternative analytical and semi-analytical models for heat flow in shallow geothermal systems. In Part III, standard and extended finite element methods suitable for heat transfer are described. Then the book concludes by introducing a three-dimensional finite element formulation capable of modeling heat flow in shallow geothermal systems using coarse meshes.

It is a pleasure to thank those who directly and indirectly contributed in achieving this work. For this, I wish to express my gratitude to the head of the Computational Mechanics Group of the Faculty of Civil Engineering and Geosciences of Delft University of Technology, Prof. dr. ir. L.J. (Bert) Sluys for his support. I am utterly indebted to my brother, Ray al Khoury of Outotec, Melbourne Australia, for closely reviewing the manuscript. I also would like to acknowledge and thank the CRC/Balkema publishing staff for their efficient handling of the book project. Finally, I would like to thank my family for their heartfelt support and encouragement.

Rafid Al-Khoury
Delft, June 2011

Part I

Preliminaries

CHAPTER 1

Introduction

In this chapter, we introduce some basic information about two main issues: geothermal energy and shallow geothermal systems. A brief description of the availability of geothermal energy and the most common technologies for its utilization is given first. Then, focus is placed on shallow geothermal systems, which constitute the core subject of this book. For this we give an introduction on the types of shallow geothermal systems and the most commonly used geothermal borehole heat exchangers. The term shallow geothermal system will be used throughout this book to indicate a geothermal heat pump, a ground-source heat pump, or a ground heat pump, consisiting of borehole heat exchangers embedded in a soil mass.

1.1   GEOTHERMAL ENERGY SYSTEMS

Geothermal energy is a form of thermal energy generated in the core of the earth, about 6000 km below the surface. It is renewable because temperatures hotter than the surface of the sun are continuously produced inside the earth as a result of radiogenic decay of naturally occurring isotopes, particularly those of potassium, uranium, and thorium. The temperature at the core of the earth reaches up to 5000°C. Far from the core and towards the surface of the earth, this temperature decreases gradually to reach around 10°C at the surface. This gradient of temperature drives a continuous flux of thermal energy from the core to the surface, forming a wide range of energy sources. Geothermal energy sources range from shallow depths, of the order of few tens to few hundred meters; to intermediate depths, of the order of few kilometers; to even deeper depths with extremely high temperatures of molten rocks. Energy systems relying on the first type of energy source are commonly known as *shallow geothermal systems*, whereas those based on the other two types of sources are commonly referred to as *deep geothermal systems*.

Geothermal energy offers a number of advantages over conventional fossil fuel resources. Primarily, the geothermal heat resources are renewable, economic, and their environmental impact in terms of CO_2 emission is significantly lower. However, there are mainly two limitations: availability and cost. Geothermal projects, especially deep, can only be established in regions with abundant geothermal resources; generally, areas with recent tectonic activity. The cost of establishing deep geothermal systems is quite high and there is a high risk of failure. Slight difference in the temperature of the extracted water from the planned can mean a large loss of investment.

Geothermal energy can be produced by drilling deep holes into the earth crust. Two systems are in use: open and closed. In the *open system,* the energy is produced by pumping cold water through one side of the reservoir; circulating it through hot fractured rock; and then collecting it in the other side, to be brought back to the surface. In the *closed system,* the energy is produced using a single borehole where cold water is pumped in one path and returned in another, using the same pipe. Figure 1.1 shows a schematic representation of these two systems.

Many geothermal systems have been developed, making use of the many advantages of the geothermal energy. They are in general classified into two categories: deep geothermal energy systems, and shallow geothermal systems. Deep geothermal systems are mainly those go as deep as few kilometers below the surface and reach *hydrothermal aquifers, dry hot rocks* or *magma.* Projects involving electricity production or direct heat production are classified in this category. Shallow geothermal systems, on the other hand, are those that do not go more than 250 m below the surface. The *ground heat pump* (GHP) is classified in this category. In this book we focus

3

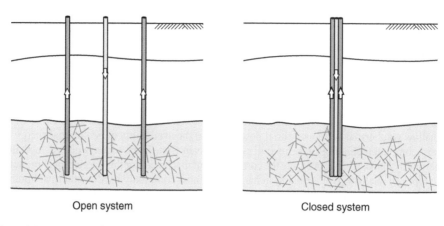

Figure 1.1   A scheme of open and closed geothermal systems.

on shallow geothermal systems, but here we give examples of different geothermal technologies used in practice.

1.1.1   *Geothermal electricity*

Geothermal electricity constitutes one of the most important, but also challenging sources of energy. It is an attractive source of energy because of its low CO_2 emission. The emission of CO_2 from existing geothermal electric plants is on average 122 kg of CO_2 per megawatt-hour of electricity, whereas CO_2 emission from a conventional oil combustion thermal power station is on average 760 kg of CO_2 per megawatt-hour of electricity. A geothermal power plant does not require fuel for its operation, but the capital cost is quite high. Drilling of wells accounts for over half the costs, and exploration of deep resources entails significant risks, with a probability of over 20% failure rate.

Technologies in use include *dry steam* power plants, *flash steam* power plants and *binary cycle* power plants. Dry steam plants are the simplest and oldest design. They require geothermal steam of 150°C or more to drive the turbines. This entails drilling wells deeper than 3 km inside the earth. Flash steam plants pull superheated high-pressured water from deep reservoirs into lower-pressure tanks and use the resulting flashed steam to drive the turbines. They require fluid temperatures of at least 180°C, usually more. This is the most common type of plants in operation today. Binary cycle power plants are the most recent development, and can accept fluid temperatures as low as 60°C to drive the turbines. Water of such a temperature can be reached from hydrothermal systems, not deeper than 3 km below the surface. Due to this relatively low temperature requirement, this type of geothermal electricity plants is recently becoming the most commonly built system. For more details see Wikipedia (b).

1.1.2   *Geothermal direct use*

This kind of energy constitutes one of the oldest ways of energy consumption, and the most popular source of geothermal energy. It has been utilized for hundreds of years when people started using natural hot springs for cooking, heating, and entertainment. In modern time, hot water has been extracted from deep geothermal reservoirs, few kilometers below the surface of the earth, and used directly for heating of buildings, greenhouses and industry. Geothermal reservoirs of low-to-moderate water temperature ranging from 40°C to 150°C are used for this purpose. This range of temperature is lower than those required for most geothermal power plants.

Geothermal direct-use systems consist of basically three parts: down hole well pumps, piping network, and heat exchangers. Different types of well pumps are in use and share many of the properties of those usually used in oil industry. Pipe networking is somewhat more complicated than those used in delivering urban cities with clean water since this network carries hot water. Care must be taken for insulating the pipes and reducing clogging of metals on the pipe joints and fittings. The heat exchangers work to extract heat from the coming water and dispose the resulting cold water back into the reservoir.

1.1.3 *Geothermal heat pumps*

Geothermal heat pumps (GHP) constitutes one of the most easy to extract and locally available in all parts of the world. It does not share the requirements of the other geothermal energy types in terms of geology or equipment. Rather, it makes use of the relatively constant temperature at shallow depths. In shallow grounds, just 10 m below the surface, the earth maintains nearly constant temperature ranging between 10°C and 20°C, depending on the region. Usually, the ground temperature at this depth is warmer than the air in winter and cooler in summer. Geothermal heat pumps are commonly used to exploit this abundant source of energy for heating and cooling of individual buildings and small compounds.

Geothermal heat pump systems consist of basically two parts: the ground heat exchanger, and the heat pump unit. The ground heat exchanger is a system of pipes, known as a loop, which is buried in the ground either vertically or horizontally. In winter, heat from the earth is extracted via a fluid, usually water or a mixture of water and antifreeze, circulating through the pipes at a certain rate and collecting heat from the earth. The heat pump extracts heat from the fluid and pumps it into the building. In summer, the process is reversed, and the heat pump extracts heat from the indoor air and transfers it to the heat exchanger. Heat removed from the indoor air during summer can also be used for heating water, which can be used for cooking and bathing.

1.2 SHALLOW GEOTHERMAL SYSTEMS

A shallow geothermal energy system is an important source of thermal energy suitable for heating and cooling of individual buildings and small compounds. It is a renewable source of energy as heat transfer due to temperature gradient between the bottom of the earth and the air is continuously taking place. The seasonal temperature variation at the ground surface is reduced to a nearly constant temperature of around 12°C at approximately 10 m below the surface. Under this depth, the temperature is known to increase with an average gradient of 3°C per 100 m depth. The continuous thermal interaction between the air and the earth makes the first 100 m sustainable and hence, suitable for supply and storage of thermal energy, though at a relatively low temperature.

There are two commonly utilized energy systems that make use of this kind of energy: the ground source heat pumps, and the *underground thermal energy storage*. In the following, we give an overview of these two systems (Mands and Sanner, 2005).

1.2.1 *Ground-source heat pumps*

The ground-source heat pump (GSHP), also known as the geothermal heat pump (mentioned above), or the *downhole heat exchanger* (DHE), is a shallow geothermal system. In practice, there are mainly two kinds of GSHP: open and closed. The open system, also known as *doublet*, extracts heat from the groundwater and carries it to the surface using at least two separate wells. One well is used for the extraction of the groundwater, and another is used for the reinjection of the groundwater. The basic advantage of this system, as compared to the closed system, is that it is around eight times more efficient. However, this can be realized if the aquifer formation is highly permeable and the groundwater is relatively low in metals, avoiding thus maintenance problems associated with scaling, clogging and corrosion.

The closed system, on the other hand, extracts heat from the earth using heat exchangers embedded in the soil mass. The heat is carried by a fluid circulating inside a heat exchanger pipe and passed to a heat pump at the surface, usually inside the building. At the heat pump the collected heat is increased and transported to the heating system. The main advantage of this system, as compared to the open, is that there is no need to an aquifer, and, in principle, only one borehole is necessary for the extraction of thermal energy. Furthermore, the *circulating fluid* is usually clean, and hence needs little maintenance.

Closed systems are installed in mainly two ways: horizontal and vertical. The horizontal systems, also known as *ground heat collectors*, and *horizontal loops*, are relatively easy to install as less complicated excavation equipment is needed. The main source of heat of this kind of systems is the surface temperature, which comes by direct exposure to the sun. Depending on the area available, the horizontal pipes are laid in different ways, including series, parallel or spiral.

The vertical system is widely used, especially in areas where the land is scarce. The heat in such a system is extracted by *borehole heat exchangers* (BHE), also known as *vertical ground heat exchangers* or *downhole heat exchangers*, which consist of plastic pipes, mainly polyethylene or polypropylene, installed in a borehole as U-tubes and fixed by filling the borehole with grout. The U-tube carries a circulating fluid, referred to in the literature as *refrigerant* or *working fluid*. The U-tube effectively forms two pipes. One pipe receives the circulating fluid from the heat pump and conveys it downward. We will be denoting this pipe as *pipe-in* throughout this book. The other pipe collects the circulating fluid at the bottom of pipe-in and brings it out to the surface, to enter the heat pump. We will be denoting this pipe as *pipe-out* throughout this book. The heat pump, usually located inside the building, extracts designed amount of heat from the fluid and pumps it back to the BHE. The circulating fluid, usually water with 20%–25% anti-freezing coolant such as Mono Ethylene Glycol, gets into contact with the surrounding soil via the U-tube material and the grout. The grout, usually bentonite-cement mix, exchanges heat with the soil and the BHE inner pipes.

Borehole heat exchangers are slender heat pipes with dimensions of the order of 30 mm in diameter for the inner pipes, 150 mm in diameter for the borehole, and 100 m in length for the borehole and the inner pipes. In practice there are different types of BHE. They mainly differ in their configurations. Here we give examples of the most commonly used configurations, namely:

- U-tubes: This BHE configuration consists of one (or more) inner pipe, which is inserted in the borehole from its middle to form a U-shape. Two main configurations are in use: single U-tube BHE and double U-tube BHE. Figure 1.2 shows a sketch of these two types of BHEs. The single U-tube BHE consists of pipe-in, where the working fluid enters; pipe-out, where the working fluid leaves; and grout, where the contact with the surrounding soil mass takes place. The double U-tube consists of two pipes-in, two pipes-out and grout.
- Coaxial: This BHE configuration consists of concentric pipes. In practice, there are mainly two coaxial configurations: *annular* (CXA) and *centered* (CXC), Figure 1.3. (In geothermics, this terminology seems to be coined by Diersch *et al.* (2011).) In CXA, pipe-out is configured inside pipe-in, forming an annular inlet and a centered outlet. Heat exchange of the inner pipes with the grout occurs along the surface area of pipe-in. In CXC, pipe-in is configured inside pipe-out, forming a centered inlet and an annular outlet. Here, heat exchange with the grout occurs via pipe-out.

The captured temperature from the earth is elevated by a heat pump and conveyed to a heating system. A heat pump is an important part of the shallow geothermal system. It is a mechanical device that transfers thermal energy from a region of a lower temperature to a region of a higher temperature. In the heating mode, the circulating fluid absorbs thermal energy from the ground. The heat pump exerts work on it, to increase its temperature, and then conveys it to the heating system of the building. In the cooling mode, the system is reversed and heat from the building is extracted and injected into the ground.

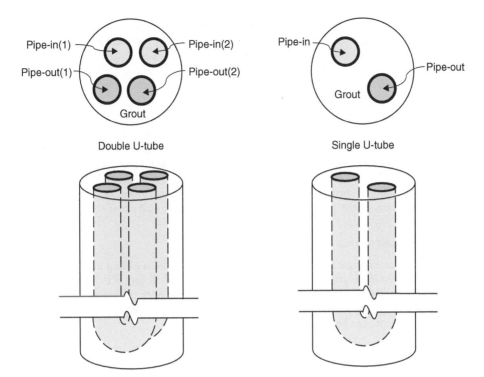

Figure 1.2   Single and double U-tubes BHE.

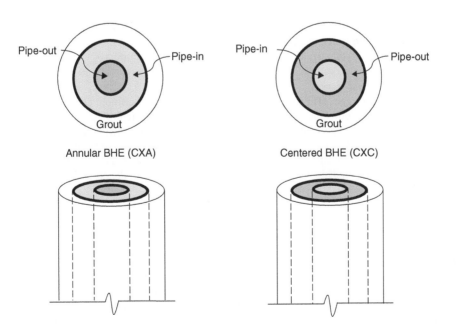

Figure 1.3   Annular (CXA) and centered (CXC) coaxial tubes.

The effectiveness of the heat pump is described in terms of a number called the *coefficient of performance* (COP) (Serway 1996). COP is defined as the ratio of the heat transferred into the system to the work required to transfer that heat, i.e.

$$\text{COP} \equiv \frac{\text{heat transferred}}{\text{work done by pump}} = \frac{Q}{W} \tag{1.1}$$

If the work needed to raise 1 kWh of heat is 0.25 kWh, the COP of the system is 4. That is the heat transferred to the building is four times greater than the work done by the motor of the heat pump. However, as the temperature from the ground decreases, it becomes more difficult for the heat pump to extract sufficient heat, and the COP drops. A good heating or cooling system should have a high coefficient of performance, typically 5 or 6 (Serway 1996). Shallow geothermal systems using geothermal heat pumps typically have COP ranging from 3.5 to 4 at the beginning of the heating season. As the ground temperature decreases due to heat extraction, the COP drops.

1.2.2   *Underground thermal energy storage*

The technology for storing energy in shallow underground depths is getting momentum as cold and hot water stored in the ground can be utilized for heating and cooling of buildings all over the year. As for the geothermal heat pumps, two systems are in use: open, known as *aquifer thermal storage*, and closed, known as *borehole storage* or *Duct thermal storage*. A comprehensive treatment of this topic can be found in Dincer and Rosen (2011).

1.2.2.1   *Aquifer thermal storage*
The aquifer thermal storage (ATES) is an open system, and hence relies on heat carried by the groundwater. It is usually built in porous aquifers with medium to high permeability and transmissivity. Typically two wells are used: one for the warm water, and another for the cold water. In winter the warm water is utilized for heating, and the resulting cold water is stored in the cold reservoir. In summer, the process is reversed. The advantage of this system is that it is environmentally safe. The extracted ground water does not mix with other types of waters and hence cannot be contaminated. Moreover, the amount of the extracted water is equal to the amount of re-injected water, and hence there is no net loss of water from underground. The only problem is that this system works only on areas with permeable aquifers.

1.2.2.2   *Duct thermal storage*
The duct thermal storage (DTES) is a closed system and somewhat more complicated than the ATES. In this system, boreholes typically 50 m to 200 m in depth are drilled and inserted in them plastic heat exchangers carrying fluid. The circulating fluid carries the thermal energy from the earth to the surface and vice versa. The advantage of this system, as compared to the ATES, is that the pipe network and fittings are much less exposed to scaling and clogging. The efficiency of this system, however, depends on the geology, the temperature of the ground, the groundwater conditions and the thermal properties of the ground.

1.3   BOOK THEME AND OBJECTIVE

This book addresses heat flow in shallow geothermal systems, with a main objective is to introduce detailed computational methodologies for developing analytical, semi-analytical and finite element models for such systems. Focus will be placed on modeling ground-source heat pump systems, in particular, the vertical borehole heat exchangers and their thermal interaction with the soil mass and air. However, the computational models will be generic and can be extended to other types of geothermal systems. For this, we will be using the term shallow geothermal system to indicate a ground-source heat pump.

CHAPTER 2

Heat transfer

In this chapter we introduce the mechanisms of heat transfer in a system in thermal contact with its surrounding or with other systems. Focus will be placed on heat conduction and convection. These two mechanisms are relevant to heat flow processes occurring in shallow geothermal systems. We give basic derivations of the involved heat equations, and introduce some important thermal parameters, including thermal material properties and relevant fluid numbers. These parameters will be extensively utilized in subsequent chapters, and hence it is useful to give a short description of their physical and mathematical definitions.

2.1 INTRODUCTION

Heat transfer is a thermodynamic process dealing with the rate at which thermal energy is transferred between a system and its surrounding, and with the mechanisms responsible for the transfer. Heat transfer involves measuring and computing of two main entities: temperature and heat flow. The temperature represents the amount of thermal energy in a system, whereas the heat flow represents the movement of thermal energy as a result of the temperature gradient. The thermal energy is related to the kinetic energy of the material molecules, as at a higher temperature, the material exhibits more kinetic energy.

The mechanisms of heat transfer, also known as modes, are classified into three main categories: *conduction*, *convection* and *radiation*. The concept of heat conduction and convection has its paradigm in *mass transfer*, where *diffusion* is used to describe the spread of particles through random *Brownian motion* from regions of higher concentration to regions of lower concentration; and *advection* is used to describe the transport mechanism of a substance by a fluid, due to the fluid bulk motion. Heat convection due to fluid flow occurs through both diffusion and advection. Heat transfer in physical systems often exhibits a complicated combination of these three modes of energy transfer. In a typical shallow geothermal system, heat flow arises as a combination of heat conduction and heat convection. Heat radiation plays no effective role in such a system, and therefore will not be treated in this book.

Heat transfer in a body or adjacent bodies in thermal contact is physically modulated by several material parameters, including *thermal conductivity*, *specific heat*, *mass density*, *thermal diffusivity*, *viscosity*, etc. These parameters have a significant impact on the amount of thermal energy lost or gained by a body when subjected to different boundary conditions. System geometry and physical properties are also important parameters that affect the heat transfer process. The significance of these parameters, including the fluid flow rate, is manifested by known fluid and thermal flow numbers, including the *Nusselt number*, the *Peclet number*, the *Reynolds number*, the *Prandtl number*, etc. Considering these parameters together makes the design of an efficient engineering gadget an involved process. Designing a geothermal heat pump is not exceptional and a proper understanding of the thermal parameters together with the heat transfer mechanisms constitutes a key to any attempt to mastering this topic.

2.2 HEAT TRANSFER MECHANISMS

As stated above, heat transfer in a shallow geothermal system involves coupling between conduction and convection. Radiation heat transfer has no or negligible effect. Hence the focus here is placed on two heat transfer mechanisms: heat conduction and heat convection.

2.2.1   *Heat conduction*

Heat conduction, also referred to as heat diffusion, is one of the most important modes of thermal energy flow. On a microscopic scale, and as required by the second law of thermodynamics, when two objects at different temperatures are placed in thermal contact, thermal energy always flows from the object with high temperature to the one with lower temperature, never in the opposite direction. Heat flow between the two regions continues until they reach the same temperature, a state known as *thermal equilibrium*. On a macroscopic scale, the heat transfer rate per unit area normal to the direction of heat flow, known as *heat flux*, is described by *Fourier's law* of heat conduction. In one-dimension, it is defined as

$$q''_x = -\lambda_x \frac{dT}{dx} \tag{2.1}$$

where dT/dx is the temperature gradient along the x-axis, and λ_x is the *thermal conductivity* of the material in the x-direction. The minus sign imposes the condition that heat must flow from the high temperature region to the lower temperature. For a constant cross sectional area, A, heat transfer is usually written as

$$q_x = -\lambda_x A \frac{dT}{dx} \tag{2.2}$$

That is $q_x = q''_x A$. In three-dimensions, Fourier's law is written as

$$\mathbf{q}'' = -\lambda \nabla T \tag{2.3}$$

where ∇ is the gradient operator, and λ is a tensor representing three-dimensional thermal conductivity of the material.

Consider a control volume $dx\,dy\,dz$ shown in Figure 2.1. Based on the first law of thermodynamics, we know that heat flow in and out of a body can be expressed as

Rate of heat flow in + Rate of energy generation

= Rate of heat flow out + Rate of internal energy storage

Using Fourier's law, the rate of heat flow across the rectangular area ABCD, is

$$q_{x1} = -\lambda_x \frac{\partial T}{\partial x} dy\, dz \tag{2.4}$$

At the other side of the control volume, EFGH, and by the use of *Taylor series*, the rate of flow is

$$\begin{aligned} q_{x2} &= q_{x1} + \frac{\partial q_{x1}}{\partial x} dx \\ &= -\lambda_x \frac{\partial T}{\partial x} dy\, dz - \frac{\partial}{\partial x}\left(\lambda_x \frac{\partial T}{\partial x}\right) dx\, dy\, dz \end{aligned} \tag{2.5}$$

The net flow of heat in the element crossing these two faces is therefore

$$q_{x1} - q_{x2} = \frac{\partial}{\partial x}\left(\lambda_x \frac{\partial T}{\partial x}\right) dx\, dy\, dz \tag{2.6}$$

Similarly, the net flow across the other four faces is

$$q_{y1} - q_{y2} = \frac{\partial}{\partial y}\left(\lambda_y \frac{\partial T}{\partial y}\right) dx\, dy\, dz$$

$$q_{z1} - q_{z2} = \frac{\partial}{\partial z}\left(\lambda_z \frac{\partial T}{\partial z}\right) dx\, dy\, dz \tag{2.7}$$

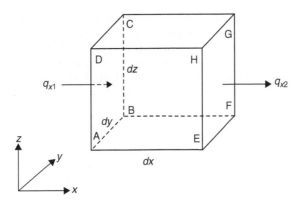

Figure 2.1   Control volume *dx dy dz*.

Summing these equations, heat flow in the element volume can then be described as

$$\left[\frac{\partial}{\partial x}\left(\lambda_x\frac{\partial T}{\partial x}\right) + \frac{\partial}{\partial y}\left(\lambda_y\frac{\partial T}{\partial y}\right) + \frac{\partial}{\partial z}\left(\lambda_z\frac{\partial T}{\partial z}\right)\right]dx\,dy\,dz \qquad (2.8)$$

If heat is generated at a rate $Q(x, y, z)\,dx\,dy\,dz$ in the element, the total rate of heat flow inside the control volume becomes

$$\left[\frac{\partial}{\partial x}\left(\lambda_x\frac{\partial T}{\partial x}\right) + \frac{\partial}{\partial y}\left(\lambda_y\frac{\partial T}{\partial y}\right) + \frac{\partial}{\partial z}\left(\lambda_z\frac{\partial T}{\partial z}\right) + Q(x, y, z)\right]dx\,dy\,dz \qquad (2.9)$$

The internal energy storage per unit volume and unit temperature for a material with density ρ (kg) and specific heat at constant pressure c_p (J/kg · K) (see below for a detailed definition) is given by

$$\frac{\partial U}{\partial t} = \rho c_p\frac{\partial T}{\partial t}dx\,dy\,dz \qquad (2.10)$$

Equating Eq. (2.9) to Eq. (2.10) and dividing throughout by $dx\,dy\,dz$, the governing partial differential equation for heat conduction in a body is obtained, as

$$\rho c_p\frac{\partial T}{\partial t} = \frac{\partial}{\partial x}\left(\lambda_x\frac{\partial T}{\partial x}\right) + \frac{\partial}{\partial y}\left(\lambda_y\frac{\partial T}{\partial y}\right) + \frac{\partial}{\partial z}\left(\lambda_z\frac{\partial T}{\partial z}\right) + Q(x, y, z) \qquad (2.11)$$

This equation, known as the *heat equation*, can be written in a more concise way, as

$$\rho c_p\frac{\partial T}{\partial t} = \nabla \cdot (\lambda\nabla T) + Q(x, y, z) \qquad (2.12)$$

For a constant thermal conductivity, and in the absence of a heat generation, the heat equation can be written as

$$\frac{1}{\alpha}\frac{\partial T}{\partial t} = \nabla^2 T \qquad (2.13)$$

in which $\alpha = \lambda/\rho c_p$ (m²/s) is the thermal diffusivity, and ∇^2 is the *Laplacian operator*. In the Cartesian coordinate system, it is described as

$$\nabla^2 = \nabla \cdot \nabla = \frac{\partial^2}{\partial x^2} + \frac{\partial^2}{\partial y^2} + \frac{\partial^2}{\partial z^2} \qquad (2.14)$$

In the cylindrical coordinate system, it is given by

$$\nabla^2 = \frac{\partial^2}{\partial r^2} + \frac{1}{r}\frac{\partial}{\partial r} + \frac{1}{r^2}\frac{\partial^2}{\partial\phi^2} + \frac{\partial^2}{\partial z^2} \qquad (2.15)$$

Depending on the system geometry and the involved initial and boundary conditions, solution of Eq. (2.12) varies between simple to quite complicated. Problems of academic interest are somewhat simple, whereas physical problems representing real-life geometry and initial and boundary conditions might be very complicated. In this book we will be solving heat equations for rather complicated physical problems using both analytical and numerical methods. However, different solution methodologies using problems of academic nature will be illustrated.

2.2.2   *Heat convection*

Heat convection is an important mode of heat transfer, basically associated with fluid flow. On a microscopic scale, heat convection occurs as a result of thermal diffusion, whereas on a macroscopic scale, it occurs as a result of bulk motion of the fluid that transports heat from region to region along the direction of its motion. Hence, heat convection occurs due to simultaneous heat diffusion and advection. However, unlike diffusion of particles where the transport is held by the Brownian motion, heat diffusion takes place due to transfer of energy by vibrations at a molecular level through the fluid.

At the macroscopic level, the convective heat flow rate per unit area normal to the direction of heat flow (convective heat flux) is described, for a one-dimensional case, as

$$q''_x = \underbrace{-\lambda_x \frac{dT}{dx}}_{\text{diffusion}} + \underbrace{\rho c_p u_x T}_{\text{advection}} \tag{2.16}$$

where u_x is the fluid velocity along the x-axis.

Convection sometimes identified by the transfer of heat of a solid body in contact with a moving fluid. The energy in this case is transferred from or to the body by the fluid along its surface area. This kind of convection involves a wide class of heat transfer applications, particularly those dealing with *heat exchangers* including geothermal borehole heat exchangers. Convective processes involving heat transfer from a boundary surface of temperature T_s exposed to a relatively low velocity fluid of temperature T_f, is described by the *Newton's law of cooling*, as

$$q'' = h(T_s - T_f) \tag{2.17}$$

in which h (W/m$^2 \cdot$ K) is the convective heat transfer coefficient. At the surface between the solid body and the fluid, $x = s$, where the velocity is zero, the exchange of energy can be described by equating Eq. (2.16) to Eq. (2.17), yielding

$$h(T_s - T_f) = -\lambda_x \frac{dT}{dx}\bigg|_{x=s} \tag{2.18}$$

If the fluid motion is induced by an external force, such as a pump or a fan, the heat flow process is called *forced convection*. If, however, the fluid motion is due to density variations with temperature, the process is called *free* or *natural convection*. The fluid motion is categorized by two different types: *laminar* and *turbulent*. In laminar motion, the flow is smooth and streamlined. Within the boundary layer, the velocity is constant, but might be varying with time. Heat flux in the laminar flow is as described in Eq. (2.16). In the turbulent motion, the flow is disturbed and the fluid velocity and temperature are not constant and exhibit irregular fluctuation with time. Commonly, the fluid velocity and temperature are represented, at any instant, as the sum of a *time-mean* value, $\bar{u}$, $\bar{T}$, and a fluctuation component u', T'. The time-mean value is taken over a time that is large compared to the period of typical fluctuation, and if $\bar{T}$ is independent of time, the time-mean is in steady-state. Following this averaging process, heat flux in a turbulent flow can be described as (Rohsenow *et al.* 1998):

$$q''_x = -\lambda_x \frac{d\bar{T}}{dx} + \rho c_p \overline{u'_x T'} \tag{2.19}$$

In heat exchanger technology, laminar heat flow is not too common because the rate of heat flow is lower than that of the turbulent flow. However, it is sometimes desirable due to the lower pumping power (Pitts and Sisson 1998). Convective heat transfer in a heat exchanger is usually described as

$$q = \dot{m}c_p(T_o - T_i) \tag{2.20}$$

where $\dot{m}$ is the flow rate, T_o and T_i are the pipe outlet and inlet temperatures, respectively. In a differential form, Eq. (2.20) can be written as

$$dq = \dot{m}c_p dT_f \tag{2.21}$$

where dT_f is the temperature variation of the fluid. Equating this equation to the Newton's law of cooling, yields

$$\frac{dT_f}{dx} = \frac{2\pi r}{\dot{m}c_p}h(T_s - T_f) \tag{2.22}$$

in which r is the pipe inner radius. Let $\varphi = T_s - T_f$, and assuming a constant surface temperature T_s, Eq. (2.22) can be written as

$$\frac{d\varphi}{dx} = -\frac{2\pi r}{\dot{m}c_p}h\varphi \tag{2.23}$$

For a pipe of length L, separating variables and integrating over the pipe length (Pitts and Sisson 1998), gives

$$\int_{\varphi_i}^{\varphi_o} \frac{d\varphi}{\varphi} = -\frac{2\pi r}{\dot{m}c_p}\int_0^L h\,dx \tag{2.24}$$

where $\varphi_i = T_s - T_i$ and $\varphi_o = T_s - T_o$. This yields

$$\ln \varphi_o - \ln \varphi_i = -\frac{2\pi rL}{\dot{m}c_p}\int_0^L \frac{1}{L}h\,dx = -\frac{2\pi rL}{\dot{m}c_p}\bar{h} \tag{2.25}$$

in which $\bar{h}$ is the average value of the convective coefficient over the pipe length. The temperature at the outlet can then be calculated as

$$T_o = T_s - (T_s - T_i)\exp\left(-\frac{2\pi rL}{\dot{m}c_p}\bar{h}\right) \tag{2.26}$$

To determine the fluid temperature at any point within the pipe, we integrate Eq. (2.25) from 0 to x to obtain

$$T_f = T_s - (T_s - T_i)\exp\left(-\frac{2\pi rx}{\dot{m}c_p}\bar{h}\right) \tag{2.27}$$

For three-dimensional problems, and in the presence of fluid flow, the temperature must be determined by solving for the corresponding three-dimensional heat equation. As for conduction, and by considering the control volume of Figure 2.1, the rate of heat flow from the left-hand side of the control volume is described as

$$q_{x1} = -\lambda_x \frac{\partial T}{\partial x}dy\,dz + \rho c u_x T\,dy\,dz \tag{2.28}$$

Using Taylor series, the heat flow rate on the right-hand side can be described as

$$q_{x2} = -\lambda_x \frac{\partial T}{\partial x}dy\,dz + \rho c u_x T\,dy\,dz - \frac{\partial}{\partial x}\left(\lambda_x \frac{\partial T}{\partial x}\right)dx\,dy\,dz - \frac{\partial}{\partial x}(\rho c u_x T)\,dx\,dy\,dz \tag{2.29}$$

Then, the net heat flow yields

$$q_{x1} - q_{x2} = \frac{\partial}{\partial x}\left(\lambda_x \frac{\partial T}{\partial x}\right)dx\,dy\,dz + \frac{\partial}{\partial x}(\rho c u_x T)\,dx\,dy\,dz \tag{2.30}$$

Similarly,

$$q_{y1} - q_{y2} = \frac{\partial}{\partial y}\left(\lambda_y \frac{\partial T}{\partial y}\right)dx\,dy\,dz + \frac{\partial}{\partial y}(\rho c u_y T)\,dx\,dy\,dz$$

$$q_{z1} - q_{z2} = \frac{\partial}{\partial z}\left(\lambda_z \frac{\partial T}{\partial z}\right)dx\,dy\,dz + \frac{\partial}{\partial z}(\rho c u_z T)\,dx\,dy\,dz$$

(2.31)

Summing Eq. (2.30) and Eq. (2.31), equating them to the rate of internal energy, Eq. (2.10), and the rate of heat generation, and dividing throughout by $dx\,dy\,dz$, the three-dimensional heat conduction-convection equation can be obtained as

$$\rho c_p \frac{\partial T}{\partial t} = \nabla \cdot (\lambda \nabla T) + \nabla(\rho c_p \mathbf{u} T) + Q(x, y, z) \tag{2.32}$$

in which $\mathbf{u}$ is the flow velocity vector.

2.3   THERMAL PARAMETERS

There are several thermal parameters that have direct influence on the rate of heat transfer and the pattern of the temperature distribution in solids and fluids. Here we give an overview of the thermal material parameters and the most important fluid flow numbers that are of direct use in heat transfer in shallow geothermal systems.

2.3.1   *Thermal conductivity*

Thermal conductivity (W/m · K) is an intrinsic property of a material which describes its ability to conduct heat. It appears primarily in Fourier's law for heat conduction. It is defined as the quantity of heat, q, transmitted through a unit thickness, L, in a direction normal to a surface of a unit area, A, due to a unit temperature gradient, ΔT, under a steady-state condition. For a continuum body, using Eq. (2.2), the thermal conductivity can be expressed as

$$\lambda = \frac{qL}{A\Delta T} \tag{2.33}$$

Thermal conductivity of solids and liquids are usually temperature dependent. For solids, the thermal conductivity increases with temperature, normally in the form of

$$\lambda = \lambda_0(1 + a\varphi) \tag{2.34}$$

where $\varphi = T - T_{ref}$, λ_0 is the conductivity at a reference temperature T_{ref}, and a is a material constant, determined experimentally. For liquids, the relationship between the thermal conductivity and the temperature is in many cases the opposite. Thermal conductivity of most liquids decreases with increasing temperature. The exception is water, which exhibits increasing in thermal conductivity up to about 150°C, after which it decreases. Water has the highest thermal conductivity of all common liquids except the so-called liquid metals (Pitts and Sisson 1998).

For a multiphase material consisting of solid, liquid and gas, the thermal conductivity of the mix is influenced by the individual conductivity of the constituents. Also, and since the thermal conductivity of the solid is different from that of the fluid (liquid or gas), the distribution of the constituents in the material influences heat conduction significantly. Porosity and connectivity of the pores play a significant role. Quantification of this property is important for designing porous materials, commonly utilized as conductors or insulators. For example, in designing a grout mix for a geothermal borehole heat exchanger, and knowing that the thermal conductivity of the air is much less than that of the solid, care must be taken when choosing the porosity of the mix and the way the pores are connected. In Chapter 3 we elaborate on the determination of the thermal conductivity in multiphase materials.

Table 2.1 Thermo-physical properties of typical shallow geothermal materials[†]

Material	λ (W/m·K)	ρ (kg/m³)	c (kJ/kg·K)	n	v (m²/s)
Ground material					
Limestone	1.2–2.15	2300–2500	0.8–0.9	0–0.2	
Sandstone	1.8–2.9	2160–2300	0.7–0.8	0.05–0.3	
Sand	0.15–4	1280–2150	0.8–1.48	0.2–0.6	
Clay	0.15–2.5	1070–1600	0.92–2.2	0.33–0.6	
Soil	0.4–0.6	1600–2050	1.8–1.9	0.3–0.5	
Pipe material and fluid					
Polyethylene	0.33	960	2.1		
Grout	0.8–1.5	1100–1400	2–2.2		
Water	0.56	1000	4.18		1.006E–6
Water + 25% Ethylene glycol	0.5	1050	3.795		4.95E–6

[†]Detailed figures are available in many books and online. See for examples Rohsenow *et al.* (1998).

Thermal conductivities of materials typically utilized in shallow geothermal systems, in round figures, are given in Table 2.1.

2.3.2 *Density*

For a homogeneous continuum material, the mass density (kg/m³) of a material is defined as the mass per unit volume, no matter how small the volume is. That is

$$\rho \equiv \lim_{\Delta V \to 0} \frac{\Delta m}{\Delta V} \tag{2.35}$$

If the material is inhomogeneous, the density becomes a function of position. The density at a point in this case can be described as

$$\rho \equiv \lim_{\Delta V \to \Delta V^\pi} \frac{\Delta m}{\Delta V} \tag{2.36}$$

where ΔV^π represents the volume of a material constituent. The mass of the material as a whole can then be expressed as

$$m = \int_V \rho(\mathbf{r}) \, dV \tag{2.37}$$

The mass density of a material varies, in general, with temperature and pressure. Depending on the physical application, this variance differs significantly, but in all cases it is more pronounced in gases. Within the context of shallow geothermal systems, the mass densities of the involved liquids and solids can be reasonably regarded as constant.

For multiphase materials, the local volume averaging is usually used. The local average mass density of a two-phase porous material is given by

$$\rho = n\rho_f + (1 - n)\rho_s \tag{2.38}$$

where ρ_f is the fluid density, ρ_s is the solid density, and n is the porosity. Densities of materials typically utilized in shallow geothermal systems, in round figures, are given in Table 2.1.

2.3.3 *Specific heat capacity*

The specific heat capacity (J/kg·K), also termed specific heat, is the amount of heat required to change a unit mass of a substance by one degree in temperature. It is expressed in terms of heat supply to a unit mass, as

$$c \equiv \frac{Q}{m \Delta T} \tag{2.39}$$

Thus, a thermal energy transferred between a mass of a substance and its surrounding, for a constant c, can be expressed as

$$Q = mc\Delta T \tag{2.40}$$

From thermodynamics, there are two definitions of the specific heat, namely

$$c_p = \left.\frac{\partial H}{\partial T}\right|_p$$
$$c_V = \left.\frac{\partial U}{\partial T}\right|_V \tag{2.41}$$

where c_p is the specific heat at constant pressure, and H is the *enthalpy* per unit mass, which is a measure of the total energy of a physical system; c_V is the *specific heat at constant volume*, and U is the internal energy per unit mass. In the first, energy added or removed from the system is transferred in two forms: one does work, and the other transforms into internal thermal energy. In the second, the total energy is transformed into internal thermal energy, with no work done. For incompressible materials, i.e. most solids and liquids, the two properties are practically equal. For gases, however, they are considerably different. Within the context of shallow geothermal systems, the specific heat of the involved solids and liquids can be denoted by c_p. Hence, for the rest of this book, we will denote, c, as the specific heat at constant pressure, c_p, unless it is otherwise necessary.

The specific heat capacity varies with temperature. If the temperature interval is not too big, such as that in shallow geothermal systems, c can be treated as constant. For example, at a constant atmospheric pressure, the specific heat of water varies by only 1% for a temperature ranging between 0°C and 100°C.

Multiplying the specific heat by the density, i.e. ρc, gives the *volumetric heat capacity*, also termed *volume specific heat capacity*. It is defined as the ability of a given volume of a substance to store thermal energy while undergoing a temperature change. For multiphase materials, local volume averaging is usually used. For a two-phase material it is given by

$$(\rho c_p)_{\text{eff}} = n(\rho c_p)_f + (1-n)(\rho c_p)_s \tag{2.42}$$

where the subscripts f and s denotes the fluid and solid respectively, and n denotes the porosity.

Specific heat capacities of materials typically utilized in shallow geothermal systems are given in Table 2.1.

2.3.4   *Thermal diffusivity*

Thermal diffusivity (m²/s) is the ratio of the thermal conductivity to the volumetric heat capacity of the material, commonly described as

$$\alpha = \frac{\lambda}{\rho c} \tag{2.43}$$

As to the thermal conductivity, density and specific heat capacity, thermal diffusivities of liquids and solids are only weakly dependent on temperature and pressure. The relationship in Eq. (2.43) shows that, due to direct proportionality of the thermal diffusivity to the thermal conductivity, materials with high thermal diffusivity rapidly adjust their temperature to that of their surroundings, because they conduct heat quickly in comparison to their volumetric heat capacity. In many engineering applications, this property is important for the design of materials that constitute a substance or an object. For shallow geothermal systems, thermal diffusivity of the involved components, such as the U-tube, the circulating fluid and the grout, is an important measure for analyzing the conductivity and resistivity of the borehole heat exchanger.

2.3.5 *Viscosity*

Viscosity is a measure of fluid resistance to deformation. This property may be thought of as an internal friction, where higher viscosity means more friction, and thus a greater force is needed to deform the material. Except superfluids, all fluids in nature have some resistance to stresses, and therefore viscous. A fluid which has no resistance to shear stresses is known as an *ideal fluid* or *inviscid fluid*. A *Newtonian fluid* is a fluid whose stress versus strain rate is linear and passes through the origin, such that

$$\tau = \mu \frac{du}{dy} \tag{2.44}$$

where τ (Pa) is the shear stress exerted by the fluid; du/dy is the velocity gradient perpendicular to the direction of shear; and μ is the *coefficient of dynamic viscosity*, or simply the *dynamic viscosity* or *absolute viscosity* (Pa · s). Water is a Newtonian fluid. By definition, the viscosity of a Newtonian fluid is a function of temperature and pressure, not the forces acting upon it. Nevertheless, viscosity of liquids depends mainly on temperature. Viscosity increases with decreasing temperature. For example, water viscosity goes from 0.28 mPa · s to 1.79 mPa · s in the temperature range from 100°C to 0°C. Following that, within the temperature range of shallow geothermal systems, the viscosity can reasonably be considered constant. Contrast to the Newtonian fluid, a *non-Newtonian* is a fluid whose viscosity varies with the applied stress. Paint is an example of a non-Newtonian fluid.

In practice, the viscosity coefficient is defined in two ways:

- Dynamic viscosity, also termed absolute viscosity, with units Pascal-second (Pa · s, Poise), is a measure of a shear resistance of a fluid placed between two plates and pushed sideway with a shear stress of one Pascal a distance equal to the thickness of the layer between the two plates in one second. It is the most commonly used coefficient of viscosity, namely in mechanics, and chemistry.
- Kinematic viscosity, is the dynamic viscosity divided by the density (m²/s, Stokes). That is

$$v = \frac{\mu}{\rho} \tag{2.45}$$

Note that 1 Stokes $= 10^{-4}$ m²/s. This viscosity is utilized in heat transfer problems and is also termed *diffusivity of momentum* because it has the same unit of the diffusivity of heat, see Eq. (2.43).

Viscosity of materials typically utilized in shallow geothermal systems are given in Table 2.1.

2.3.6 *Porosity*

Porosity is defined as the ratio of the volume of pores (voids) to the total volume of the material. Mathematically it is expressed as

$$n = \frac{V_V}{V_T} \tag{2.46}$$

where V_V is the volume of the pores and V_T is the total volume of the material. Apparently, porosity represents the storage capacity of the material. Together with the hydraulic conductivity (permeability), the porosity represents an important factor that controls the flow and storage of fluids in porous media. In shallow geothermal systems, the porosity plays an important role in defining the amount of heat carried by the groundwater. More porosity, and permeability, means more heat carried by the water to the geothermal heat pumps. Even in the absence of the liquid, i.e. the pores are filled with air, the porosity plays a role in defining the thermal conductivity, the mass density and the average volume heat capacity of the material. See for example Eq. (2.38) and Eq. (2.42).

2.3.7   *Reynolds number*

In fluid mechanics, Reynolds number, Re, is a dimensionless number defined as

$$\text{Re} = \frac{\text{inertial force}}{\text{viscous force}} = \frac{\rho u^2/L}{\mu u/L^2}$$
$$= \frac{\rho u L}{\mu} = \frac{u L}{v} = \frac{Q L}{v A} \tag{2.47}$$

where u is the mean velocity, L is a characteristic length, μ is the dynamic viscosity, v is the kinematic viscosity, ρ is the density, Q is the volumetric flow rate, and A is the cross sectional area. The characteristic length of a tube is its diameter. Hence, for fluid flow in a pipe or a tube, the Reynolds number becomes

$$\text{Re} = \frac{\rho u D_h}{\mu} \tag{2.48}$$

where D_h is the hydraulic diameter of the pipe. Reynolds number is used as a measure of flow, whether laminar, transient or turbulent. Typically, the following classification is in use:

$$\begin{array}{ll} \text{Re} < 2300 & \text{laminar} \\ 2300 < \text{Re} < 4000 & \text{transient} \\ \text{Re} > 4000 & \text{turbulent} \end{array} \tag{2.49}$$

2.3.8   *Prandtl number*

Prandtl number, Pr, is a dimensionless number defined as the ratio of kinematic viscosity, v, to the thermal diffusivity, α, as

$$\text{Pr} = \frac{v}{\alpha} = \frac{c_p \mu}{\lambda} \tag{2.50}$$

where c_p is the specific heat, μ is the dynamic viscosity, and λ is the thermal conductivity. In heat transfer problems, the Prandtl number is a measure of diffusion with respect to the fluid velocity. For $\text{Pr} = 1$, the temperature profile coincide with the velocity profile. This is approximately the case for most gases ($0.6 < \text{Pr} < 1$). For liquids, however, it is ranging between large values, for viscous oils, to small values of the order of 0.01, for liquid metals that have high thermal conductivities (Pitts and Sisson 1998). When Pr is small, heat of the fluid diffuses quickly as compared to the velocity. Prandtl number of water at 0°C is 12.99 and at 20°C is 6.96 (Rohsenow *et al.* 1998).

2.3.9   *Peclet number*

Peclet number is defined as the ratio of the rate of advection to the rate of diffusion. In the context of heat transfer, the Peclet number is equivalent to the product of the Reynolds number and the Prandtl number, that is

$$\text{Pe} = \frac{u L}{\alpha} = \text{Re} \cdot \text{Pr} \tag{2.51}$$

where u is the fluid velocity, L is a characteristic length, and $\alpha = \lambda/\rho c_p$ is the thermal diffusivity. The Peclet number is used to characterize the mode of the heat flow. At low Pe, the heat flow is dominated by conduction, whereas at high Pe, the heat flow is dominated by convection.

2.3.10   *Nusselt number*

In heat transfer, Nusselt number is the ratio of convective to conductive heat transfer across a boundary, defined as

$$\text{Nu} = \frac{h L}{\lambda_f} \tag{2.52}$$

where L is a characteristic length, h is the convective heat transfer coefficient, and λ_f is the fluid thermal conductivity. A Nusselt number close to unity is a characteristic of a laminar flow. A larger Nusselt number corresponds to more active convection, with turbulent flow typically in the range between 100 and 1000. The characteristic length is the length or thickness of the boundary layer in the direction of growth. For a cylindrical pipe, the characteristic length is the outer diameter of the pipe. The Nusselt number in this case, for a fully developed laminar pipe flow, is described as

$$\mathrm{Nu} = \frac{hD_h}{\lambda_f} \tag{2.53}$$

where D_h is the hydraulic diameter. For forced convection, which is what we will be dealing with in this book, the Nusselt number is generally a function of the Reynolds number, Re, and the Prandtl number, Pr, usually written as

$$\mathrm{Nu} = a\mathrm{Re}^m \mathrm{Pr}^n \tag{2.54}$$

where a, m and n are empirical constants. Empirical correlations for a wide variety of geometries are available in the literature. Here we give some examples of Nusselt numbers, as reported in Wikipedia (c):

Laminar flow: For fully-developed internal laminar flow in cylindrical tubes, the Nusselt numbers are constant. For convection with constant surface heat flux, the Nusselt number is:

$$\mathrm{Nu}_D = 4.36 \tag{2.55}$$

For convection with constant surface temperature, the Nusselt number is:

$$\mathrm{Nu}_D = 3.66 \tag{2.56}$$

Turbulent flow: Gnielinski correlation for turbulent flow in tubes is:

$$\mathrm{Nu}_D = \frac{(f/8)(\mathrm{Re} - 1000)\,\mathrm{Pr}}{1 + 12.7(f/8)^{1/2}(\mathrm{Pr}^{2/3} - 1)}, \quad \begin{array}{l} 0.5 < \mathrm{Pr} < 200 \\ 3000 < \mathrm{Re} < 5E6 \end{array} \tag{2.57}$$

where f is the *Darcy friction factor* defined, for smooth tubes, as

$$f = (0.79 \ln(\mathrm{Re}) - 1.64)^{-2} \tag{2.58}$$

Dittus-Boelter correlation for turbulent flow in tubes is:

$$\mathrm{Nu}_D = 0.023\mathrm{Re}^{4/5}\mathrm{Pr}^n, \quad \begin{array}{l} 0.7 < \mathrm{Pr} < 160 \\ \mathrm{Re} \geq 10000 \\ L/D \geq 10 \end{array} \tag{2.59}$$

where $n = 0.4$ for heating, and $n = 0.3$ for cooling. Note that Dittus-Boelter correlation is explicit, but less accurate, especially when there is a large temperature difference between the bulk fluid and the surface. In practice it is used for smooth pipes.

Sieder-Tate correlation for turbulent flow in tubes is:

$$\mathrm{Nu}_D = 0.027\mathrm{Re}^{4/5}\mathrm{Pr}^{1/3}\left(\frac{\mu}{\mu_s}\right)^{0.14}, \quad \begin{array}{l} 0.7 < \mathrm{Pr} < 16700 \\ \mathrm{Re} \geq 10000 \\ L/D \geq 10 \end{array} \tag{2.60}$$

where μ is the fluid viscosity at the bulk fluid temperature, and μ_s is the fluid viscosity at the heat transfer surface temperature. This correlation is implicit non-linear, since the fluid viscosity is a function of the fluid Nusselt number. However, it is more accurate and valid for a wider range of applications.

CHAPTER 3

Heat transfer in porous media

This chapter gives a brief overview of heat transfer in porous media. We derive the macroscopic energy balance equation in porous media based on the averaging theory. We also discuss two ways of modeling heat flow in a two-phase material consisting of a solid matrix and a fluid. In one case, we assume that there is no local thermal equilibrium between the two phases, and hence the balance of energy is represented by two coupled energy equations for each individual constituent. In the other, we assume that there is a local thermal equilibrium, and we show that the balance energy can be represented by a single energy equation, using effective thermal parameters.

3.1 INTRODUCTION

A porous medium is a medium that is composed of a porous material constituting particles of different thermodynamic properties, such as densities and thermal conductivities. Such a material is also referred to as a multiphase material. Typical soil mass, for instance, consists of three phases: solid, water and air. The water and the air are enclosed in the pores (voids) of the material, whereas the solid particles constitute the skeleton of the material, often called the solid matrix. Within the context of shallow geothermal systems, the soil mass surrounding the borehole heat exchangers, and the filling grout are typical sorts of porous materials.

Transport in porous media is a subject of wide interest and has emerged as a separate field of study, usually denoted as *poromechanics*. In this field, the balance and field equations are derived on the basis of the classical continuum mechanics, but emphasis is placed on the contribution of the individual constituents and their interaction. An important aspect in modeling multiphase materials is the correct mathematical description of the thermodynamic processes at boundary surfaces between the individual constituents. Currently, no theory exists which adequately accounts for the interaction between different constituents at the microscopic level and their effects at the macroscopic level. However, important modeling theories have been introduced to account for these effects using different level of complexity. There are at least two modeling approaches. The first approach is phenomenological, where the interaction between the phases is done at the macroscopic level, using some empirical or experimental relationships. The most well know theory in this field is the Biot theory of consolidation (Biot 1941).

The second approach is based on the *mixture theory*, first introduced by Bowen (1976). The classical mixture theory starts from the macroscopic level and the individual phases constituting a material are postulated as overlying continua in such a way that each thermodynamic property, such as density and energy, is defined spatially everywhere. The interaction between the phases is defined at the microscopic level by introducing an additional term in the field equations to account for the exchange of quantities among the constituents. The constituent interaction is scaled up to the macroscopic level using some sort of averaging procedure. One known averaging procedure is the *spatial convolution method*, reported in Coussy *et al.* (1998). Here, the macroscopic quantity is obtained by convolving the microscopic quantity by some weighting function describing the presence or absence of a constituent in a point in space. However, most models based on the classical mixture theory do not include the thermodynamic interaction at the interface between the involved constituents, and as a result certain essential features of multiphase systems are easily overlooked. In many cases this leads to misinterpretation of the macroscopic quantities from their corresponding microscopic ones. Furthermore, the involved microscopic parameters may not be tractable by usual macroscopic experiments.

A more physically sound scaling up of the microscopic quantities to their corresponding macroscopic counterparts can be achieved by the *hybrid mixture theory*, also known as the *averaging theory*. This approach, reported and competently employed by Hassanizadeh and Gray (1979 and 1990), provides both physical and thermodynamic description of multiphase processes. In this approach the macroscopic quantities are obtained from formal local volume averaging of the corresponding microscopic quantities. As a result of the formal interpretation, the interfacial effects are explicitly accounted for, including the possibility of exchange of mass, momentum and energy between the constituents. The constitutive relationships used in this approach are usually formulated on the basis of the *entropy inequality*. The averaging process is conducted over a *Representative Elementary Volume* (REV), which is much larger than the constituent individual volumes, and at the same time much smaller than the volume of the physical system. However, for highly heterogeneous materials this approach is not valid, as in the averaging theory, REV is assumed to be invariant in space and time.

Within the context of this book, which deals mainly with relatively slow heat flow between the multiphase constituents, the local thermodynamic equilibrium is assumed to hold, as the time scale of the modeled phenomena is substantially larger than the relaxation time required to reach equilibrium locally. That is the temperatures of the constituents at a spatial point are equal. Accordingly, modeling procedure based on continuum and phenomenological approaches is to a high extent sufficient. However, we here derive the fundamental energy field equation on the basis of the averaging theory for two reasons. First, we show that the averaged field equations at the macroscopic level leads to the same field equation obtained by the continuum approach, outlined in Chapter 2 for a one-phase material, and in Chapter 12 for a fully saturated material. Second, we introduce some basic outlines for developing more complicated models for heat flow in porous media.

3.2 ENERGY FIELD EQUATION: FORMAL REPRESENTATION

Formal up-scaling of the microscopic level to the macroscopic level makes the averaging theory fundamentally sound and applicable, as its variables are based on the thermodynamics principles and are experimentally measurable. The averaging theory has been described by many authors including Hassanizadeh and Gray (1979 and 1990) and Lewis and Schrefler (2000). Here we follow, in general terms, the lines described by Lewis and Schrefler (2000). We derive the macroscopic field equations of a multiphase medium, dealing mainly with the transport of the energy field.

At the microscopic level, the conservation equation of phase π of a multiphase system can be described as

$$\frac{\partial}{\partial t}(\rho\psi) + \text{div}\,(\rho\psi\dot{\mathbf{r}}) - \text{div}\,\mathbf{j} - \rho b = \rho G \tag{3.1}$$

in which ψ is a generic thermodynamic variable, which might be mass, linear momentum or energy (enthalpy), ρ is the mass density of phase π, $\dot{\mathbf{r}}$ is the local value of the velocity field at a fixed point in space, $\mathbf{j}$ is the flux vector, b is the external supply of ψ, and G is the net production of ψ. At the interface between two constituents π and α, the discontinuity condition holds:

$$[\rho\psi(\mathbf{w} - \dot{\mathbf{r}}) + \mathbf{j}]|_{\pi} \cdot \mathbf{n}^{\pi\alpha} + [\rho\psi(\mathbf{w} - \dot{\mathbf{r}}) + \mathbf{j}]|_{\alpha} \cdot \mathbf{n}^{\alpha\pi} = 0 \tag{3.2}$$

where $\mathbf{w}$ is the interface velocity, and $\mathbf{n}^{\pi\alpha}$ is the unit normal vector pointing out of π phase and into α phase, with $\mathbf{n}^{\pi\alpha} = -\mathbf{n}^{\alpha\pi}$. This equation indicates that it is possible to transport mass, momentum and energy between the adjacent constituents. Since, in this book we are interested mainly on energy transport, the focus will be placed on specializing Eq. (3.1) and Eq. (3.2) to describe heat flow.

The microscopic conservation equation, Eq. (3.1), cannot be solved directly due to the geometrical complexity of the porous media, especially those related to geological and geotechnical

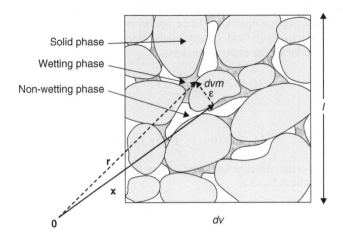

Figure 3.1   REV of a multiphase medium.

materials that we are dealing with. Here arises the importance of the averaging technique, where the transport of energy at the microscopic level is transformed to the macroscopic level by integrating the involved microscopic quantities over a representative elementary volume (REV) of volume dv and area da. The multiphase domain is postulated as a superposition of all π phases and each spatial point is simultaneously occupied by the material points of all phases. However, the state of motion of each phase is described independently.

Consider a multiphase domain occupying a total volume V, and bounded by a surface A. The constituents $\pi, \pi = 1, 2, \ldots, k$, have the partial volume V_π. Each point within the total volume V is considered as the centroid of REV with volume dv. The volume of a microscopic constituent π is denoted as dv_m, described by the position vectors $\mathbf{x}$ (the center of REV) and $\mathbf{r}$ Figure 3.1. The volume of this constituent within REV, referred to as the average volume element dv_π, is defined as

$$dv_\pi(\mathbf{x}, t) = \int_{dv} \gamma_\pi(\mathbf{r}, t) \, dv_m \tag{3.3}$$

where $\gamma_\pi(\mathbf{r}, t)$ is a phase distribution function, defined as

$$\gamma_\pi(\mathbf{r}, t) = \begin{cases} 1 & \text{for } \mathbf{r} \in dv_\pi \\ 0 & \text{for } \mathbf{r} \in dv_\alpha, \quad \alpha \neq \pi \end{cases} \tag{3.4}$$

This distribution function describes step discontinuity at the interface between different constituents. The same is valid for the microscopic area element da_m. Following these definitions, the concept of volume fraction, η_π, which is an important parameter in multiphase systems, can be introduced:

$$\eta_\pi(\mathbf{x}, t) = \frac{dv_\pi}{dv} = \frac{1}{dv} \int_{dv} \gamma_\pi(\mathbf{r}, t) \, dv_m \tag{3.5}$$

with

$$\sum_{\pi=1}^{k} \eta_\pi = 1 \tag{3.6}$$

The volume of fractions of a porous material consisting of a solid matrix, s, and water, w, are defined as:

- solid phase: $\eta_s = 1 - n$, where $n = dv_w/dv$ is the porosity
- water phase: $\eta_w = n S_w$, where S_w is the saturation. $S_w = 1$ denotes a fully saturated material.

Also, following the above definitions, we are able to define the volume average, the mass average and the area average operators. These operators can be applied to functions, such as $\zeta(\mathbf{r}, t)$, which is a microscopic field variable, indicating for instance the internal energy. The volume average operator for the phase average is

$$\langle \zeta \rangle_\pi (\mathbf{x}, t) = \frac{1}{dv} \int_{dv} \zeta(\mathbf{r}, t) \gamma_\pi(\mathbf{r}, t) \, dv_m \tag{3.7}$$

Knowing that

$$\int_{dv_\pi} \zeta(\mathbf{r}, t) \, dv_m = \int_{dv} \zeta(\mathbf{r}, t) \gamma_\pi(\mathbf{r}, t) \, dv_m \tag{3.8}$$

the intrinsic phase average, which refers to the volume actually occupied by the constituent at the micro level, can then be defined as

$$\langle \zeta \rangle_\pi^\pi (\mathbf{x}, t) = \frac{1}{dv_\pi} \int_{dv_\pi} \zeta(\mathbf{r}, t) \, dv_m$$

$$= \frac{1}{dv_\pi} \int_{dv} \zeta(\mathbf{r}, t) \gamma_\pi(\mathbf{r}, t) \, dv_m \tag{3.9}$$

Using Eq. (3.5) gives

$$\langle \zeta \rangle_\pi (\mathbf{x}, t) = \eta_\pi(\mathbf{x}, t) \, \langle \zeta \rangle_\pi^\pi (\mathbf{x}, t) \tag{3.10}$$

The mass average operator, with a microscopic mass density $\rho(\mathbf{r}, t)$ as a weighting function, is expressed as

$$\bar{\zeta}_\pi (\mathbf{x}, t) = \frac{\int_{dv} \rho(\mathbf{r}, t) \zeta(\mathbf{r}, t) \gamma_\pi(\mathbf{r}, t) \, dv_m}{\int_{dv} \rho(\mathbf{r}, t) \gamma_\pi(\mathbf{r}, t) \, dv_m} \tag{3.11}$$

For a constant microscopic mass density, it follows that

$$\langle \zeta \rangle_\pi (\mathbf{x}, t) = \eta_\pi(\mathbf{x}, t) \, \bar{\zeta}_\pi (\mathbf{x}, t) \tag{3.12}$$

The area average operator is

$$\bar{\bar{\zeta}}_\pi (\mathbf{x}, t) = \frac{1}{da} \int_{da} \zeta(\mathbf{r}, t) \cdot \mathbf{n} \, \gamma_\pi(\mathbf{r}, t) \, da_m \tag{3.13}$$

with $\mathbf{n}$ the outward normal unit vector of an area element da_m. Due to this averaging process, a deviation of a microscopic function $\zeta_\pi(\mathbf{r}, t)$ from the mass-averaged quantity $\bar{\zeta}_\pi(\mathbf{x}, t)$ arises, and can be described as

$$\tilde{\zeta}_\pi (\mathbf{x}, t) = \zeta_\pi(\mathbf{r}, t) - \bar{\zeta}_\pi(\mathbf{x}, t) \tag{3.14}$$

Based on the above definitions, the macroscopic balance equations of a porous medium can be obtained by multiplying Eq. (3.1) with the distribution function $\gamma_\pi(\mathbf{r}, t)$, integrating the resulting product over the REV volume element dv, and then over the total volume V. For example, the description of the phase average, Eq. (3.7), over the total volume and area is given by

$$\int_V \langle \zeta \rangle_\pi (\mathbf{x}, t) \, dV = \int_V \left[\frac{1}{dv} \int_{dv} \zeta(\mathbf{r}, t) \gamma_\pi(\mathbf{r}, t) \, dv_m \right] dV$$

$$\int_A \bar{\bar{\zeta}}_\pi (\mathbf{x}, t) \, dA = \int_A \left[\frac{1}{da} \int_{da} \zeta(\mathbf{r}, t) \cdot \mathbf{n} \, \gamma_\pi(\mathbf{r}, t) \, da_m \right] dA \tag{3.15}$$

Following this, and by making use of the volume, mass and area operators, Eqs. (3.7)–(3.13), the averaged macroscopic conservation equation can be obtained as (Lewis and Schrefler 2000)

$$\frac{\partial}{\partial t} (\langle \rho \rangle_\pi \bar{\Psi}_\pi) + \text{div} \, (\langle \rho \rangle_\pi \bar{\Psi}_\pi \bar{\mathbf{v}}_\pi) - \text{div} \, \mathbf{j}_\pi - \langle \rho \rangle_\pi \bar{b}_\pi = \langle \rho \rangle_\pi \bar{G}_\pi \tag{3.16}$$

where the bar (‾) denotes averaged quantity, $\mathbf{j}_\pi$ is the flux vector associated with $\overline{\psi}_\pi$, $\overline{b}_\pi$ is the external body force associated with $\overline{\psi}_\pi$, $\langle\rho\rangle_\pi$ is the volume-averaged value of mass density, and $\overline{\mathbf{v}}_\pi$ is the mass averaged velocity of the π phase, defined as

$$\overline{\mathbf{v}}_\pi = \overline{\mathbf{r}}_\pi(\mathbf{x}, t) = \frac{1}{\langle\rho\rangle_\pi dv} \int_{dv} \rho(\mathbf{r}, t)\dot{\mathbf{r}}(\mathbf{r}, t)\gamma_\pi(\mathbf{r}, t)\,dv_m \tag{3.17}$$

Note that the differential form of Eq. (3.16) is derived under certain smoothness conditions of the integral form.

Now we specialize the microscopic generic balance equation, Eq. (3.1), and its corresponding macroscopic balance equation, Eq. (3.16), for deriving the energy balance equation in porous media. For the energy balance, the generic microscopic conserved variable ψ and its associated parameters are denoted by

$$\psi = E + \frac{1}{2}\dot{\mathbf{r}}\cdot\dot{\mathbf{r}}$$
$$\mathbf{j} = \mathbf{t}_m\dot{\mathbf{r}} - \mathbf{q} \tag{3.18}$$
$$b = \mathbf{g}\cdot\dot{\mathbf{r}} + Q$$
$$G = 0$$

where E is the specific intrinsic energy, $\mathbf{t}_m$ is the stress tensor, $\mathbf{q}$ is the heat flux vector, $\mathbf{g}$ is the gravitational vector, and Q is the intrinsic heat source. Substituting Eq. (3.18) into Eq. (3.1) and ignoring the contribution of the mass transport and the linear momentum, the conservation law of energy can be expressed as

$$\rho\frac{DE}{Dt} = -\operatorname{div}\mathbf{q} + \rho Q \tag{3.19}$$

where the operator D/Dt is the *material time derivative* defined as

$$\frac{D}{Dt} = \frac{\partial}{\partial t} + \operatorname{grad}\cdot\mathbf{v} \tag{3.20}$$

with $\mathbf{v}$ the field velocity vector. The average quantities at the macroscopic level are:

$$\overline{\psi}_\pi = \overline{E}_\pi + \frac{1}{2}\overline{\mathbf{v}}_\pi\cdot\overline{\mathbf{v}}_\pi$$
$$\overline{\mathbf{j}} = \mathbf{t}_m\overline{\mathbf{v}}_\pi - \mathbf{q}_\pi \tag{3.21}$$
$$\overline{b} = \mathbf{g}\cdot\overline{\mathbf{v}}_\pi + Q_\pi$$

Substituting Eq. (3.21) into Eq. (3.16), and considering the conditions in Eq. (3.19), the conservation law of energy can be expressed as

$$\rho_\pi\frac{D_\pi\overline{E}_\pi}{Dt} = -\operatorname{div}\overline{\mathbf{q}}_\pi + \rho_\pi Q_\pi \tag{3.22}$$

The macroscopic form of this energy equation is similar to that obtained from the mixture theory and other macroscopic continuum theories. The essential difference, however, is in the thermal interaction at the interface between the constituents of the porous material. The average internal heat flux, using Eqs. (3.13) and (3.14), is described as

$$\tilde{\mathbf{q}}_\pi = \frac{1}{da}\int_{da}[\mathbf{t}_m\tilde{\dot{\mathbf{r}}} - \mathbf{q} - \rho(E + \tilde{\dot{\mathbf{r}}}_\pi\cdot\tilde{\dot{\mathbf{r}}}_\pi)\tilde{\dot{\mathbf{r}}}_\pi]\cdot\mathbf{n}\,\gamma_\pi(\mathbf{r}, t)\,da_m \tag{3.23}$$

in which $\tilde{\dot{\mathbf{r}}}_\pi = \dot{\mathbf{r}}_\pi - \overline{\mathbf{v}}_\pi$. Note that we used $\tilde{\mathbf{q}}$ instead of $\overline{\mathbf{q}}$ to include the deviation of the microscopic function, see Eq. (3.14). The constitutive relationship for the heat flux is described by Fourier's law, such that

$$\tilde{\mathbf{q}}_\pi = -\lambda_\pi\operatorname{grad}T_\pi \tag{3.24}$$

where λ_π is the thermal conductivity of π phase, and T_π is its temperature.

The internal energy $\overline{E}_\pi$ is a thermodynamic state function of pressure and temperature, expressed, in a differential form, as

$$d\overline{E}_\pi = \left(\frac{\partial \overline{E}_\pi}{\partial T_\pi}\right)_{\overline{p}} dT_\pi + \left(\frac{\partial \overline{E}_\pi}{\partial \overline{P}_\pi}\right)_T d\overline{P}_\pi \qquad (3.25)$$

For incompressible materials such as liquids and solids, the dependence on pressure is negligible. Hence the internal energy becomes

$$d\overline{E}_\pi = c_p^\pi dT_\pi \qquad (3.26)$$

where c_p^π is the specific heat, defined as

$$c_p^\pi = \left(\frac{\partial \overline{E}_\pi}{\partial T_\pi}\right)_{\overline{p}} \qquad (3.27)$$

Using the intrinsic phase-averaged densities of Eq. (3.22), dividing by the volume fraction η_π and neglecting the density variation due to temperature and momentum, the final form of the energy field equation yields

$$\rho_\pi c_p^\pi \frac{D_\pi T_\pi}{Dt} = -\text{div } \tilde{\mathbf{q}}_\pi + \rho_\pi Q_\pi \qquad (3.28)$$

where $\rho_\pi c_p^\pi$ is the volume heat capacity of π phase. This equation is a fundamental representation of the energy equation in a multiphase medium. In its macroscopic form it is very much similar to those obtained by other derivation procedures. However, the details of handling the thermal interactions between the involved constituents are significantly different. In engineering practice, the thermal interactions are treated intuitively. In many applications, models based on continuum heat flow balance and constitutive relationships, together with engineering intuition, give reasonably accurate results.

3.3   HEAT FLOW IN A TWO-PHASE SOIL MASS: ENGINEERING REPRESENTATION

The energy field equation, Eq. (3.28), can be utilized to describe heat conduction and convection in a multiphase porous medium constituting solid, liquid and gas. In deriving this equation we have ignored the mass transport and the energy dissipation due to momentum because heat transfer in shallow geothermal systems does not really depend on these aspects. Considering the heat flow processes and the materials involved in such a system, we can still apply further simplifications by considering that the soil mass is fully saturated: a two-phase material consisting of a solid matrix and water.

Thermal parameters of multiphase constituents may be significantly different from each other. For instance, for a fully saturated sand, the thermal conductivity of the sand reaches up to $4 \text{ W/m} \cdot \text{K}$, whereas for water it is $0.56 \text{ W/m} \cdot \text{K}$. Similarly, the specific heat of the sand is on the order of 1 kJ/kg · K, but for the water it is 4.18 kJ/kg · K. Therefore, at the microscopic level, it is expected that there is a temperature gradient between the solid particles and the water particles. The significance of this gradient is very much influenced by the boundary conditions. For a highly transient case or where heat generation exists in one of the phases, the temperature gradient is significant, and must be taken into consideration. In this case, the material is said to be in *local thermal non-equilibrium*. However, for ordinary transient conductive or conductive-convective cases, the microscopic temperature gradient is generally much smaller than that at the macroscopic level, and hence can be ignored. In this case the material is said to be in *local thermal equilibrium*. Hereafter we discuss both cases.

3.3.1 *Local thermal non-equilibrium*

When phases representing the constituents of a porous material are in local thermal non-equilibrium, due to either very fast transient processes or where there is a significant heat generation in any of the phases, heat transfer is described by thermal coupling between the solid particles and the fluid particles. As stated earlier, there are basically two approaches to model this kind of problems. One is based on the mixture theory, and another is based on phenomenological interpretation.

The first approach was discussed in Section 3.2. According to Eq. (3.28), the macroscopic field equation for a two-phase soil material can be described as

$$\rho_s c_p^s \frac{D_s T_s}{Dt} = -\text{div } \tilde{\mathbf{q}}_s + \rho_s Q_s$$

$$\rho_w c_p^w \frac{D_w T_w}{Dt} = -\text{div } \tilde{\mathbf{q}}_w + \rho_w Q_w$$

(3.29)

where the subscripts s and w represent the solid and water phases respectively. These macroscopic energy equations can also be obtained from any other mixture theory or those obtained from the continuum theory, but they differ in the thermal interaction between the two constituents within the pores, namely $\tilde{\mathbf{q}}_s$ and $\tilde{\mathbf{q}}_w$. Quantification of this interaction at the microscopic level in complicated materials, such as in natural soil masses, is rather difficult and in many cases not necessary. In these cases, the phenomenological approach is more realistic.

Phenomenological models for heat flow in multiphase media are based on the continuum theory of heat transfer, as each phase is represented by appropriate effective total thermal parameters. Several models with different complexities are in use for describing heat flow in porous media. One group postulates that the porous medium is consisted of parallel rods or sheets, where local thermal equilibrium is not valid. This way of modeling is described by Nield and Bejan (1999), where the energy equation of the two phases is expressed as

$$(1-n)(\rho c)_s \frac{\partial \langle T \rangle_s}{\partial t} = (1-n)\nabla \cdot (\lambda_s \nabla \langle T \rangle_s) + b_{sw}(\langle T \rangle_s - \langle T \rangle_w)$$

(3.30)

$$n(\rho c)_w \frac{\partial \langle T \rangle_w}{\partial t} + (\rho c)_w \mathbf{v} \cdot \nabla \langle T \rangle_w = n\nabla \cdot (\lambda_w \nabla \langle T \rangle_w) + b_{ws}(\langle T \rangle_w - \langle T \rangle_s)$$

(3.31)

where n is the porosity, and $\langle T \rangle_s$ and $\langle T \rangle_w$ are the average temperatures of the solid and water respectively; λ_s, $(\rho c)_s$, λ_w, $(\rho c)_w$ are the thermal conductivity and volumetric heat capacities of the solid and water respectively; $\mathbf{v}$ is the *Darcy velocity* vector, b_{sw} and b_{ws} are the interfacial heat transfer coefficients, determined experimentally and usually treated as equal. Note that the averaging here is intuitive since it is not derived on the basis of formal scaling-up procedure, as that for the averaging theory. This model is simple and reflects the concept of two macroscopic continua in thermal contact.

Another group of models postulate that there is a more complicated thermal interaction between the constituents that must be considered in order to obtain a reasonable representation. Non-equilibrium models based on local volume averaging and intuitive thermal interactions have been introduced by Whitaker and Kaviany and their co-workers (Kaviany 1995). They proposed several formulations for heat flow in porous media, of the form:

$$(\rho c_p)_s \frac{\partial \langle T \rangle_s}{\partial t} + (\rho c_p)_s \mathbf{v}_{ss} \cdot \nabla \langle T \rangle_s + (\rho c_p)_s \mathbf{v}_{sw} \cdot \nabla \langle T \rangle_w$$
$$+ \nabla \cdot \lambda_{ss} \cdot \nabla \langle T \rangle_s + \nabla \cdot \lambda_{sw} \cdot \nabla \langle T \rangle_w = \frac{A_{ws}}{V_s} h_{ws}(\langle T \rangle_w - \langle T \rangle_s)$$

(3.32)

$$(\rho c_p)_w \frac{\partial \langle T \rangle_w}{\partial t} + (\rho c_p)_w \mathbf{v}_{ww} \cdot \nabla \langle T \rangle_w + (\rho c_p)_w \mathbf{v}_{ws} \cdot \nabla \langle T \rangle_s$$
$$+ \nabla \cdot \lambda_{ww} \cdot \nabla \langle T \rangle_w + \nabla \cdot \lambda_{ws} \cdot \nabla \langle T \rangle_s = \frac{A_{ws}}{V_w} h_{ws}(\langle T \rangle_s - \langle T \rangle_f)$$

(3.33)

where λ_{ss}, λ_{ff}, λ_{sf} and λ_{fs} are the total thermal conductivity tensors, describing the individual phase thermal conductivities and their interactions, and h_{fs} is the interfacial heat transfer coefficient, describing the convective thermal interaction between the two phases. Details on the determination of these parameters can be found in Rohsenow *et al.* (1998). This model has been further elaborated and more complicated averaging process has been introduced. See for example Quintrad *et al.* (1997).

3.3.2   *Local thermal equilibrium*

In most practical cases, the temperature gradient at the microscopic (pore) level is less than that at the macroscopic level (REV), and both are much less than that occurring at the megascopic level, the physical system. That is:

$$\Delta T_{micro} < \Delta T_{REV} \; << \; \Delta T_{system} \tag{3.34}$$

This indicates that the local temperature gradient between the phases is negligible, and hence allows for assuming that the solid and fluid phases within a representative elementary volume are in local thermal equilibrium. This means that the average temperatures of the two phases are equal, such that

$$T_s = T_w = T \tag{3.35}$$

where the standard averaging notation $\langle \cdots \rangle$ is ignored for simplicity. Applying this condition to the energy field equations, Eqs. (3.32) and (3.33), it can readily be realized that all thermal interaction terms must be discarded. Thus, the macroscopic energy field equation for the solid phase and the fluid phase can be rewritten, respectively, as

$$(\rho c_p)_s \frac{\partial T}{\partial t} + (\rho c_p)_s \mathbf{v}_s \cdot \nabla T + \nabla \cdot \lambda_s \cdot \nabla T = 0 \tag{3.36}$$

$$(\rho c_p)_w \frac{\partial T}{\partial t} + (\rho c_p)_w \mathbf{v}_w \cdot \nabla T + \nabla \cdot \lambda_w \cdot \nabla T = 0 \tag{3.37}$$

Adding Eq. (3.36) to Eq. (3.37), and omitting the convective term of the solid, $(\rho c_p)_s \mathbf{v}_s \cdot \nabla T$, the system macroscopic field equation can be obtained, as

$$(\rho c_p)_{\text{eff}} \frac{\partial T}{\partial t} + (\rho c_p)_w \mathbf{v}_w \cdot \nabla T + \nabla \cdot \lambda_{\text{eff}} \cdot \nabla T = 0 \tag{3.38}$$

in which $(\rho c_p)_{\text{eff}}$ and $\lambda_{\text{eff}} \equiv \lambda_s + \lambda_w$ are the local averaged effective heat capacity and thermal conductivity of the porous medium, respectively; $\mathbf{v}_w$ in is the fluid velocity which is described by the generalized Darcy's law, expressed as

$$\mathbf{v}_w = -\frac{k_r}{\mu_w} \mathbf{k} \cdot (\text{grad } P \; - \; \rho_w \mathbf{g}) \tag{3.39}$$

where $\mathbf{k}$ is the intrinsic permeability tensor, k_r is the relative permeability, ranging from 0 to 1, μ_w is the dynamic viscosity, and $\mathbf{g}$ is the gravity vector; k_r/μ_w is referred to as the mobility of the fluid phase.

The effective heat capacity is usually obtained by a simple volume averaging, such that

$$(\rho c_p)_{\text{eff}} = n (\rho c_p)_w + (1 - n)(\rho c_p)_s \tag{3.40}$$

where n is the porosity. However, the effective thermal conductivity is more complicated and proved to be dependent on many factors, among which is the relative magnitude of λ_s to that of λ_w, and the thermal resistance between the particles (Rohsenow *et al.* 1998). Commonly, it is defined as

$$\lambda_{\text{eff}} = n\lambda_w + (1 - n)\lambda_s + \frac{\lambda_w - \lambda_s}{V} \int_{A_{ws}} \mathbf{n}_{ws} \mathbf{b}_w \, dA \tag{3.41}$$

where $\mathbf{n}_{ws}$ is the outward normal unit vector on the interface between the two phases, and $\mathbf{b}_w$ is a transformation function of the gradient of local volume-averaged temperature. Determination of $\mathbf{b}_w$ can be found in Kaviany (1995). Note that this form of the effective thermal conductivity can be obtained by applying the volume averaging theorem on the energy equations, Eqs. (3.36) and (3.37), where the transformation of the temperature can be obtained by (Slattery 1969)

$$\langle \nabla T \rangle = \nabla \langle T \rangle + \frac{1}{V} \int_{A_{sw}} \mathbf{n} T \, dA \tag{3.42}$$

In the literature, there are several formulations describing the effective thermal conductivity of two phase materials. Krupiczka (1967), for $0.2 \leq n \leq 0.6$, gives

$$\lambda_{\text{eff}} = \lambda_w \left(\frac{\lambda_s}{\lambda_w} \right)^{0.28 - 0.757 \log n - 0.057 \log(\lambda_s / \lambda_w)} \tag{3.43}$$

Bromberg and Shirtliffe (1978) proposed a simpler relationship of the form

$$\lambda_{\text{eff}} = \lambda_s \left(1 + 4 \frac{n S_w \rho_w}{(1 - n)\rho_s} \right) \tag{3.44}$$

For shallow geothermal systems where the temperature gradient is not significant, a simple volume averaging, similar to the effective heat capacity, can be used. That is

$$\lambda_{\text{eff}} = n \lambda_w + (1 - n)\lambda_s \tag{3.45}$$

will suffice.

CHAPTER 4

Heat transfer in borehole heat exchangers

In this chapter, we establish a consistent procedure for deriving mathematical models capable of simulating transient and steady-state heat transfer in commonly utilized geothermal boreholes heat exchangers. We begin by formulating the heat equation of a multiple conductive-convective continuum. Then we generalize the procedure to establish heat equations of typical borehole heat exchangers, namely, single U-tube borehole heat exchanger (1U), double U-tube (2U), coaxial with annular (CXA) and coaxial with centered inlet (CXC).

4.1 INTRODUCTION

The mechanism of heat transfer in a borehole heat exchanger (BHE) is somewhat complicated as it involves conductive and convective processes occurring in a multiple component medium. A typical borehole heat exchanger consists of one or more plastic U-tubes, surrounded by a bentonite-cement grout. The U-tubes carry a circulating (working) fluid, usually water with 20%–25% anti-freezing coolant. Since these materials have different thermal parameters, the mechanism of heat transfer in each component differs significantly from the others, and as a result, affects the mechanism of heat transfer in the whole borehole heat exchanger.

Another important factor that affects the mechanism of heat transfer in the borehole heat exchanger is the geometry of the involved components and their thermal interactions. As shown in Figure 1.2 and Figure 1.3, Chapter 1, each component is in direct contact with one or more components, and in an indirect contact with some others. The importance of the thermal interactions in the whole process depends on the size of the borehole and the distances between the pipes, known as the *shank space*.

Due to this relatively complicated heat transfer mechanism, several models with different complexities have been introduced in the literature. In Chapter 8, we will introduce some of the current models, especially those relevant to our treatment of the problem. In this chapter, we derive mathematical models for typical borehole heat exchangers. These models will be utilized in subsequent chapters to describe the initial and boundary value problems of shallow geothermal systems.

To derive the heat equation of a borehole heat exchanger, we need to consider the particularity of its geometry. Two geometrical characteristics need to be addressed: slenderness and multiple channels. Typical borehole heat exchangers are quite slender with diameter to length ratio of the order of 1/700 or less. This entails that the main stream of heat flow is along its axial dimension. The radial heat flow is negligible. Therefore, it is reasonably accurate to utilize a one-dimensional heat equation to describe heat flow in the involved BHE components.

A typical borehole heat exchanger is composed of one or more pipe(s)-in and pipe(s)-out, and grout. Each BHE component acts as a channel for transferring heat along its axial dimension, and at the same time exchanging heat along its contact surface area with other components. In this respect, there is a similarity between heat flow in the borehole heat exchanger and those dealing with heat flow in parallel rods, sheets, tubes or multiple-diffusivity continuum, where there is no local equilibrium between components, as postulated by Aifantis (1979), and Nield and Bejan (1999), among others. In what follows we derive the heat equation of a multiple component system. Then we tailor the equation to describe particular borehole heat exchanger geometry.

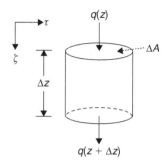

Figure 4.1   Control volume of a single component.

4.2   HEAT EQUATION OF A MULTIPLE COMPONENT SYSTEM

Heat equation in a domain can be derived from the law of conservation of energy and Fourier's law. We first derive the heat equation of a single component system, and then expand it to a multiple component system.

Consider a small element of a single component system bounded in space by $z \leq \zeta \leq z + \Delta z$, and in time by $t \leq \tau \leq t + \Delta t$, Figure 4.1. According to Fourier's law, the rate of flow of heat across a surface ΔA, in one-dimension, is described as

$$q(z) = -\lambda \frac{\partial T(z, t)}{\partial z} \Delta A \qquad (4.1)$$

where λ is the thermal conductivity and T is the temperature. At $z + \Delta z$ the rate of heat flow is

$$q(z + \Delta z) = -\lambda \frac{\partial T(z + \Delta z, t)}{\partial z} \Delta A \qquad (4.2)$$

In the absence of an external work, the amount of heat (internal energy) in the element at time t is

$$H(t) = \rho c_p T(z, t) \Delta A \Delta z \qquad (4.3)$$

where ρ is the mass density of the material and c_p is the specific heat capacity. At time $t + \Delta t$ the amount of heat is

$$H(t + \Delta t) = \rho c_p T(z, t + \Delta t) \Delta A \Delta z \qquad (4.4)$$

Assuming a constant cross section area, A, and using the *first fundamental theorem of calculus*, which states that the sum of infinitesimal changes in a quantity over time (or over some other quantity) adds up to the net change in the quantity (Wikipidia (a)), the change in the internal energy in time, $t \leq \tau \leq t + \Delta t$, can be expressed as

$$H(t + \Delta t) - H(t) = \rho c_p A \int_z^{z+\Delta z} [T(\zeta, t + \Delta t) - T(\zeta, t)] \, d\zeta$$

$$= \rho c_p A \int_t^{t+\Delta t} \int_z^{z+\Delta z} \frac{\partial T}{\partial \tau} \, d\zeta \, d\tau \qquad (4.5)$$

Physically, this expression represents heat gain in the system. In the same way, the change in the internal energy in space, $z \leq \zeta \leq z + \Delta z$, may be expressed as

$$q(z + \Delta z) - q(z) = \lambda A \int_t^{t+\Delta t} \left[\frac{\partial T}{\partial z}(z + \Delta z, \tau) - \frac{\partial T}{\partial z}(z, \tau) \right] d\tau$$

$$= \lambda A \int_t^{t+\Delta t} \int_z^{z+\Delta z} \frac{\partial^2 T}{\partial \zeta^2} \, d\zeta \, d\tau \qquad (4.6)$$

Physically, this expression represents heat flux in the system. Following the conservation of energy, the heat flux, Eq. (4.6), must be equal to the heat gain, Eq. (4.5), giving

$$\int_{t}^{t+\Delta t} \int_{z}^{z+\Delta z} \left[\rho c_p \frac{\partial T}{\partial \tau} - \lambda \frac{\partial^2 T}{\partial \zeta^2} \right] d\zeta \, d\tau = 0 \tag{4.7}$$

If we assume that this expression is true for all time and space of interest, the integrand must vanish identically, yielding

$$\rho c_P \frac{\partial T}{\partial t} - \lambda \frac{\partial^2 T}{\partial z^2} = 0 \tag{4.8}$$

This equation is the transient heat conduction equation in one dimension. Note that, as shown in Chapter 2, we could derive this equation using Taylor series, instead of the fundamental theorem of calculus.

If convection is also involved, we need to include an advection term. This can be done by replacing the derivative of temperature with time with the material derivative (also known as convective, advective and Lagrangian derivative). This kind of derivatives is taken along a path moving with velocity u and is often used to describe the time rate of change of a scalar or a vector quantity. The material derivative of a scalar field, φ, is described as

$$\frac{D\varphi}{Dt} = \frac{\partial \varphi}{\partial t} + u \cdot \nabla \varphi \tag{4.9}$$

Replacing the time derivative (Euler derivative, $\partial/\partial t$) in Eq. (4.8) by a material derivative, the heat equation might then be expressed as

$$\rho c_P \frac{DT}{Dt} - \lambda \frac{\partial^2 T}{\partial z^2} = 0 \tag{4.10}$$

Using Eq. (4.9), we obtain

$$\rho c_P \left(\frac{\partial T}{\partial t} + u_z \frac{\partial T}{\partial z} \right) - \lambda \frac{\partial^2 T}{\partial z^2} = 0 \tag{4.11}$$

This equation can also be written as

$$\rho c_P \frac{\partial T}{\partial t} + \frac{\partial q}{\partial z} = 0 \tag{4.12}$$

in which

$$q = -\lambda \frac{\partial T}{\partial z} + \rho c_P u_z T \tag{4.13}$$

In three-dimensions, Eq. (4.12) can be written as

$$\rho c_p \frac{\partial T}{\partial t} + \operatorname{div} \mathbf{q} = 0 \tag{4.14}$$

in which

$$\mathbf{q} = -\lambda \operatorname{grad} T + \rho c_p \mathbf{u} T \tag{4.15}$$

Now we expand the heat equation of a single component system to a multiple component. We begin with the balance of energy in a control volume consisting of multiple pipe components, Figure 4.2. Designate $T_1, T_2, \ldots, T_n$ as temperatures in a borehole heat exchanger consisting of n components, $q_1, q_2, \ldots, q_n$ as their heat fluxes and $Q_1, Q_2, \ldots, Q_n$ as source items. Using Eq. (4.12), the energy balance for each individual pipe component, α, can be described as

$$\rho c_\alpha \frac{\partial T_\alpha}{\partial t} + \frac{\partial q_\alpha}{\partial z} = Q_\alpha, \quad \alpha = 1, 2, \ldots, n \tag{4.16}$$

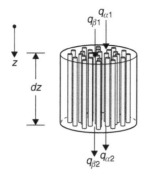

Figure 4.2   Control volume of a multiple pipe component.

where $c_\alpha = c_p^\alpha$ is the specific heat in α component (solid or fluid) and Q_α is its heat source. The total energy balance for the whole system is

$$\rho c \frac{\partial T}{\partial t} + \frac{\partial q}{\partial z} = 0 \qquad (4.17)$$

with

$$T = \sum_{\alpha=1}^{n} T_\alpha, \quad q = \sum_{\alpha=1}^{n} q_\alpha \qquad (4.18)$$

Comparing Eq. (4.16) with Eq. (4.17) it can readily be deduced that thermal equilibrium can only be attained by the constraint (Aifantis 1979):

$$\sum_{\alpha=1}^{n} Q_\alpha = 0 \qquad (4.19)$$

As will be apparent later, this constraint is important for describing the thermal interaction between the pipe components. Following this, Eq. (4.13) can be modified, and the conductive-convective heat flow in pipe component α in contact with pipe component β can be described as

$$q_\alpha = -\sum_{\beta=1}^{n} \lambda_{\alpha\beta} \frac{\partial T_{\alpha\beta}}{\partial z} + \sum_{\beta=1}^{n} (\rho c u)_{\alpha\beta} T_{\alpha\beta} \qquad (4.20)$$

in which $\lambda_{\alpha\beta}$ represents the thermal conductivity of component α in contact with component β, and similarly $T_{\alpha\beta}$ represents the temperature of component α in contact with component β. The first term on the right hand side of Eq. (4.20) describes heat conduction, and the second term describes heat convection. Obviously, for a U-tube borehole heat exchanger consisting of pipe-in, pipe-out and grout, heat conduction and convection occurs in pipe-in and pipe-out, but in the grout, only conduction takes place.

We are now left with introducing the constitutive relationship for the source term Q_α. In analogy to Newton's law of cooling, Q_α can be described by a linear function of temperature difference between the pipe components, such that

$$Q_\alpha = \sum_{\beta=1}^{n} b_{\alpha\beta}(T_{\alpha\alpha} - T_{\alpha\beta}) \qquad (4.21)$$

where $b_{\alpha\beta}$ is a thermal coefficient describing the thermal interaction between pipe component α and pipe component β. As it will be shown in Chapter 5, quantification of $b_{\alpha\beta}$ can be done using the analogy between Fourier's law and *Ohm's law*.

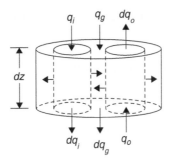

Figure 4.3 Control volume of a single U-tube BHE.

Equating Eq. (4.20) to Eq. (4.21), the heat equation of a multiple component system can be obtained of the form

$$-\sum_{\beta=1}^{n} \lambda_{\alpha\beta} \frac{\partial T_{\alpha\beta}}{\partial z} + \sum_{\beta=1}^{n} (\rho c u)_{\alpha\beta} T_{\alpha\beta} = \sum_{\beta=1}^{n} b_{\alpha\beta}(T_{\alpha\alpha} - T_{\alpha\beta}) \qquad (4.22)$$

Notice the similarity of the this model with the phenomenological models of porous media presented in Chapter 3, see Eqs. (3.32)–(3.33).

4.3   HEAT EQUATION OF A BOREHOLE HEAT EXCHANGER

The theory presented so far is suitable for modeling thermal systems consisting of multiple channels or components. This theory can be adapted to model all known borehole heat exchangers. To give an example, here we tailor the theory to model a single U-tube borehole heat exchanger consisting of three pipe components: pipe-in, pipe-out and grout. This entails that, in Eqs. (4.20) and (4.21), the number of components is $n = 3$.

Consider a control volume of a typical single U-tube BHE of length dz, consisting of pipe-in, pipe-out, and grout, Figure 4.3. Each pipe component transfers heat across its cross sectional area and exchange fluxes with other components across its surface area. If we designate index 1 to correspond to pipe-in, 2 to correspond to pipe-out, and 3 to correspond to grout to Eq. (4.20), we obtain

$$q_1 = -\lambda_{11} \frac{\partial T_{11}}{\partial z} + (\rho c u)_{11} T_{11} - \lambda_{12} \frac{\partial T_{12}}{\partial z} + (\rho c u)_{12} T_{12} - \lambda_{13} \frac{\partial T_{13}}{\partial z}$$

$$q_2 = -\lambda_{21} \frac{\partial T_{21}}{\partial z} + (\rho c u)_{21} T_{21} - \lambda_{22} \frac{\partial T_{22}}{\partial z} + (\rho c u)_{22} T_{22} - \lambda_{23} \frac{\partial T_{23}}{\partial z} \qquad (4.23)$$

$$q_3 = -\lambda_{31} \frac{\partial T_{31}}{\partial z} - \lambda_{32} \frac{\partial T_{32}}{\partial z} - \lambda_{33} \frac{\partial T_{33}}{\partial z}$$

Similarly, Eq. (4.21) yields

$$Q_1 = b_{11}(T_{11} - T_{11}) + b_{12}(T_{11} - T_{12}) + b_{13}(T_{11} - T_{13})$$

$$Q_2 = b_{21}(T_{22} - T_{21}) + b_{22}(T_{22} - T_{22}) + b_{23}(T_{22} - T_{23}) \qquad (4.24)$$

$$Q_3 = b_{31}(T_{33} - T_{31}) + b_{32}(T_{33} - T_{32}) + b_{33}(T_{33} - T_{33})$$

Note that the terms describing the thermal interaction within the channel itself cancel out automatically. As the pipe components of a borehole heat exchanger are physically continuous, the coupled thermal parameters in Eq. (4.23) must be neglected. That is

$$\lambda_{12} = \lambda_{13} = \lambda_{21} = \lambda_{23} = \lambda_{31} = \lambda_{32} = 0$$
$$(\rho c u)_{12} = (\rho c u)_{21} = 0$$
(4.25)

We adapt the geothermal notation:

$$T_{11} = T_i$$
$$T_{22} = T_o$$
$$T_{33} = T_g$$
$$\lambda_{11} = \lambda_{22} = \lambda_{ref}$$
$$\lambda_{33} = \lambda_g$$
$$(\rho c u)_{11} = \rho_{ref} c_{ref} u_{ref}$$
$$(\rho c u)_{22} = -\rho_{ref} c_{ref} u_{ref}$$
(4.26)

where "ref" refers to the circulating (working, refrigerant) fluid, and i, o, g refer to pipe-in, pipe-out and grout respectively. The minus sign in the last line of Eq. (4.26) indicates that the velocity in pipe-out is opposite in direction to that in pipe-in. Applying Eqs. (4.25) and (4.26) into the heat flux equation, Eq. (4.23), we obtain

$$q_i = -\lambda_{ref}\frac{\partial T_i}{\partial z} + \rho_{ref}\, c_{ref}\, u_{ref} T_i$$
$$q_o = -\lambda_{ref}\frac{\partial T_o}{\partial z} - \rho_{ref}\, c_{ref}\, u_{ref} T_o$$
$$q_g = -\lambda_g\frac{\partial T_g}{\partial z}$$
(4.27)

In the same way, the thermal interaction terms, Eq. (4.24), can be modified to give

$$Q_i = b_{io}(T_i - T_o) + b_{ig}(T_i - T_g)$$
$$Q_o = b_{oi}(T_o - T_i) + b_{og}(T_o - T_g)$$
$$Q_g = b_{gi}(T_g - T_i) + b_{go}(T_g - T_o)$$
(4.28)

Using Eq. (4.16), the heat equation of a single U-tube borehole heat exchanger can then be expressed as

$$\rho_{ref}\, c_{ref}\frac{\partial T_i}{\partial t} + \frac{\partial}{\partial z}\left(-\lambda_{ref}\frac{\partial T_i}{\partial z} + \rho_{ref}\, c_{ref}\, u_{ref}\, T_i\right) = b_{io}(T_i - T_o) + b_{ig}(T_i - T_g)$$
$$\rho_{ref}\, c_{ref}\frac{\partial T_o}{\partial t} + \frac{\partial}{\partial z}\left(-\lambda_{ref}\frac{\partial T_o}{\partial z} - \rho_{ref}\, c_{ref}\, u_{ref}\, T_o\right) = b_{oi}(T_o - T_i) + b_{og}(T_o - T_g)$$
$$\rho_g c_g\frac{\partial T_g}{\partial t} + \frac{\partial}{\partial z}\left(-\lambda_g\frac{\partial T_g}{\partial z}\right) = b_{gi}(T_g - T_i) + b_{go}(T_g - T_o)$$
(4.29)

Important to note that the terms on the left-hand side represent heat flow and storage per unit volume, and on the right-hand side represent heat flow per unit area, along the surface area. In a formal format, Eq. (4.29) can be written as

$$
\int_{V_i} \left[\rho_{ref} c_{ref} \frac{\partial T_i}{\partial t} + \frac{\partial}{\partial z} \left(-\lambda_{ref} \frac{\partial T_i}{\partial z} + \rho_{ref} c_{ref} u_{ref} T_i \right) \right] dV_i
$$

$$
= \int_{S_{io}} b_{io}(T_i - T_o) \, dS_{io} + \int_{S_{ig}} b_{ig}(T_i - T_g) \, dS_{ig}
$$

(4.30)

$$
\int_{V_o} \left[\rho_{ref} c_{ref} \frac{\partial T_o}{\partial t} + \frac{\partial}{\partial z} \left(-\lambda_{ref} \frac{\partial T_o}{\partial z} - \rho_{ref} c_{ref} u_{ref} T_o \right) \right] dV_o
$$

$$
= \int_{S_{oi}} b_{oi}(T_o - T_i) \, dS_{oi} + \int_{S_{og}} b_{og}(T_o - T_g) \, dS_{og}
$$

(4.31)

$$
\int_{V_g} \left[\rho_g c_g \frac{\partial T_g}{\partial t} + \frac{\partial}{\partial z} \left(-\lambda_g \frac{\partial T_g}{\partial z} \right) \right] dV_g
$$

$$
= \int_{S_{gi}} b_{gi}(T_g - T_i) \, dS_{gi} + \int_{S_{go}} b_{go}(T_g - T_o) \, dS_{go}
$$

(4.32)

where dV_i, dV_o, dV_g are pipe-in, pipe-out and grout partial volumes, and dS_{io}, dS_{ig} are the corresponding contact surface areas between pipe-in and pipe-out and pipe-in and grout, etc.

4.4 HEAT EQUATIONS OF SOME TYPICAL BOREHOLE HEAT EXCHANGERS

In this section, and on the basis of the above modeling methodology, we formulate the heat equations of four commonly used borehole heat exchangers, namely, single U-tube (1U), double U-tube (2U), coaxial pipe with annular (CXA), and coaxial pipe with centered inlet (CXC). We will not use the subscript ref, representing the circulating fluid, for simplicity of notation.

4.4.1 *Heat equations of a single U-tube borehole heat exchanger (1U)*

In Eq. (4.29) we derived, in general terms, the heat equations of a single U-tube borehole heat exchanger. For a vertical single U-tube borehole heat exchanger, it is reasonable to state that thermal interactions within the borehole occur along the contact surface area between pipe-in and grout, and pipe-out and grout. No direct interaction exists between pipe-in and pipe-out themselves, leading to $b_{io} = b_{oi} = 0$. Letting $b_{ig} = b_{gi}$ and $b_{og} = b_{go}$, heat equations of a single U-tube borehole heat exchanger can be expressed as

Pipe-in:

$$
\rho c \frac{\partial T_i}{\partial t} - \lambda \frac{\partial^2 T_i}{\partial z^2} + \rho c u \frac{\partial T_i}{\partial z} = b_{ig}(T_i - T_g)
$$

(4.33)

Pipe-out:

$$
\rho c \frac{\partial T_o}{\partial t} - \lambda \frac{\partial^2 T_o}{\partial z^2} - \rho c u \frac{\partial T_o}{\partial z} = b_{og}(T_o - T_g)
$$

(4.34)

Grout:

$$
\rho c_g \frac{\partial T_g}{\partial t} - \lambda_g \frac{\partial^2 T_g}{\partial z^2} = b_{ig}(T_g - T_i) + b_{og}(T_g - T_o)
$$

(4.35)

where dV's and dS's are not included for clarity of notation. We will pursue this style for the rest of this book, unless otherwise necessary.

4.4.2 *Heat equations of a double U-tube borehole heat exchanger (2U)*

Following the procedure for deriving the governing heat equations of a single U-tube BHE, the governing heat equations for a double U-tube BHE can be formulated. The number of pipe

components of this heat exchanger is two pipes-in (denoted as $i1$ and $i2$), two pipes-out (denoted as $o1$ and $o2$) and grout (denoted as g), forming five coupled partial differential equations of the form:
Pipes-in:

$$\rho c \frac{\partial T_{i1}}{\partial t} - \lambda \frac{\partial^2 T_{i1}}{\partial z^2} + \rho c u \frac{\partial T_{i1}}{\partial z} = b_{ig1}(T_{i1} - T_g)$$

$$\rho c \frac{\partial T_{i2}}{\partial t} - \lambda \frac{\partial^2 T_{i2}}{\partial z^2} + \rho c u \frac{\partial T_{i2}}{\partial z} = b_{ig2}(T_{i2} - T_g)$$

(4.36)

Pipes-out:

$$\rho c \frac{\partial T_{o1}}{\partial t} - \lambda \frac{\partial^2 T_{o1}}{\partial z^2} - \rho c u \frac{\partial T_{o1}}{\partial z} = b_{og1}(T_{o1} - T_g)$$

$$\rho c \frac{\partial T_{o2}}{\partial t} - \lambda \frac{\partial^2 T_{o2}}{\partial z^2} - \rho c u \frac{\partial T_{o2}}{\partial z} = b_{og2}(T_{o2} - T_g)$$

(4.37)

Grout:

$$\rho c_g \frac{\partial T_g}{\partial t} - \lambda_g \frac{\partial^2 T_g}{\partial z^2} = b_{ig1}(T_g - T_{i1}) + b_{ig2}(T_g - T_{i2}) + b_{og1}(T_g - T_{o1}) + b_{og2}(T_g - T_{o2}) \quad (4.38)$$

4.4.3   Heat equations of a coaxial borehole heat exchanger with annular (CXA)

CXA heat pipe exchanger consists of three pipe components: pipe-in, pipe-out and grout, forming three coupled partial differential equations, as
Pipe-in:

$$\rho c \frac{\partial T_i}{\partial t} - \lambda \frac{\partial^2 T_i}{\partial z^2} + \rho c u \frac{\partial T_i}{\partial z} = b_{ig}(T_i - T_g) + b_{io}(T_i - T_o) \quad (4.39)$$

Pipe-out:

$$\rho c \frac{\partial T_o}{\partial t} - \lambda \frac{\partial^2 T_o}{\partial z^2} - \rho c u \frac{\partial T_o}{\partial z} = b_{io}(T_o - T_i) \quad (4.40)$$

Grout:

$$\rho c_g \frac{\partial T_g}{\partial t} - \lambda_g \frac{\partial^2 T_g}{\partial z^2} = b_{ig}(T_g - T_i) \quad (4.41)$$

where a new thermal parameter, b_{io}, representing the thermal interaction between pipe-in and pipe-out is introduced. In CXA, pipe-in acts as the intermediate medium that transfers heat between the three pipe components, see Figure 1.3, Chapter 1.

4.4.4   Heat equations of a coaxial borehole heat exchanger with centered inlet (CXC)

Similar to CXA, CXC consists of three pipe components: pipe-in, pipe-out and grout. The difference is that in CXC, pipe-out is the intermediate medium that transfers heat between the three components, see Figure 1.3, Chapter 1. The governing heat equations of a CXC borehole heat exchanger can then be described as
Pipe-in:

$$\rho c \frac{\partial T_i}{\partial t} - \lambda \frac{\partial^2 T_i}{\partial z^2} + \rho c u \frac{\partial T_i}{\partial z} = b_{io}(T_i - T_o) \quad (4.42)$$

Pipe-out:

$$\rho c \frac{\partial T_o}{\partial t} - \lambda \frac{\partial^2 T_o}{\partial z^2} - \rho c u \frac{\partial T_o}{\partial z} = b_{io}(T_o - T_i) + b_{og}(T_o - T_g) \quad (4.43)$$

Grout:

$$\rho c_g \frac{\partial T_g}{\partial t} - \lambda_g \frac{\partial^2 T_g}{\partial z^2} = b_{og}(T_g - T_o) \quad (4.44)$$

CHAPTER 5

Thermal resistance

This chapter focuses on determining the thermal resistance, which is a measure of a material ability to resist heat. This parameter is particularly important for heat flow in shallow geothermal systems. It has a significant impact on the amount of heat flow between the borehole heat exchanger components, and between the borehole and the surrounding soil mass. Therefore, it is an important parameter for designing the borehole heat exchangers and evaluating the cost-effectiveness of the ground-source heat pumps. In this chapter, we discuss the analogy between the electric circuits and the thermal circuits, and outline different methods, which are commonly utilized for the determination of the thermal resistances in shallow geothermal systems.

5.1 INTRODUCTION

Thermal resistance, also known as *heat resistance*, is a measure of a material ability to resist heat. It is the reciprocal of thermal conductance. This property is important in many engineering applications as it is used as a means for enhancing the energy efficiency of an object or a gadget. In some applications, the thermal resistance of a material is required to be high, as in electrical cables and equipments. In many other applications, the material is required to be low in thermal resistivity, as in cooking pans and other related tools. In yet other applications, the material must exhibit a balanced resistivity. In geothermal systems, for instance, the thermal resistance of pipe-in is required to be as low as possible, to allow for more heat to be collected along the path of the fluid flow; whereas in pipe-out, where the fluid flows out of the pipe, the opposite applies. The grout is required to conduct heat when the fluid goes down along pipe-in, and resists heat when the fluid goes up along pipe-out. Therefore, in designing a borehole heat exchanger, a careful optimization of the thermal resistivity is essential for obtaining efficient geothermal systems.

Accurate determination of the thermal resistance is important, especially for designing heat exchangers. For this purpose, several methods have been devised and, in general, can be classified in three categories: experimental; analytical or numerical; and thermal circuit. Hereafter we give an overview of these three methods, focusing on the determination of thermal resistances in shallow geothermal systems. But first, we start with some preliminary issues, including the analogy between electric circuits and thermal circuits.

5.2 FOURIER'S LAW VS. OHM'S LAW

Ohm's law states that an electric current through a conductor between two points is directly proportional to the potential difference or voltage across the points, and inversely proportional to the resistance between them. Mathematically, it is described as

$$I = \frac{\Delta V}{R_e} \tag{5.1}$$

where I (amperes) is the current through the conductor, ΔV (volts) is the potential difference measured across the conductor, and R_e (ohms) is the electric resistance of the conductor.

Fourier's law states that the rate of heat conduction through a homogeneous domain is directly proportional to the cross sectional area of the domain and the temperature difference across its

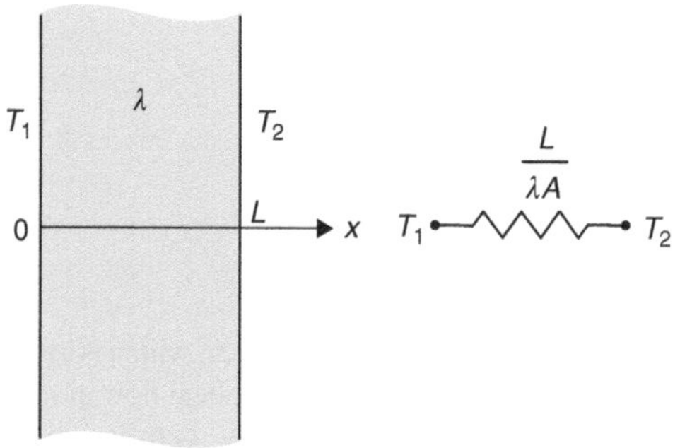

Figure 5.1   A plane wall and its thermal circuit.

boundaries along the path of heat flow. Mathematically, Fourier's law, as we shall see later, can be expressed as

$$q = \frac{\Delta T}{R_t} \tag{5.2}$$

where q (W) is the rate of heat flow, ΔT (K) is the temperature difference measured across the system, and R_t (K/W) is the thermal resistance. In what follows, we use R to indicate R_t, unless otherwise necessary.

Comparing Eq. (5.1) to Eq. (5.2), we readily notice that there is a similarity between Fourier's law and Ohm's law, and thus there is an analogy between heat flow and electrical current flow. In this, the electric current is resembled by the heat flow; the voltage difference is resembled by the temperature difference; and the electric resistance is resembled by the thermal resistance. However, unlike the electric current, the heat flow is categorized in terms of the mode by which heat transfers in the medium. That is whether it is conductive or convective. Hence, two types of thermal resistances can be identified: conductive thermal resistance and convective thermal resistance.

5.2.1   Conductive thermal resistance

Based on the analogy between Fourier's law and Ohm's law we derive here the conductive thermal resistance of a plane wall and a cylindrical pipe.

5.2.1.1   Plane wall
Consider a one-dimensional steady-state heat flow in a homogeneous plane wall with surface area A, length L, and a constant thermal conductivity, λ, Figure 5.1. The wall surfaces are kept at constant temperatures, T_1 and T_2. The boundary value problem of this case can be described as

$$\frac{d^2 T}{dx^2} = 0 \tag{5.3}$$

$$T(0) = T_1, \quad T(L) = T_2$$

Integrating Eq. (5.3) twice, and imposing the boundary conditions, yields

$$T = T_1 + (T_2 - T_1)\frac{x}{L} \tag{5.4}$$

Fourier's law for heat conduction in a homogeneous one-dimensional domain is

$$q = -\lambda A \frac{dT}{dx} \tag{5.5}$$

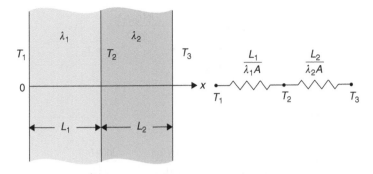

Figure 5.2   A composite plane wall and its thermal circuit.

Substituting Eq. (5.4) into Eq. (5.5), gives

$$q = -\frac{\lambda A}{L}(T_2 - T_1) \tag{5.6}$$

This equation can be rearranged to give

$$q = \frac{T_1 - T_2}{L/\lambda A} = \frac{T_1 - T_2}{R} = \frac{\text{temperature difference}}{\text{thermal resistance}} \tag{5.7}$$

where R is the *coefficient of thermal resistance*. This relationship is similar to Ohm's law, Eq. (5.1), and thus, as mentioned earlier, heat flow in a one-dimensional domain can be described using the analogy with an electrical circuit, as shown on the right-hand side sketch of Figure 5.1.

This principle can readily be extended to a composite system. Consider a two-layer composite plane wall shown in Figure 5.2, with surface area A, layer thicknesses L_1 and L_2, and heat conductivities λ_1 and λ_2. The outer left, inner and outer right planes are held at constant temperatures, T_1, T_2 and T_3 respectively. In the steady-state, and as a result of energy conservation, the heat flow rate entering the left plane is equal to the heat flow rate leaving the right plane, i.e.

$$q = \frac{T_1 - T_2}{L_1/\lambda_1 A} \quad \text{and} \quad q = \frac{T_2 - T_3}{L_2/\lambda_2 A} \tag{5.8}$$

Eliminating T_2, gives

$$q = \frac{T_1 - T_3}{(L_1/\lambda_1 A) + (L_2/\lambda_2 A)} \tag{5.9}$$

This equation can also be written as

$$q = \frac{T_1 - T_3}{R_1 + R_2} \tag{5.10}$$

where R_1 and R_2 is the first and second layer thermal resistances in series, respectively. (Later in this chapter, we will discuss the thermal resistances in series and parallel configurations.) This concept can be generalized to a composite wall consisting of n layers, giving

$$q = \frac{T_1 - T_n}{(L_1/\lambda_1 A) + (L_2/\lambda_2 A) + \cdots + (L_n/\lambda_n A)}$$
$$= \frac{T_1 - T_n}{R_1 + R_2 + \cdots + R_n} \tag{5.11}$$

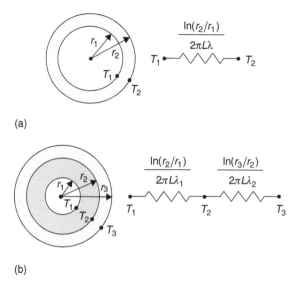

(a)

(b)

Figure 5.3   Cylindrical pipes with their thermal circuits. (a) single-layer. (a) two-layers.

5.2.1.2   *Cylindrical pipes*

Consider a single-layer cylindrical pipe of length L, inside radius r_1, outside radius r_2, and a constant thermal conductivity, λ. The inner and outer surface temperatures are held at T_1 and T_2, respectively, Figure 5.3(a). For this case, the steady-state heat conduction can be described using a cylindrical coordinate system as

$$\nabla^2 T = \frac{\partial^2 T}{\partial r^2} + \frac{1}{r}\frac{\partial T}{\partial r} + \frac{1}{r^2}\frac{\partial^2 T}{\partial \theta^2} + \frac{\partial^2 T}{\partial z^2} = 0 \tag{5.12}$$

where we assume that there is no heat supply in the system. For a long and axial symmetric pipe, the heat equation reduces to

$$\frac{\partial^2 T}{\partial r^2} + \frac{1}{r}\frac{\partial T}{\partial r} = \frac{1}{r}\frac{\partial}{\partial r}\left(r\frac{\partial T}{\partial r}\right) = 0 \tag{5.13}$$

The boundary conditions in this case are:

$$\begin{aligned} T(r_1) &= T_1 \\ T(r_2) &= T_2 \end{aligned} \tag{5.14}$$

Integrating Eq. (5.13) twice and applying the boundary conditions, Eq. (5.14), gives

$$T = T_1 + \frac{T_1 - T_2}{\ln(r_2/r_1)}\ln\left(\frac{r}{r_1}\right) \tag{5.15}$$

Using Fourier's law, the heat flow rate gives

$$q = \frac{2\pi L\lambda}{\ln(r_2/r_1)}(T_1 - T_2) \tag{5.16}$$

In analogy to the composite plane wall, heat flow rate in a composite cylindrical pipe, shown in Figure 5.3(b), can be described as

$$q = \frac{T_1 - T_2}{(1/2\pi L\lambda_1)\ln(r_2/r_1) + (1/2\pi L\lambda_2)\ln(r_3/r_2)} \tag{5.17}$$

Similarly, for an n-layered cylindrical pipe, we obtain

$$q = \frac{T_1 - T_n}{(1/2\pi L\lambda_1)\ln(r_2/r_1) + (1/2\pi L\lambda_2)\ln(r_3/r_2) + \cdots + (1/2\pi L\lambda_n)\ln(r_{n+1}/r_n)}$$

$$= \frac{T_1 - T_n}{R_1 + R_2 + \cdots + R_n} \tag{5.18}$$

Notice that heat flow in a cylindrical pipe is dependent on the ratio between the outer and inner diameters of the pipe, entailing that the pipe thickness is included. This has important implication in the design of cylindrical heat exchangers, such as the borehole heat exchangers.

5.2.2   Convective thermal resistance

In case where heat flow occurs as a result of heat advection processes, the convective thermal resistance must be included. For this, Newton's law of cooling can be utilized to describe heat convection at the boundary between a conductive surface and its surrounding convective fluid. Newton's law of cooling can be expressed as

$$q = hA\Delta T \tag{5.19}$$

where h (W/m$^2 \cdot$ K) is the convective heat transfer coefficient, A (m^2) is the area normal to the direction of flow, and ΔT (K) is the temperature difference between the solid surface and the fluid. In the following, we derive the conductive-convective thermal resistance for plane walls and cylindrical systems.

5.2.2.1   Plane wall
Consider a homogeneous plane wall of length L and having a constant thermal conductivity, λ. The wall surface temperatures are held at T_1 and T_2, and on one surface the wall is exposed to a fluid of T_{f1} temperature, and on another, the wall is exposed to a fluid of T_{f2} temperature, Figure 5.4. As indicated earlier, for a steady-state, the conservation of energy entails that heat flow rate entering the left plane is equal to the heat flow rate leaving the right plane. Equating Eq. (5.6) and Eq. (5.19), and rearranging gives

$$q = \frac{T_{f1} - T_1}{1/\overline{h}_1 A} \qquad q = \frac{T_1 - T_2}{L/\lambda A} \qquad q = \frac{T_{f2} - T_2}{1/\overline{h}_2 A} \tag{5.20}$$

where the over-bar on h denotes an average value of the convective heat transfer coefficient. Similar to the conductive thermal resistance, Eq. (5.7), $1/\overline{h}A$ can be described as the thermal resistance due to the convective boundary. Eliminating T_1 and T_2 from Eq. (5.20) gives

$$q = \frac{T_{f1} - T_{f2}}{(1/\overline{h}_1 A) + (L/\lambda A) + (1/\overline{h}_2 A)} \tag{5.21}$$

This equation can also be written as

$$q = \frac{T_{f1} - T_{f2}}{R_1 + R_2 + R_3} \tag{5.22}$$

where R_1, R_2 and R_3 are the thermal resistances in series, see the right-hand side of Figure 5.4. As for the conductive thermal resistance, this can readily be extended to a multilayer system.

5.2.2.2   Cylindrical pipe
Consider a cylindrical single-layer pipe with a constant thermal conductivity and uniform inner and outer surface temperatures, T_1 and T_2. The pipe is of length L, inside radius r_1, and outside

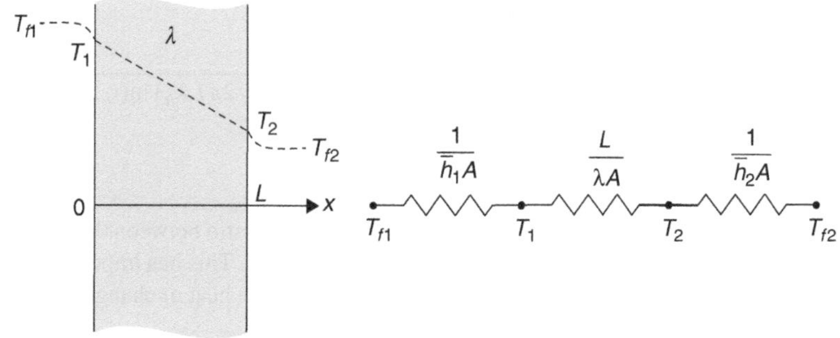

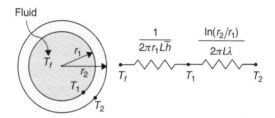

Figure 5.4   Conductive-convective heat flow in a plane wall and its thermal circuit.

Fluid

Figure 5.5   Single-layer pipe conducting a fluid.

radius r_2; and carries a fluid of temperature T_f, Figure 5.5. The conservation of energy entails that the rate of heat flow in the conductive part is equal to the convective part, i.e.

$$q = \frac{T_1 - T_2}{(1/2\pi L\lambda)\ln(r_2/r_1)} \quad \text{and} \quad q = A\bar{h}(T_f - T_1) \tag{5.23}$$

in which $A = 2\pi r_1 L$ is the inner surface area of the pipe. Eliminating T_1 gives

$$q = \frac{T_f - T_2}{1/2\pi r_1 L\bar{h} + (1/2\pi\lambda L)\ln(r_2/r_1)}$$
$$= \frac{T_f - T_2}{\sum R} \tag{5.24}$$

where

$$\sum R = \frac{1}{2\pi r_1 L\bar{h}} + \frac{\ln(r_2/r_1)}{2\pi\lambda L} \tag{5.25}$$

In the literature, see for example Pitts and Sisson (1998), the thermal resistance is defined by the overall heat transfer coefficient, which is the reciprocal of the sum of the thermal resistances $b = 1/(A\sum R)$, where A is usually denoted by the outer surface area $2\pi r_2 L$. Thus the reciprocal of the thermal resistance of Eq. (5.24) can be expressed as

$$b_o = \frac{1}{r_2/(r_1\bar{h}) + (r_2/\lambda)\ln(r_2/r_1)} \tag{5.26}$$

where the subscript o denotes that it is based on the outside surface area of the pipe.

For a composite cylindrical pipe exposed to fluid temperatures from the inside and the outside surfaces, T_{f1} and T_{f2}, respectively, the rate of heat flow is expressed as

$$q = \frac{T_{f1} - T_{f2}}{\sum R_{\text{conv}} + \sum R_{\text{cond}}} \tag{5.27}$$

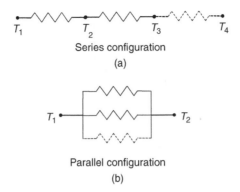

Figure 5.6   Series and parallel configurations.

in which R_{conv} denotes the convective thermal resistance, and R_{cond} denotes the conductive thermal resistance.

5.3   SERIES AND PARALLEL CONFIGURATIONS

Thermal interaction between components of a composite system occurs in different ways, depending on its configuration. Similar to electric circuits, there are mainly two configurations: series and parallel. Components connected in series are aligned along a single path, that is, all components carry the same heat flux, Figure 5.6(a). The *total thermal resistance*, R, of a series configuration is expressed as

$$R = R_1 + R_2 + \cdots + R_n \tag{5.28}$$

where R_1 represents the thermal resistance of component 1, etc. The total thermal resistance is also known as the *equivalent thermal resistance*.

Components connected in parallel are not in line with each other, and in this case the heat flux is divided over the parallel paths of its flow, Figure 5.6(b). The total thermal resistance of a parallel configuration is expressed as

$$\frac{1}{R} = \frac{1}{R_1} + \frac{1}{R_2} + \cdots + \frac{1}{R_n} \tag{5.29}$$

For only two parallel components, R_1 and R_2, for instance, the unreciprocated total resistance is

$$R = \frac{R_1 R_2}{R_1 + R_2} \tag{5.30}$$

Note that in the series configuration, the total resistance is always larger than the value of the individual resistances, whereas in the parallel configuration the total resistance is always less than the value of the smallest resistance in the system.

In practice, thermal systems constituting composite layers or constituents might be thermally connected in series and parallel. Figure 5.7 shows a typical series-parallel thermal circuit, where an equal heat flux flows through R_1, and R_4, connected in series, and beyond R_1, the flux branches into two resistors, R_2 and R_3, connected in parallel. The equivalent thermal resistance, in this case, can readily be expressed as

$$R = R_1 + \frac{1}{\dfrac{1}{R_2} + \dfrac{1}{R_3}} + R_4 \tag{5.31}$$

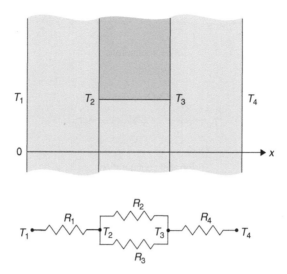

Figure 5.7   Series-parallel connection.

5.4   THERMAL RESISTANCE OF A BOREHOLE HEAT EXCHANGER

Heat transfer in a borehole heat exchanger and its surrounding soil mass is influenced by a combination of different factors, including the initial and boundary conditions; the thermal and physical parameters of the involved materials and geometry; and the thermal resistances between the BHE and the soil mass, and between the different components of the BHE. The initial and boundary conditions involve mainly long term initial soil temperature, short term variation of air temperature, and a transient fluid temperature entering pipe-in from the heat pump. The thermal and physical properties affecting heat transfer are mainly thermal conductivities, thermal capacities, fluid flow velocity and the physical dimensions of the geometry. The thermal resistance, and due to the slenderness of the borehole heat exchanger, affects the transfer of thermal energy between the different components of the BHE, and between the BHE and the surrounding soil mass. It is treated as one of the important parameters for designing borehole heat exchangers. Determination of the BHE thermal resistances can be done in three different methods: experimental; analytical or numerical; and thermal circuit.

5.4.1   *Experimental methods*

Basically, accurate BHE thermal resistance can best be determined experimentally. One of the most popular experiments in this field is the *thermal response test* (TRT), introduced by Morgensen (1983). This test is usually conducted before designing a shallow geothermal system. It is mainly an *in-situ* determination of the soil thermal conductivity and the borehole thermal resistance. In the TRT experiment, temperatures at the inlet and outlet of a BHE subjected to a constant heat flux are recorded at specific time intervals for around 72 hours.

In principle, *in-situ* determination of thermal parameters must be conducted iteratively by minimizing a system *objective function*, which describes the difference between measured data and results obtained from an analytical or a numerical model describing heat flow in the system. This iterative algorithm is known as *parameter identification*, which involves solving for *forward* and *inverse* models. The forward model is a mathematical expression which describes the physical system and capable of predicting the values of some response quantities, the temperature for instance, from given values of the model parameters. The inverse model is a mathematical optimization capable of inferring the values of the model parameters from measured values of the response quantities.

Despite the elegance of the experimental method, it has many drawbacks. The accuracy of this method depends on several factors, including the quality of measured data, the adequacy of the forward model to describe the physical processes, and the accuracy, efficiency and robustness of the optimization algorithm for solving the objective function. Measured data is usually contaminated with noise, which might be either physical, due to deviations from the theoretical model; or electronics, due to the inevitable noise in the electrical equipment.

The choice of the forward model is an important entity in the parameter identification procedure. Basically, two factors determine whether a forward model is suitable or not: accuracy and computational efficiency. Simplified models, such as the simplified version of the Kelvin's line-source model, is easy to handle, but in many cases far from being adequate for describing the involved thermal processes in the system. On the other hand, complicated models, such as those based on the finite element method or the finite difference method, are more versatile, but computationally inefficient for utilization in iterative algorithms. The iterative nature of the parameter identification procedure necessitates an appraisal between accuracy and computational efficiency.

The choice of the optimization technique (also referred to as minimization) is another important entity in the parameter identification procedure. Minimization is a mathematical technique especially devised for solving non-linear equations, the objective function in our case. Basically, with more unknown parameters, it becomes difficult to solve the system objective function, and above that, it becomes less likely that a unique solution will be obtained. Therefore, any attempt for conducting parameter identification of thermal systems must reduce the number of the unknown parameters to the minimum.

In geothermics, and in order to circumvent these difficulties, parameter identification of the soil thermal conductivity and the borehole thermal resistance is usually conducted based on the infinite line source model (ILS) and some simple least square curve fitting. In Chapter 8, we derive the ILS model, but here we state its application for the TRT. For sufficiently long time, Carslaw and Jaeger (1959) have shown that temperature in a soil mass subjected to heat flux per unit length, q, can be approximated as

$$T(r,t) \simeq T_0 + \frac{q}{4\pi\lambda}\left(\ln\frac{4\alpha t}{r^2} - \gamma\right) \tag{5.32}$$

in which T_0 is the initial temperature, λ is the soil thermal conductivity, $\alpha = \lambda/\rho c$ is the thermal diffusivity, and $\gamma = 0.5772\ldots$, is Euler's constant. At the borehole surface, $r = r_b$, the average borehole temperature can be described as (Eskilson 1987, and Wagner and Clauser 2005)

$$\overline{T} = T|_{r=r_b} + q\,R_b$$
$$= \underbrace{\frac{q}{4\pi\lambda}\ln(t)}_{a} + \underbrace{\left[qR_b + \frac{q}{4\pi\lambda}\left(\ln\frac{4\alpha}{r_b^2} - \gamma\right)\right] + T_0}_{b} \tag{5.33}$$

where R_b is the thermal resistance between the BHE fluid and the borehole wall, and $\overline{T} = (T_{in} - T_{out})/2$, is the average temperature between the inlet of pipe-in, T_{in}, and the outlet of pipe-out, T_{out}. Eq. (5.33) can be written as

$$\overline{T} = a\ln(t) + b \tag{5.34}$$

which represents a line in a semi-log scale with slope, a. Thus, the soil thermal conductivity can be calculated from the slope of the line as

$$\lambda = \frac{q}{4\pi a}$$
$$= \frac{q}{4\pi}\frac{\Delta\ln(t)}{\Delta\overline{T}} \tag{5.35}$$

Once the thermal conductivity is calculated, the borehole thermal resistance is commonly calculated using Eq. (5.33), though a system of at least two simultaneous equations must be solved using an optimization algorithm.

Apparently, this *in-situ* parameter identification technique is simple, but has a sever drawback in simulating the complex nature of heat flow inside the BHE and the soil mass. The model overlooks the geometry of the BHE, the vertical heat flow variation in the BHE and the soil mass, the heat convection in the BHE, and the transient nature of the TRT. Despite these shortcomings, this technique is commonly utilized in practice, see for example Eskilson (1987) and Yu *et al.* (2004). Schiavi (2009) studied this technique using the finite element method and concluded that, within certain limitations of operating time and number of unknown parameters, this technique is reasonably accurate.

TRT data is also studied using numerical methods. Wagner and Clauser (2005) used the finite difference method for devising the forward model; and Bozzoli *et al.* (2011) utilized the finite element method for formulating the forward model and the Gauss linearization method for the inverse model. The numerical methods are versatile and several parameters can be determined from the TRT data. However, and as indicated earlier, they are computationally inefficient for utilization in iterative schemes. Due to the disproportionate geometry of the shallow geothermal systems, and the inherited difficulties in simulating convective problems, extremely fine meshes or grids are necessary. This makes utilization of numerical models for parameter identification of geothermal systems less attractive, giving reasons for the utilization of the ILS model. However, this problem can be circumvented if use is made of a computationally efficient forward model capable of simulating the physical processes using a coarse mesh. Al-Khoury *et al.* (2005 and 2006) made a first attempt to reduce the finite element mesh size considerably by simulating a pseudo three-dimensional heat flow in a BHE using a line element. The derivation of this model will be elaborated in Chapter 12.

In-situ experiments are relatively expensive since they require heavy equipment and specialized software for parameter identification. As an alternative, but not a substitute, laboratory experiments are in use for parameter identification of, mainly, the grout thermal conductivity and the borehole thermal resistance. A known experiment for the determination of the thermal resistance is introduced by Paul N.D. in 1996 (reported by Lamarche *et al.* 2010). Paul measured the effective grout thermal resistance of a single U-tube BHE for three different shank space configurations, and introduced an empirical relationship of the form

$$R_g = \frac{1}{S\lambda_g} \tag{5.36}$$

where the subscript g denotes the grout, and S is a *shape factor* defined as

$$S = a(r_b/r_o)^b \tag{5.37}$$

in which r_b is the borehole radius, r_o is the outer radius of the pipe, and a and b are geometrical constants, depending on the shank space, determined from curve fitting of the experimental data. In this model the temperatures in pipe-in and pipe-out are assumed equal. This technique is practical but suffers from two main shortcomings: first, the forward model does not simulate the complex nature of the problem, and second, the calculated parameters are typically fit for the tested samples. Different samples of the same material with different geometry might produce different parameters.

5.4.2 *Analytical and numerical methods*

In practice, experimental determination of thermal resistances of the borehole heat exchanger, whether *in-situ* or laboratory, are not conducted on a daily basis, and it is common to rely on analytical or numerical methods, or on those based on simple thermal circuits. Several analytical expressions have been formulated to calculate the equivalent thermal resistance in shallow geothermal systems. Typically, the equivalent thermal resistance is calculated as the sum of two main resistances: the *borehole resistance* and the *internal resistance*, such that

$$R = R_{\text{BHE-soil}} + R_{\text{BHE}} \tag{5.38}$$

where the borehole resistance, $R_{\text{BHE-soil}}$, describes the thermal resistance between the grout and the soil mass, and the internal resistance, R_{BHE}, describes the thermal resistance between the inner components of the BHE. They are devised by solving the basic relationship

$$R = \frac{T_1 - T_2}{q} \tag{5.39}$$

in which T_1 and T_2 are the temperatures of any two components in thermal contact. The determination of these temperatures constitutes the main task in this line of work, and for this, several techniques have been employed. Bannet *et al.* (1987), (reported by Lamarche *et al.* 2010), utilized the *multipole expansion* to calculate the temperature in the system. The multipole expansion is a mathematical technique which is capable of representing fields, thermal field in our case, at distant points in terms of sources at the origin. The multipole expansion is expressed as a sum of truncated series, where only few terms are sufficient for convergence. Based on this method, Bannet *et al.* calculated the borehole thermal resistance up to a *j*th order. The first order expression is given by

$$R_{\text{BHE-soil}} = \frac{1}{4\pi\lambda_g} \left[\ln\left(\frac{\gamma_1 \gamma_2^{1+4\sigma}}{2(\gamma_2^4 - 1)^\sigma} \right) - \frac{\gamma_3^2 [1 - \{4\sigma/\gamma_2^4 - 1\}]^2}{1 + \gamma_3^2 (1 + [16\sigma/(\gamma_2^2 - 1\gamma_2^2)^2])} \right] \tag{5.40}$$

where

$$\gamma_1 = r_b/r_p, \quad \gamma_2 = r_b/s, \quad \gamma_3 = r_p/2s, \quad \sigma = (\lambda_g - \lambda_s)/(\lambda_g + \lambda_s) \tag{5.41}$$

in which r_b is the borehole radius, r_p is the pipe radius, and s is the shank spacing. The second term on the right-hand-side represents the correction of the thermal field due to the far distant pipes.

An easier formulation based on the line-source model is introduced by Hellstrom (1991), (reported in Lamarche *et al.* 2010). The borehole resistance derived from this model is given by

$$R_{\text{BHE-soil}} = \frac{1}{4\pi\lambda_g} \left[\ln\gamma_1 - \ln\gamma_2 + \sigma\ln\left(\frac{\gamma_2^4}{\gamma_2^4 - 1} \right) \right] \tag{5.42}$$

On the other hand, there are several expressions for calculating the thermal resistance that are derived on the basis of numerical methods. Sharqawy *et al.* (2009) have developed an expression for the borehole resistance based on 2D finite element simulations. They proposed:

$$R_{\text{BHE-soil}} = \frac{1}{2\pi\lambda_g} \left[\frac{-1.49}{\gamma_2} + 0.656\ln\gamma_1 + 0.436 \right] \tag{5.43}$$

The internal resistance has also been derived on the basis of the multipole expansion and the line-source model. Hellstrom (1991), (reported in Lamarche *et al.* 2010), suggested two expressions. One on the basis of the line-source model, of the form:

$$R_{\text{BHE}} = \frac{1}{\pi\lambda_g} \ln\left(\frac{2\gamma_1 (\gamma_2^2 + 1)^\sigma}{\gamma_2 (\gamma_2^2 - 1)^\sigma} \right) \tag{5.44}$$

and another, based on the multipole expansion, of the form:

$$R_{\text{BHE}} = \frac{1}{\pi\lambda_g} \ln\left(\frac{2\gamma_1(\gamma_2^2 + 1)^\sigma}{\gamma_2(\gamma_2^2 - 1)^\sigma} \right) - \frac{1}{\pi\lambda_g} \left[\frac{\gamma_3^2 [1 + \{4\sigma\gamma_2^2/(\gamma_2^4 - 1)\}]^2}{1 - \gamma_3^2 + \{8\sigma\gamma_2^2\gamma_3^2(\gamma_2^4 + 1)/(\gamma_2^4 - 1)^2\}} \right] \tag{5.45}$$

These expressions take into consideration pipe-in to pipe-out thermal interactions. Details of these expressions are best covered by Lamarche *et al.* 2010. They conducted extensive comparison of results obtained from these expressions with 2D and 3D finite element analyses. They found that the expressions based on the multipole expansion were best matching the finite element results.

Most algorithms for calculating the equivalent thermal resistance are based on thermal conduction processes between the borehole and a dry soil. The effect of groundwater flow is not taken into consideration. This effect has been studied by Sutton *et al.* (2003), and developed a ground resistance model for vertical borehole heat exchangers on the basis of the moving line heat source model (Carslaw and Jaeger, 1959). The borehole resistance derived from this model is given, for a transient case, by

$$R_{\text{BHE-soil}} = \frac{1}{4\pi\lambda_{\text{eff}}} \exp\left(\frac{\text{Pe}}{2}\cos\vartheta\right)\Gamma\left(0, \frac{1}{4\text{Fo}}; \frac{\text{Pe}^2}{16}\right) \tag{5.46}$$

and for a steady-state case, by

$$R_{\text{BHE-soil}} = \frac{1}{2\pi\lambda_{\text{eff}}} \exp\left(\frac{\text{Pe}}{2}\cos\vartheta\right)K_0\left(\frac{\text{Pe}}{2}\right) \tag{5.47}$$

in which λ_{eff} is the effective thermal conductivity of a fully saturated sand, Pe is the Peclet number, Fo is the Fourier number, and ϑ is the angular coordinate of the cylindrical system. Γ is a generalized incomplete gamma function, defined (Chaudhry, and Zubair, 1994) by

$$\Gamma(a, x; b) = \int_x^\infty \xi^{a-1} \exp\left(-\xi - \frac{b}{\xi}\right)d\xi \tag{5.48}$$

Apparently, these thermal resistances are not symmetric with respect to the angular coordinate ϑ, due to the presence of the groundwater flow.

Despite the rigor of the analytical expressions, they are derived assuming simplified geometry and boundary conditions. The theoretical models describing heat flow in the borehole heat exchanger and the soil mass are either based on the line source formulation, or based on the direct coupling between pipe-in and pipe-out, regardless of the grout and its interaction with the soil mass. The boundary conditions are in most cases simple, assuming mainly constant heat fluxes. These simplifications certainly degrade the rigor of the analytical expressions, and this might explain why up until now there is no consensus among geothermal engineers as to the most appropriate expression for describing the thermal resistances of shallow geothermal systems.

5.4.3   *Thermal circuit methods*

Following the shortcomings of the experimental and analytical or numerical methods, it might be more effective to utilize some simple formulations based on the analogy between Fourier's law and Ohm's law, outlined above. The usual practice is to lump the effects of the system constituents into an effective coefficient representing the sum of the local thermal resistances between thermally interacting components. Using the analogy with electric circuits, and depending on the governing heat equations, it is possible to assemble any combination of local thermal resistances. Thermal circuits of different configurations have been established to describe the effective thermal resistance in the borehole heat exchangers and their interaction with the surrounding soil mass. They basically fall into two configurations: Delta configuration, and Y configuration.

5.4.3.1   *Delta configuration*
In electrical engineering, this configuration is also referred to as the Pi configuration, Figure 5.8. In this configuration the thermal resistance between any two points is a series-parallel combination of the three resistors. As such, the equivalent resistance can be described as $R_{12}\|(R_1 + R_2)$, and can mathematically be expressed as

$$\frac{1}{R} = \frac{1}{R_{12}} + \frac{1}{R_1 + R_2} \tag{5.49}$$

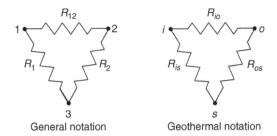

Figure 5.8   Delta configuration.

or

$$R = \frac{R_{12}(R_1 + R_2)}{R_1 + R_2 + R_{12}} \tag{5.50}$$

where R_1 is the thermal resistance between point 1 and point 3, R_2 is the thermal resistance between point 2 and point 3, and R_{12} is the thermal resistance between point 1 and point 2. Using geothermal notations, where we denote pipe-in by a subscript i, pipe-out by a subscript o, grout by a subscript g, and soil by a subscript s, Eq. (5.50) can be expressed as

$$R_b = \frac{R_{io}(R_{is} + R_{os})}{R_{is} + R_{os} + R_{io}} \tag{5.51}$$

in which R_b is the borehole equivalent thermal resistance, R_{is} is pipe-in – soil thermal resistance, and R_{os} is pipe-out – soil thermal resistance, and R_{io} is pipe-in – pipe-out thermal resistance. See the right-hand side sketch of Figure 5.8. These local thermal resistances can be determined either analytically or numerically, described in Sub-section 5.4.2, or using the technique of the next sub-section.

This configuration has been utilized in geothermal models of the form similar to Eskilson and Claesson (1988) model. This model assumes that there is a direct thermal interaction between pipe-in and soil, pipe-out and soil, and pipe-in and pipe-out. We discuss this model in Chapter 8.

5.4.3.2   *Y configuration*
In electrical engineering, this configuration is referred to as the Wye configuration or the T configuration, Figure 5.9. In this configuration, the thermal resistance between any two of the three components is a series combination of the three resistors. Using the analogy to Figure 5.7 and Eq. (5.31), the equivalent thermal resistance of this circuit can be visualized as a series connection of resistor R_3 to a parallel connection of resistors R_1 and R_2, $R_3 + (R_1 \| R_2)$. Mathematically, the equivalent resistance be expressed as

$$R = R_3 + R_{12}$$

$$= R_3 + \frac{1}{\dfrac{1}{R_1} + \dfrac{1}{R_2}} \tag{5.52}$$

in which R_{12} is the equivalent thermal resistance of the parallel circuit between R_1 and R_2. In geothermal notation, this thermal resistance can be written as

$$R_b = R_{\text{BHE-soil}} + R_{\text{BHE}}$$

$$= R_{gs} + \frac{1}{\dfrac{1}{R_{ig}} + \dfrac{1}{R_{og}}} \tag{5.53}$$

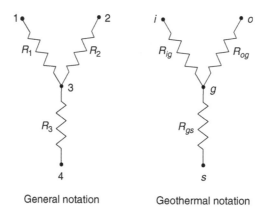

General notation   Geothermal notation

Figure 5.9   Y-configuration.

where R_b is the borehole thermal resistance R_{ig} is pipe-in – grout thermal resistance, R_{og} is pipe-out – grout thermal resistance, and R_{gs} is grout-soil thermal resistance. Here, the equivalent thermal resistance between the borehole and the soil can be visualized as a series-parallel config-uration, where pipe-in and pipe-out are parallel in their interactions with the grout, and the grout is in series with them and with the soil. This configuration has been utilized in heat transfer models of the form similar to Al-Khoury *et al.* model (2005 and 2006), see Chapter 4. This model entails that there is a thermal interaction between pipe-in and grout, pipe-out and grout, and between the grout and the soil. No direct thermal interaction between pipe-in and pipe-out exists.

Determination of R_{gs} of Eq. (5.53) can be done by lumping the areas of the inner pipes to create an equivalent area of the form:

$$A_{eq} = A_i + A_o \tag{5.54}$$

The radius of the equivalent area is hence

$$r_{eq} = \sqrt{r_i^2 + r_o^2} \tag{5.55}$$

where r_i is pipe-in radius and r_o is pipe-out radius, as shown in Figure 5.10. The grout-soil thermal resistance can then be determined using the simple thermal circuit of Figure 5.3(a) and Eq. (5.16), yielding

$$R_{gs} = \frac{\ln(r_g/r_{eq})}{2\pi L \lambda_g} \tag{5.56}$$

in which r_g is the radius of the grout (borehole) and λ_g is its thermal conductivity.

Determination of R_{ig} can be obtained in terms of the conductive-convective thermal resis-tance between the working fluid and the pipe material, and between the pipe material and the grout. Using the thermal circuit of Figure 5.5 and Eq. (5.25), R_{ig} (and similarly R_{og}) can be described as

$$R_{ig} = \frac{1}{2\pi r_{\text{inner}}\, L \bar{h}} + \frac{\ln(r_{\text{outer}}/r_{\text{inner}})}{2\pi L \lambda_p} \tag{5.57}$$

in which L is the pipe length, r_{inner} and r_{outer} are the inner and the outer radius of the pipe respectively, and λ_p the thermal conductivity of the pipe material, see Figure 5.11. Using

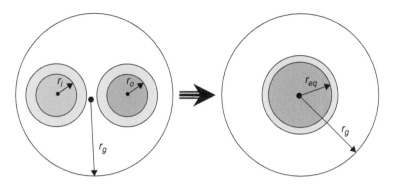

Figure 5.10   Lumping the inner pipes.

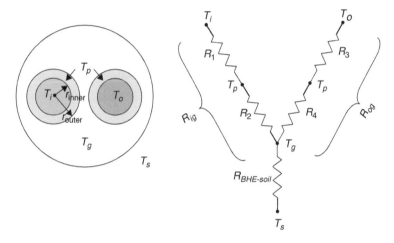

Figure 5.11   BHE-soil Y-configuration thermal circuit.

Eqs. (5.56) and (5.57), the total borehole thermal resistance, for a single U-tube BHE can then be described as

$$R_b = R_{gs} + R_{BHE}$$

$$= \frac{\ln(r_g/r_{eq})}{2\pi L\lambda_g} + \cfrac{1}{\left[\cfrac{1}{2\pi r_{inner} L\bar{h}} + \cfrac{\ln(r_{outer}/r_{inner})}{2\pi L\lambda_p}\right]_{ig}} + \cfrac{1}{\left[\cfrac{1}{2\pi r_{inner} L\bar{h}} + \cfrac{\ln(r_{outer}/r_{inner})}{2\pi L\lambda_p}\right]_{og}}$$

$$(5.58)$$

Using the representation in Eq. (5.26), the reciprocal of the equivalent thermal resistance of a borehole can be expressed as

$$b_o = \frac{1}{AR_b} \qquad (5.59)$$

Part II

Analytical and semi-analytical modeling

CHAPTER 6

Eigenfunction expansions and Fourier transforms

In this chapter, we introduce the eigenfunction expansions for solving Sturm-Liouville problems, typically related to heat conduction applications. The objective of this chapter is to establish a methodology for solving heat conduction in shallow geothermal systems that will be addressed in subsequent chapters. We first define the concept of initial and boundary value problems. Then introduce the notion of the eigenfunctions and eigenvalues. After that we give a brief description of Fourier series, Fourier transforms and Fourier integrals. Later, examples describing the solution of heat conduction in finite, semi-infinite and infinite media will be presented.

6.1 INTRODUCTION

Eigenfunction expansions are very important techniques for solving boundary value problems, since they yield, out of rather complicated functions, series of simple terms. Nearly all functions representing physical processes can be decomposed into a set of basis (characteristic) functions. Fourier transforms are one of the most powerful and commonly used eigenfunction expansions. Prior to Fourier's work on "The analytical theory of heat" in 1822, the heat equation could only be solved under simple boundary conditions, mainly under sinusoidal heat source conditions. Fourier showed that any arbitrary function can be expressed by summing over its basis functions. By this, he revolutionized the solution of initial and boundary value problems arising from Sturm-Liouville problems of heat conduction, fluid flow, transport, etc.

In the following, we introduce the basic concepts and applications of Fourier series, integrals and transforms in connection with heat equations. Full account of these techniques can be found in Kreyszig (1993); and Zill and Gullen (2000). But first, we define the concept of initial and boundary value problems, and give a short description of the Sturm-Liouville problem.

6.2 INITIAL AND BOUNDARY VALUE PROBLEMS

Initial and boundary value problems arise in many physical and engineering fields dealing with differential equations restrained by initial and boundary conditions. Solution of an initial value problem entails solving a differential equation for a specific initial condition. In the same way, solution of a boundary value problem entails solving a differential equation for some imposed boundary conditions. The basic difference between an initial value problem and a boundary value problem is in the way the constraints (conditions) are imposed. In the initial value problem, the condition is imposed at all values of the independent variable, and that value is at one point in time or space, usually in time, at $t = 0$ for instance. On the other hand, in the boundary value problem, the conditions are imposed at the boundaries of the independent variable, usually in space, for instance, at $x = 0$ and/or $x = L$ in a domain bounded by $0 \leq x \leq L$. A transient heat conduction problem in an infinite domain, for example, is an initial value problem, since there are no boundaries involved. Whereas, a steady-state heat conduction problem with two boundary conditions forms a boundary value problem. While a transient heat conduction problem with an initial condition and one or more boundary conditions forms an initial and boundary value problem. Heat flow in a typical geothermal system represents an initial and boundary value problem. In this book, the term boundary value problem will be utilized to describe initial and boundary value problems, unless it is necessary to specify.

A typical initial condition of a heat transfer problem is

$$T(x,0) = f(x) \tag{6.1}$$

which states that at time $t = 0$ the temperature distribution in the domain is $f(x)$.

Boundary conditions arising from boundary value problems of heat transfer applications are mainly of three types:

1. Boundary condition of the first kind (Dirichlet, essential). This boundary condition corresponds to a prescribed temperature at a point and/or along a boundary surface, and in general it is a function of time and space, such that

$$T(x_i, t) = T_i(x_i, t) \tag{6.2}$$

where T_i is a prescribed function of position x_i, and time t. In many cases, including shallow geothermal systems, the prescribed temperature is either a function of time, space or both. If the temperature at the boundary surface vanishes, i.e. $T = 0$, the boundary condition is said to be *homogeneous of the first kind*. On the other hand, if a constant temperature, T_0, is prescribed, it is said to be *non-homogeneous*. However, in many cases, such a boundary condition might be made homogeneous if the temperature in the system is calculated in excess of T_0.

2. Boundary condition of the second kind (Neumann, natural): This boundary condition corresponds to a prescribed heat flux, q_{ij}, at a point and/or along a boundary surface. It is described as the normal derivative of temperature, and it may be a function of time and space, of the form

$$\frac{\partial T(x_i, t)}{\partial n} = q_{ij}(x_i, t) \tag{6.3}$$

where $\partial/\partial n$ denotes differentiation along the outward normal to the boundary surface x_i. If the normal derivative of temperature at the boundary surface vanishes, $\partial T/\partial n = 0$, the boundary condition is said to be *homogeneous of the second kind*, otherwise it is *non-homogeneous*. An insulated boundary condition is a homogeneous of the second kind.

3. Boundary condition of the third kind (Cauchy, mixed): This boundary condition corresponds to a linear combination of the first and second kind boundary conditions prescribed on a point and/or along a boundary surface. Well known boundary condition of this kind with physical significance is the Newton's law of cooling, which states that heat transfer along a boundary surface is proportional to the temperature difference between the body and its surrounding medium, expressed as

$$-\lambda \frac{\partial T}{\partial n} = h(T - T_a) \tag{6.4}$$

in which λ is the thermal conductivity, T_a is the temperature of the surrounding medium, and h is the heat transfer coefficient, which depends on the physical properties of the surrounding medium. Physically, Eq. (6.4) describes heat dissipation by convection from the boundary surface of a body into the surrounding medium. If the surrounding temperature is zero, this boundary condition becomes homogeneous.

The above initial and three types of boundary conditions cover most of the applications of heat transfer in shallow geothermal systems.

6.3 STURM-LIOUVILLE PROBLEM

Sturm-Liouville problem is a boundary value problem of great importance to engineers and scientists, expressed as

$$\left[R(x)u'\right]' + [P(x) + kQ(x)]u = 0, \quad a \le x \le b$$

$$A_1 u(a) + B_1 u'(a) = 0 \tag{6.5}$$

$$A_2 u(b) + B_2 u'(b) = 0$$

where $u' = du/dx$, $R(x)$, $P(x)$, and $Q(x)$ are continuous in $[a, b]$, A_1, A_2, B_1 and B_2 are real and arbitrary constants, and k is any arbitrary constant. The power of this problem is that Bessel's, Legendre's and other equations of engineering significance, such as heat equations of physical systems, can be evaluated in the form of series. To illustrate this, we consider a special case of the Sturm-Liouville problem. A boundary value problem in standard form might be represented by a second-order linear differential equation of the form

$$u'' + P(x, k)u' + Q(x, k)u = F(x), \quad a \le x \le b \tag{6.6}$$

$$A_1 u(a) + B_1 u'(a) = C_1$$
$$\tag{6.7}$$
$$A_2 u(b) + B_2 u'(b) = C_2$$

Obviously, this problem is non-homogeneous. The solution of a non-homogeneous problem is unique if and only if the associated homogeneous problem has a unique solution. The associated homogeneous problem of Eqs. (6.6) and (6.7) is

$$u'' + P(x, k)u' + Q(x, k)u = 0, \quad a \le x \le b \tag{6.8}$$

$$A_1 u(a) + B_1 u'(a) = 0$$
$$\tag{6.9}$$
$$A_2 u(b) + B_2 u'(b) = 0$$

Let $u_1(x)$ and $u_2(x)$ be two linearly independent solutions of the differential equation, Eq. (6.8). Then, the solution of this equation might be expressed as

$$u(x) = D_1(x)u_1 + D_2(x)u_2 \tag{6.10}$$

Substituting Eq. (6.10) into the boundary conditions in Eq. (6.9), yields

$$\begin{bmatrix} A_1 u_1(a) + B_1 u_1'(a) & A_1 u_2(a) + B_1 u_2'(a) \\ A_2 u_1(b) + B_2 u_1'(b) & A_2 u_2(b) + B_2 u_2'(b) \end{bmatrix} \begin{bmatrix} D_1 \\ D_2 \end{bmatrix} = 0 \tag{6.11}$$

Clearly, $u(x) = 0$ is always a solution of the problem, but it is of no practical use. This solution is commonly termed, trivial. The non-trivial solution (i.e. solution not identically equal to zero) of Eq. (6.11) exists if and only if the determinant of the matrix equals zero, i.e.

$$\begin{vmatrix} A_1 u_1(a) + B_1 u_1'(a) & A_1 u_2(a) + B_1 u_2'(a) \\ A_2 u_1(b) + B_2 u_1'(b) & A_2 u_2(b) + B_2 u_2'(b) \end{vmatrix} = 0 \tag{6.12}$$

Zeros of the determinant (roots of the resulting polynomial function) may exist for certain values, $k_n, n = 1, 2, \ldots$, but not for other values of k. Those value of k_n for which the non-trivial solution does exist, are referred to as *eigenvalues*. The corresponding non-trivial solutions are called *eigenfunctions*. Problems dealing with eigenfunction expansions are called the eigenfunction problems.

Eigenfunctions of Sturm-Liouville problems have remarkable properties, but one of the most significant properties is *orthogonality*. Functions $u_1, u_2, \ldots$ defined on some interval $a \le x \le b$ are called orthogonal with respect to a weight function $w(x) > 0$ if

$$\int_a^b w(x)u_m(x)u_n(x)\, dx = 0, \quad m \ne n \tag{6.13}$$

For $w(x) = 1$ the functions $u_1, u_2, \ldots$ are still orthogonal. For example, $u_m(x) = \sin mx$, $m = 1, 2, \ldots$ form an orthogonal set on the interval $-\pi \le x \le \pi$ for $m \ne n$ (Kreyszig 1993) since

$$\int_{-\pi}^{\pi} u_m(x)u_n(x)\, dx = \int_{-\pi}^{\pi} \sin mx \sin nx\, dx$$

$$\tag{6.14}$$

$$= \frac{1}{2}\int_{-\pi}^{\pi} \cos(m - n)x\, dx - \frac{1}{2}\int_{-\pi}^{\pi} \cos(m + n)x\, dx = 0$$

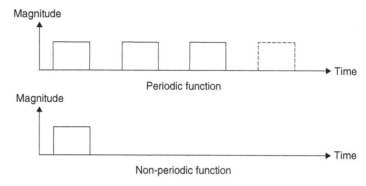

Figure 6.1   Periodic and non-periodic rectangular functions.

Another important property of the Sturm-Liouville functions is that for each eigenvalue there exists one and only one linearly independent eigenfunction. As a result of the linearity, the principle of *superposition* can be utilized. These two properties (orthogonality and superposition) make the eigenfunction expansions significant for developing numerical and semi-numerical algorithms. They yield functions in discrete forms, which are simple to manipulate and considerably suitable for computer implementations.

A wide class of functions can be represented by infinite series of some discrete basis functions. If $f(x)$ is piecewise smooth on $[a, b]$ and if $e_n(x)$ is the set of all corresponding eigenfunctions, it is possible to express $f(x)$ in a discrete form, such as

$$f(x) = \sum_{n=1}^{\infty} c_n e_n(x) \tag{6.15}$$

where

$$c_n = \frac{\int_a^b w(x) f(x) e_n(x) \, dx}{\int_a^b w(x) e_n^2(x) \, dx} \tag{6.16}$$

in which the function $w(x)$ is continuous and positive on $[a, b]$.

Functions of physical interest can be either periodic or non-periodic. A function is said to be periodic in time if its values are repeated throughout the domain at an interval of, for instance, P. For a time function $f(t)$, this implies that, $f(t + P) = f(t)$ where P is the period at which the function repeats itself. Similarly, a function may be periodic in space with a period length (wavelength), λ, implying $g(x + \lambda) = g(x)$. The trigonometric functions are one of the most important examples of periodic functions, which repeat themselves over intervals of length 2π.

A non-periodic function is a function that ends at a certain interval, and does not repeat its values. The rectangular step function is one of the known functions in applied physics and engineering. Figure 6.1 shows typical periodic and non-periodic rectangular functions.

Standard methods dealing with such functions are Fourier's methods. It is well established that Fourier series are suitable for decomposing periodic functions, and Fourier integrals are suitable for decomposing non-periodic functions. From a theoretical point of view, the Fourier series and integrals are particular cases of the Fourier transforms. These techniques will be discussed hereafter.

6.4   FOURIER SERIES

Fourier series are by far the most important eigenfunction expansions for many engineering applications. Fourier series of a function $f(x)$ constituting an orthogonal set of eigenfunctions of a Sturm-Liouville problem, $e_0, e_1, \ldots$, on an interval $a \leq x \leq b$, can be represented by

$$f(x) = \sum_{n=0}^{\infty} c_n e_n(x) = c_0 e_0(x) + c_1 e_1(x) + \cdots \tag{6.17}$$

in which $c_n, n = 0, 1, 2, \ldots$ are the Fourier constants, which can be determined using Eq. (6.16) or using the orthogonality property of the involved eigenfunctions. Different eigenfunctions can be utilized, depending on the specific problem. Hereafter, we introduce the most common Fourier series.

6.4.1 *Fourier trigonometric series*

Fourier trigonometric series are series of sine and cosine terms representing general periodic functions. In the temporal domain, the eigenfunctions are expressed as $\sin(\omega t)$, and $\cos(\omega t)$, where t refers to time and $\omega = 2\pi f = 2\pi/P$ refers to the radial frequency, with f being the frequency, and P the period. Similarly, in the spatial domain, the eigenfunctions are expressed as $\sin(kx)$ and $\cos(kx)$, where $k = 2\pi/\lambda$, representing the wavenumber with λ the wavelength. The first expansion takes on physical significance in the time-frequency relationships, and the second expansion takes on the space-wavenumber relationships.

Eigenfunctions of the Sturm-Liouville problem (Bronson and Gabriel 2006):

$$\frac{d^2 u(x)}{dx^2} + k^2 u(x) = 0; \quad 0 \le x \le L$$
$$u(0) = 0, \quad u(L) = 0$$

(6.18)

are the sine functions

$$e_n(x) = \sin \frac{n\pi x}{L}, \quad n = 1, 2, 3, \ldots$$

(6.19)

and their corresponding eigenvalues are $k = n\pi/L$. Similarly, the eigenfunctions of the Sturm-Liouville problem:

$$\frac{d^2 u(x)}{dx^2} + k^2 u(x) = 0; \quad 0 \le x \le L$$
$$\frac{du(0)}{dx} = 0, \quad \frac{du(L)}{dx} = 0$$

(6.20)

are the cosine functions

$$e_n(x) = \cos \frac{n\pi x}{L}, \quad n = 0, 1, 2, 3, \ldots$$

(6.21)

and their corresponding eigenvalues are $k = n\pi/L$. Substituting Eq. (6.19) and Eq. (6.21) into Eq. (6.17), and adding, leads to

$$f(x) = a_0 + \sum_{n=1}^{\infty} a_n \cos \frac{n\pi x}{L} + \sum_{n=1}^{\infty} b_n \sin \frac{n\pi x}{L}$$

(6.22)

where we replaced $c_n, n = 0, 1, 2, \ldots$ by $a_n, n = 0, 1, 2, \ldots$ and $b_n, n = 1, 2, \ldots$. Each term of the series in Eq. (6.22) has a period L. Hence, if the series converges, the summation results to the original $f(x)$ with period L. Practically speaking, there are many sufficient conditions for the Fourier series to converge to the right solution. One of the known conditions is whether the function is piecewise continuous, i.e. differentiable over its interval of interest. Even for a discontinuous function, the Fourier series converges to the right solution if the function has left and right derivatives at the discontinuity point, for instance x. Upon convergence, the sum of the Fourier terms produces $f(x)$, except at the discontinuity, the sum gives the average of the left- and right-hand limits of $f(x)$. In most practical cases, the Fourier series converges fairly quickly and the first few terms are adequate to produce accurate results.

The Fourier coefficients a_0, a_n, b_n $(n = 1, 2, 3, \ldots)$ can be determined using Eq. (6.16) with $w(x) = 1$, $a = 0$ and $b = L$, so that, for the sine series,

$$\int_a^b w(x) e_n^2(x)\, dx = \int_0^L \sin^2 \frac{n\pi x}{L} dx = \frac{L}{2}$$

(6.23)

Substituting this result into Eq. (6.16), gives

$$b_n = \frac{2}{L} \int_0^L f(x) \sin \frac{n\pi x}{L} dx \tag{6.24}$$

In the same way, the coefficients of the cosine series can be determined to give

$$a_0 = \frac{1}{L} \int_0^L f(x) dx$$

$$a_n = \frac{2}{L} \int_0^L f(x) \cos \frac{n\pi x}{L} dx \tag{6.25}$$

Alternatively, these coefficients can be determined by making use of the orthogonality property of the sine and cosine functions, expressed as

$$\int_0^L \sin \frac{m\pi x}{L} \sin \frac{n\pi x}{L} dx = \begin{cases} L/2 & n = m \neq 0 \\ 0 & n \neq m \end{cases}$$

$$\int_0^L \cos \frac{m\pi x}{L} \cos \frac{n\pi x}{L} dx = \begin{cases} L/2 & n = m \neq 0 \\ 0 & n \neq m \end{cases} \tag{6.26}$$

$$\int_0^L \sin \frac{m\pi x}{L} \cos \frac{n\pi x}{L} dx = 0 \qquad n \neq m$$

To determine a_0, we integrate both sides of Eq. (6.22) from 0 to L, yielding

$$a_0 = \frac{1}{L} \int_0^L f(x) dx \tag{6.27}$$

The coefficient of the cosine terms, a_n, can be determined by multiplying Eq. (6.22) by $\cos mx$, where m is any positive integer. This gives Eq. (6.25). b_n can be determined by multiplying Eq. (6.22) by $\sin mx$, to give Eq. (6.24).

Fourier series, Eq. (6.22), with its coefficients, Eqs. (6.24)–(6.25), is referred to as the *complete Fourier series*. This form can be simplified to satisfy particular symmetry conditions using the notion of even and odd functions. The cosine function is an even function. Fourier series of such a function is

$$f(x) = a_0 + \sum_{n=1}^{\infty} a_n \cos \frac{n\pi x}{L}, \qquad 0 < x < L \tag{6.28}$$

This series is called the *Fourier cosine series*. The sine function, on the other hand, is an odd function, and the Fourier series of such a function is

$$f(x) = \sum_{n=1}^{\infty} b_n \sin \frac{n\pi x}{L}, \qquad 0 < x < L \tag{6.29}$$

This series is called the *Fourier sine series*.

6.4.2   *Complex Fourier series*

Using Euler's formula:

$$e^{ix} = \cos(x) + i \sin(x)$$

$$e^{-ix} = \cos(x) - i \sin(x) \tag{6.30}$$

the Fourier series in Eq. (6.22) can be expressed in a compact form as

$$f(x) = \sum_{n=-\infty}^{\infty} c_n e^{inx} \tag{6.31}$$

in which $i = \sqrt{-1}$ is the imaginary number. Multiplying both sides of Eq. (6.31) by e^{-imx} and integrating over $[0, 2\pi]$ gives

$$\int_0^{2\pi} f(x)e^{-imx}dx = \sum_{n=-\infty}^{\infty} c_n \int_0^{2\pi} e^{inx} e^{-imx}dx \tag{6.32}$$

Making use of the orthogonality property, all terms on the right-hand side vanish except for $n = m$, where the integrand on the right-hand side is $e^0 = 1$, and the integral equals 2π. This readily gives

$$c_n = \frac{1}{2\pi} \int_0^{2\pi} f(x)e^{-inx}\,dx, \qquad n = 0, \pm 1, \pm 2, \ldots \tag{6.33}$$

This form of Fourier series is called the *complex form*, also known as the *exponential form*. The Fourier complex coefficient c_n is related to the trigonometric coefficients a_n and b_n, Eqs. (6.24)–(6.25), by

$$\begin{aligned} a_n &= c_n + c_{-n} \\ b_n &= i(c_n - c_{-n}) \end{aligned}; \qquad n = 0, 1, 2, \ldots \tag{6.34}$$

The complex form, Eq. (6.31), for a function periodic in $[0, L]$ is

$$f(x) = \sum_{n=-\infty}^{\infty} c_n e^{2in\pi x/L}$$

$$c_n = \frac{1}{L} \int_0^L f(x)e^{-2in\pi x/L}dx \tag{6.35}$$

6.4.3 Fourier-Bessel series

Trigonometric Fourier series is an important method for expanding a function in terms of an orthogonal set of trigonometric functions. However, the eigenfunction expansion is not limited to orthogonal sets of trigonometric functions. One important expansion procedure is the Fourier-Bessel series. It arises from solving a Sturm-Liouville problem in cylindrical coordinates, of the form

$$\frac{d^2u}{dr^2} + \frac{1}{r}\frac{du}{dr} + \left(k^2 - \frac{v^2}{r^2}\right)u = 0, \quad a \le r \le b$$

$$\alpha_1 u(a) + \beta_1 u'(a) = 0, \quad \alpha_1^2 + \beta_1^2 > 0 \tag{6.36}$$

$$\alpha_2 u(b) + \beta_2 u'(b) = 0, \quad \alpha_2^2 + \beta_2^2 > 0$$

This equation is called *Bessel's differential equation* of order v with a particular solution based on the Bessel function of order v of the first kind, $J_v(kr)$, and the second kind, $Y_v(kr)$.

As for the trigonometric functions, the Bessel functions, $J_v(k_n r)$, $n = 1, 2, 3, \ldots$, for a non-negative integer v, form an orthogonal set on the interval $[0, R]$ with respect to a weight function $w(r) = r$, that is

$$\int_{a=0}^{b=R} J_v(k_m r)J_v(k_n r)r\,dr = 0, \quad m \ne n \tag{6.37}$$

Thus the Bessel function can be expanded in terms of any set of functions that is orthogonal. Based on Eq. (6.15) and Eq. (6.16), a *Fourier-Bessel series* can be defined as

$$f(r) = \sum_{n=1}^{\infty} c_n J_v(k_n r) \tag{6.38}$$

with series coefficients, c_n, for a periodic function $[0, R]$, given by

$$c_n = \frac{\int_0^R J_v(k_n r) f(r) r \, dr}{\int_0^R J_v^2(k_n r) r \, dr} \tag{6.39}$$

Eigenfunctions of the Sturm-Liouville problem, Eq. (6.36), are determined based on the boundary conditions. If in Eq. (6.36), $\alpha_1 = \beta_1 = \beta_2 = 0$, the eigenfunction is

$$e_n(r) = J_v(k_n r), \quad n = 1, 2, 3, \cdots \tag{6.40}$$

and the series coefficient is

$$c_n = \frac{2}{R^2 J_{v+1}^2(k_n R)} \int_0^R J_v(k_n r) f(r) r \, dr \tag{6.41}$$

The eigenvalues, k_n, of this case is determined by solving for the roots of $J_v(k_n R) = 0$. If $\alpha_1 = \beta_1 = 0$, $\alpha_2 = h \geq 0$, the eigenfunction is

$$e_n(r) = h J_v(k_n r) + k_n R J_v'(k_n r), \quad n = 1, 2, 3, \cdots \tag{6.42}$$

and the series coefficient is

$$c_n = \frac{2k_n^2}{(k_n^2 R^2 - v^2 + h^2) J_v^2(k_n R)} \int_0^R J_v(k_n r) f(r) r \, dr \tag{6.43}$$

The eigenvalues, k_n, of this case is determined by solving for the roots of $h J_v(k_n R) + k_n R J_v'(k_n R) = 0$.

If, however, $\alpha_1 = \beta_1 = \alpha_2 = 0$, the eigenfunction is

$$e_n(r) = J_0'(k_n r), \quad n = 1, 2, 3, \cdots \tag{6.44}$$

the Fourier-Bessel series reads

$$f(r) = c_1 + \sum_{n=2}^{\infty} c_n J_0(k_n r) \tag{6.45}$$

and the coefficients are

$$c_1 = \frac{2}{R^2} \int_0^R f(r) r \, dr$$

$$c_n = \frac{2}{R^2 J_0^2(k_n R)} \int_0^R J_0(k_n r) f(r) r \, dr, \quad n = 2, 3, \cdots \tag{6.46}$$

In this case, the eigenvalues, k_n, are obtained from the roots of

$$J_0'(k_n R) = J_1(k_n R) = 0 \tag{6.47}$$

As for the trigonometric Fourier series, the Fourier-Bessel series is convergent if $f(r)$ and its derivatives $f'(r)$ are piecewise continuous on the interval $[0, R]$ at any point where $f(r)$ is continuous. At a point where there is a discontinuity, the series converges to the average between the left- and right-hand sides, $[f(x^-) + f(x^+)]/2$.

The Fourier-Bessel series is especially important for solving heat flow problems in shallow geothermal systems, since in such systems, axial-symmetry is usually utilized to describe the geometry of the borehole heat exchanger and its interaction with the soil mass. In Chapter 9 we present the application of this series for modeling heat conduction in the soil mass.

6.5 FOURIER INTEGRAL

In the previous section we used Fourier series to represent a periodic function defined over a finite interval $[0, L]$. If, however, the function is non-periodic and defined over either an infinite interval, $(-\infty, \infty)$, or a semi-infinite interval, $[0, \infty)$, we ought to resort to one of the integral transforms. Here, we focus on the Fourier integrals, one of the most important techniques for solving partial differential equations.

Fourier integral of a function $f(x)$ defined over the interval $(-\infty, \infty)$ can be obtained simply by extending the period of a periodic function to infinity and integrating its Fourier series. Fourier integrals can be derived in different ways, but here we follow the derivation given by Kreyszig (1993) to extend the method of Fourier series to Fourier integrals.

Consider a periodic trigonometric function $f_L(x)$ of period $2L$ with its Fourier series

$$f_L(x) = a_0 + \sum_{n=1}^{\infty}(a_n\cos\alpha_n x + b_n\sin\alpha_n x), \quad \alpha_n = \frac{n\pi}{L} \tag{6.48}$$

The Fourier coefficients of this series are

$$a_0 = \frac{1}{2L}\int_{-L}^{L} f_L(x)\,dx$$

$$a_n = \frac{1}{L}\int_{-L}^{L} f_L(x)\cos\alpha_n x\,dx, \quad n = 1, 2, \dots \tag{6.49}$$

$$b_n = \frac{1}{L}\int_{-L}^{L} f_L(x)\sin\alpha_n x\,dx, \quad n = 1, 2, \dots$$

If we insert a_0, a_n and b_n into Eq. (6.48) and denote the variable of integration by v instead of x, the Fourier series, Eq. (6.48), becomes

$$f_L(x) = \frac{1}{2L}\int_{-L}^{L} f_L(v)\,dv + \frac{1}{L}\sum_{n=1}^{\infty}\left[\cos\alpha_n x \int_{-L}^{L} f_L(v)\cos\alpha_n v\,dv \right.$$
$$\left. + \sin\alpha_n x \int_{-L}^{L} f_L(v)\sin\alpha_n v\,dv\right] \tag{6.50}$$

To convert the summation over infinity in Eq. (6.50) to integration over infinity, we use this relationship:

$$\Delta\alpha = \alpha_{n+1} - \alpha_n = \frac{(n+1)\pi}{L} - \frac{n\pi}{L} = \frac{\pi}{L} \tag{6.51}$$

which entails

$$\frac{1}{L} = \frac{\Delta\alpha}{\pi} \tag{6.52}$$

Substituting Eq. (6.52) into Eq. (6.50) yields

$$f_L(x) = \frac{1}{2L}\int_{-L}^{L} f_L(v)dv + \frac{1}{\pi}\sum_{n=1}^{\infty}\left[(\cos\alpha_n x)\Delta\alpha \int_{-L}^{L} f_L(v)\cos\alpha_n v\,dv\right.$$
$$\left. + (\sin\alpha_n x)\Delta\alpha \int_{-L}^{L} f_L(v)\sin\alpha_n v\,dv\right] \tag{6.53}$$

We now let $L \to \infty$, and assume the resulting non-periodic function:

$$f(x) = \lim_{L \to \infty} f_L(x) \tag{6.54}$$

is absolutely integrable, i.e. the integral

$$\int_{-\infty}^{\infty} |f(x)| dx \tag{6.55}$$

does exist. Then $1/L \to 0$, and the first term on the right-hand side of Eq. (6.53) approaches zero. Also, as $\Delta\alpha \to 0$, it becomes $d\alpha$, and the summation in Eq. (6.53) becomes an integral from 0 to ∞, namely

$$f(x) = \int_0^{\infty} [A(\alpha)\cos\alpha x + B(\alpha)\sin\alpha x]\, d\alpha \tag{6.56}$$

where

$$A(\alpha) = \frac{1}{\pi} \int_{-\infty}^{\infty} f(v)\cos\alpha v\, dv$$
$$B(\alpha) = \frac{1}{\pi} \int_{-\infty}^{\infty} f(v)\sin\alpha v\, dv \tag{6.57}$$

The similarity between Eqs. (6.56)–(6.57) and the corresponding Fourier series, Eqs. (6.48)–(6.49), is apparent. When $f(x)$ is an even function on the interval $(-\infty, \infty)$, the product $f(x)\cos\alpha x$ is also an even function. The Fourier integral of an even function can then be obtained by letting $B(\alpha) = 0$ in Eq. (6.56) to give

$$f(x) = \frac{2}{\pi} \int_0^{\infty} A(\alpha)\cos\alpha x\, d\alpha$$
$$A(\alpha) = \int_0^{\infty} f(v)\cos\alpha v\, dv \tag{6.58}$$

This pair of Fourier integrals is called the *Cosine integral* or the *Fourier cosine transform*. Similarly, when $f(x)$ is an odd function on the interval $(-\infty, \infty)$, then the product $f(x)\sin\alpha x$ is also an odd function. The Fourier integral of an odd function can then be obtained by letting $A(\alpha) = 0$ in Eq. (6.56) to get

$$f(x) = \frac{2}{\pi} \int_0^{\infty} B(\alpha)\sin\alpha x\, d\alpha$$
$$B(\alpha) = \int_0^{\infty} f(v)\sin\alpha v\, dv \tag{6.59}$$

This pair of Fourier integrals is called the *Sine integral* or the *Fourier sine transform*. We note that the pairs in Eq. (6.58) and Eq. (6.59) are symmetric.

In analogy to the complex form of the Fourier series, Eq. (6.31), the Fourier integral, Eq. (6.56), has an equivalent complex form, also known as exponential form, expressed as

$$f(x) = \frac{1}{2\pi} \int_{-\infty}^{\infty} C(\alpha)e^{-i\alpha x}\, d\alpha$$
$$C(\alpha) = \int_{-\infty}^{\infty} f(v)e^{i\alpha v}\, dv \tag{6.60}$$

It is important to note that the Fourier integrals imply replacing series summation over integer multiplies of $\alpha_n = n\pi/L$ by integrals of transcendental functions over infinity. One way to solve this kind of integrations is by using the *contour integration*. In this, the contour is chosen so that it follows the part of the complex plane and encloses the integration poles where singularities are located. Another way to solve such an integral is to utilize numerical integration procedures. Both kind of integrations, as will be shown in Chapter 7, are awkward due to the difficulty in locating the involved poles and branch points. Later, we shall see the necessity to overcome this problem when dealing with complicated infinite and semi-infinite geometry.

6.6   FOURIER TRANSFORM

Fourier transform is basically a Fourier integral, but in a complex form. The concept of the Fourier transform entails transforming a given function of a certain variable to a new function of a different variable, and appears in the form of complex integral. It is found that for physical problems, it is simpler to transform the boundary value problem, defined by a partial differential equation $F(x)$, to another function, $\hat{F}(\alpha)$ for instance, by multiplying $F(x)$ by a new function $K(\alpha, x)$, called *kernel*, and integrating with respect to x from 0 to ∞, such that

$$\hat{F}(\alpha) = \int_0^\infty F(x)K(\alpha, x)\,dx \tag{6.61}$$

This function is the *integral transform* of $F(x)$ by the forward transform kernel $K(\alpha, x)$. The transformed function must be relatively easy to solve in order for the transformation to be useful. After solving the transformed function, the original function must then be retrieved by reversing the logic, such that

$$F(x) = \int_a^b \hat{F}(\alpha)H(\alpha, x)\,d\alpha \tag{6.62}$$

This function is the *inverse transform* of $\hat{F}(\alpha)$ by the inverse transform kernel $H(\alpha, x)$. Depending on which method to be applied, the kernel might have different forms. Among the most commonly used kernels are the Laplace kernel, e^{-sx}, the Mellin transform kernel, $x^{\alpha-1}$, and the Fourier kernel. The Laplace transform is considered as the most important integral transform in engineering. This will be covered in Chapter 7. The next in order of importance is the Fourier transform. Here, we are not aiming at the formal introduction of integral transforms, but shall confine our attention to the Fourier kernels and transforms which are very convenient and sufficiently wide to embrace the initial and boundary value problems that we are dealing. For general in-depth accounts on the fundamental issues and applications of the Fourier transform method, the reader is referred to Sneddon (1951).

In the special case in which the inverse transform kernel, $H(\alpha, x)$ is equal to the forward transform kernel, $K(\alpha, x)$, the kernel is referred to as the *Fourier kernel*. In this case, the inverse Fourier transform, Eq. (6.62), becomes

$$F(x) = \int_0^\infty \hat{F}(\alpha)K(\alpha, x)\,d\alpha \tag{6.63}$$

so that the relationship between the function and its integral transform is symmetrical. This symmetry constitutes an important advantage of the Fourier transform. We will be using this pair of transformation, and utilize this notation:

$$F(t) \Leftrightarrow \hat{F}(\omega)$$

$$F(x) \Leftrightarrow \hat{F}(k) \tag{6.64}$$

where the hat ($\hat{\ }$) denotes the transformed function, and the symbol $\Leftrightarrow$ means "can be transformed into" and the transform can go in either directions.

Fourier transform is a universally accepted tool of modern analysis. Yet, there are no common expressions of the Fourier integrals and their inversion formula. A frequently used Fourier transform pairs are (Zill and Cullen 2000):

Fourier complex transform, infinite domain:

Forward:

$$\mathcal{F}[F(t)] = \hat{F}(\omega) = \int_{-\infty}^{\infty} F(t)e^{-i\omega t}\, dt \tag{6.65}$$

Inverse:

$$\mathcal{F}^{-1}[\hat{F}(\omega)] = F(t) = \frac{1}{2\pi} \int_{-\infty}^{\infty} \hat{F}(\omega)e^{i\omega t}\, d\omega \tag{6.66}$$

Fourier sine transform, semi-infinite domain:

Forward:

$$\mathcal{F}_s[F(t)] = \hat{F}(\omega) = \int_{0}^{\infty} F(t)\sin(\omega t)\, dt \tag{6.67}$$

Inverse:

$$\mathcal{F}_s^{-1}[\hat{F}(\omega)] = F(t) = \frac{2}{\pi} \int_{0}^{\infty} \hat{F}(\omega)\sin(\omega t)\, d\omega \tag{6.68}$$

Fourier cosine transform, semi-infinite domain:

Forward:

$$\mathcal{F}_c[F(t)] = \hat{F}(\omega) = \int_{0}^{\infty} F(t)\cos(\omega t)\, dt \tag{6.69}$$

Inverse:

$$\mathcal{F}_c^{-1}[\hat{F}(\omega)] = F(t) = \frac{2}{\pi} \int_{0}^{\infty} \hat{F}(\omega)\cos(\omega t)\, d\omega \tag{6.70}$$

The similarity between the Fourier transforms and the corresponding Fourier series presented in Section 6.4 is apparent. For example, if $f(t)$ is an odd function in $[-P/2, P/2]$, then

$$F(t) = \sum_{n=1}^{\infty} b_n \sin n\omega t \tag{6.71}$$

with

$$b_n = \frac{2}{P} \int_{-P/2}^{P/2} F(t)\sin n\omega t\, dt \tag{6.72}$$

This equation can be written as

$$b_n = \frac{2}{P}\hat{F}(\omega); \quad \hat{F}(\omega) = \int_{-P/2}^{P/2} F(t)\sin n\omega t\, dt \tag{6.73}$$

Then Eq. (6.71) can be expressed as

$$F(t) = \frac{2}{P} \sum_{n=1}^{\infty} \hat{F}(\omega)\sin n\omega t \tag{6.74}$$

$$= \mathcal{F}^{-1}\{\hat{F}(\omega)\}$$

Since our goal is to apply the Fourier transforms to boundary value problems dealing with partial differential equations, it is important to examine the transform of derivatives. If $F(t)$ is continuous and absolutely integrable over the interval $(-\infty, \infty)$ and $F'(t)$ is piecewise continuous over every finite interval, then the transform can be expressed as

$$\mathcal{F}[F'(t)] = \hat{F}'(\omega) = \int_{-\infty}^{\infty} F'(t)e^{i\omega t}\, dt \tag{6.75}$$

Integration by parts gives

$$\hat{F}'(\omega) = F(t)e^{i\omega t}\Big|_{-\infty}^{\infty} - i\omega \int_{-\infty}^{\infty} F(t)e^{i\omega t}\, dt$$

$$= -i\omega \int_{-\infty}^{\infty} F(t)e^{i\omega t}\, dt \qquad\qquad (6.76)$$

$$= -i\omega\, \hat{F}(\omega)$$

Similarly, if $F(t)$ and $F'(t)$ are continuous and absolutely integrable over the interval $(-\infty, \infty)$, and $F''(t)$ is piecewise continuous over every finite interval, then by applying integration by parts twice, the Fourier transform of this second derivative can be expressed as

$$\hat{F}''(\omega) = -\omega^2 \hat{F}(\omega) \qquad\qquad (6.77)$$

It is important to note that the sine and cosine transforms are not suitable for transforming the derivatives of odd orders. This can be easily handled by the complex transform, which explains the suitability of using the complex transform for solving wide range of boundary value problems.

As for a trigonometric Fourier series defined over a finite interval and has a Fourier transform over an infinite interval as a counterpart, the Fourier–Bessel series has a counterpart over an infinite interval, namely the *Hankel transform*. The Hankel transform is an integral transform, known also as the *Fourier-Bessel transform*. A fuller account on Hankel transform can be found in Ozisik (1968). Here, we introduce frequently used Hankel transform pairs in finite and semi-infinite regions.

Finite region: The Hankel transform pair of function $F(r)$ in the region $[a, b]$ is given by
Forward:

$$\mathcal{H}[F(r)] = \hat{F}(k_m) = \int_{r'=a}^{b} r' \cdot K_v(k_m, r') \cdot F(r')\, dr' \qquad\qquad (6.78)$$

Inverse:

$$F(r) = \sum_{m=1}^{\infty} K_v(k_m, r) \cdot \hat{F}(k_m) \qquad\qquad (6.79)$$

in which the kernels $K_v(k_m, r)$ and the eigenvalues k_m depend on the type of the boundary conditions and the space variable r, i.e. $0 \le r \le b$ or $a \le r \le b$.

Semi-infinite: The Hankel transform pair of function $F(r)$ in the region $[0, \infty)$ is given by
Forward:

$$\mathcal{H}[F(r)] = \hat{F}(k) = \int_{r'=0}^{\infty} r' \cdot J_v(k\, r') \cdot F(r')\, dr' \qquad\qquad (6.80)$$

Inverse:

$$F(r) = \int_{k=0}^{\infty} k \cdot J_v(k\, r) \cdot \hat{F}(k)\, dk \qquad\qquad (6.81)$$

where J_v is the Bessel function of the first kind of order v.

6.7 DISCRETE FOURIER TRANSFORM

So far, we discussed continuous Fourier transforms, where the form of the transformed functions is assumed to be known analytically over the whole interval. This technique is powerful, but has the drawback that it is not suitable for practical situations, where the function is in effect a signal made of discrete points (samples). In this case, we rely on the discrete Fourier transform (DFT) to solve the problem. Here we give the definition and the basic properties of the discrete Fourier transform. A fuller account can be found in Doyle (1997), Brigham (1988) and Lee (2009).

The sine, cosine and exponential functions have the remarkable property that they are also orthogonal over a series of discrete, equally spaced points over the orthogonality interval. In analogy to the continuum transform, such a property allows us to decompose discrete functions given by discrete data into trigonometric or exponential series. For a given data function, $f(x)$ of $2N$ samples, the discrete form of the trigonometric Fourier series can be expressed as

$$F(x_n) = \frac{a_0}{2} + \sum_{k=1}^{N-1} a_k \cos \frac{2n\pi k}{N} + \sum_{k=1}^{N-1} b_k \sin \frac{2n\pi k}{N} \tag{6.82}$$

The orthogonality property of these sines and cosines gives

$$\sum_{k=0}^{N} \sin \frac{2\pi}{N+1} nk \sin \frac{2\pi}{N+1} mk = \begin{cases} (N+1)/2 & n = m \neq 0 \\ 0 & n \neq m \end{cases}$$

$$\sum_{k=0}^{N} \cos \frac{2\pi}{N+1} nk \cos \frac{2\pi}{N+1} mk = \begin{cases} (N+1)/2 & n = m \neq 0 \\ 0 & n \neq m \end{cases} \tag{6.83}$$

$$\sum_{k=0}^{N} \sin \frac{2\pi}{N+1} nk \cos \frac{2\pi}{N+1} mk = 0$$

Accordingly, and using Eq. (6.16), the Fourier coefficients can be determined as

$$a_k = \frac{1}{N} \sum_{n=0}^{N-1} F(x_n) \cos \frac{2n\pi k}{N}, \quad k = 0, 1, 2, \ldots, N-1 \tag{6.84}$$

$$b_k = \frac{1}{N} \sum_{n=0}^{N-1} F(x_n) \sin \frac{2n\pi k}{N}, \quad k = 1, 2, 3, \ldots, N-1 \tag{6.85}$$

The forgoing trigonometric series can also be represented in a complex form, and the transform pair of a function $f(x)$ defined at discrete samples x_n can be expressed as

$$F(x_n) = \sum_{k=0}^{N-1} \hat{F}(x_k) e^{i2n\pi k/N} \tag{6.86}$$

$$\hat{F}(x_k) = \frac{1}{N} \sum_{n=0}^{N-1} F(x_n) e^{-i2n\pi k/N} \tag{6.87}$$

6.8 FAST FOURIER TRANSFORM

The essence of the discrete Fourier transform is that it is suitable for utilization in computational methods. The Fast Fourier Transform (FFT) computes the DFT in an exceptionally efficient manner. FFT was first introduced by Cooley and Tukey (1965) (see also Brigham 1988), who introduced a particular way of rearranging the terms of the discrete Fourier transform and managed to reduce drastically the number of the required numerical operations. Since then, the fast Fourier transform has revolutionized the spectral analysis method, and encouraged the utilization of the Fourier transform for solving large system of equations, including coupled partial differential equations, as those typically involved in boundary value problems of shallow geothermal systems.

When N is a product of integers, the forward function $\hat{F}(x_k)$ of Eq. (6.87) is proved to be closely interdependent. This interdependence was utilized intelligently to substantially reduce the amount of computation. To illustrate the computational efficiency of FFT, we consider the number of complex multiplication and additions of a signal data with N samples. Using DFT, the direct calculation requires N^2 multiplications and $N(N-1)$ additions. For N a power of 2, the FFT algorithm of Cooley and Tukey cuts the number of multiplications to $N/2 \log_2 N$ and additions to $N \log_2 N$. That is, for $N = 1024 = 2^{10}$ samples, the FFT achieves a computational reduction by a factor of over 200.

In all subsequent analysis, we will assume that any time function can be represented by the following transform pair (Doyle 1997):

Forward:

$$\hat{F}_n = \hat{F}(\omega_n) \approx \Delta P \sum_{m=0}^{N-1} F_m \, e^{-i\omega_n t_m} = \Delta P \sum_{m=0}^{N-1} F_m \, e^{-i2\pi nm/N} \tag{6.88}$$

Inverse:

$$F_m = F(t_m) \approx \frac{1}{P} \sum_{n=0}^{N-1} \hat{F}_n \, e^{i\omega_n t_m} = \frac{1}{P} \sum_{n=0}^{N-1} \hat{F}_n \, e^{i2\pi nm/N} \tag{6.89}$$

in which $\Delta P = P/N$, is the sampling rate, with P the period of the time signal, and N the number of samples.

FFT is a powerful tool for signal analysis of measured and computed data. If a measured data function, $f(t)$, is non-periodic, which is normally the case, it is still possible to construct an FFT approximation. In this case, the given $f(t)$ may then be imagined to extend periodically beyond the interested interval. This can be done by multiplying the signal function by a data window. For a rectangular window with period $P/2 \leq t \leq P/2$, the window function might be of the form

$$w(t) = \begin{cases} 1 & |t| \leq P/2 \\ 0 & |t| > P/2 \end{cases} \tag{6.90}$$

The Fourier transform of this function is

$$\hat{W}(\omega) = P \frac{\sin(\omega P/2)}{\omega P/2} \tag{6.91}$$

When multiplying the measured signal, by the window signal in the time domain, the Fourier transform of this product is convolved, using the convolution theorem, such that

$$\mathcal{F}\{f(t)w(t)\} = \hat{F}(\omega) * \hat{W}(\omega) \tag{6.92}$$

In signal analysis, this process is referred to as "windowing" where a measured signal is truncated and only small portion of the "imagined" infinite trace is visible through the window. Examples of commonly used time windows are the *rectangle* and the *Hanning* window. In windowing, it is important to sample and truncate the signal properly, otherwise, upon the inverse transform, the reconstructed signal might exhibit significant disturbances. Commonly encountered phenomena related to wrong signal analysis are *aliasing* and *leakage*.

6.8.1 *Aliasing*

This phenomenon occurs when a discrete signal is truncated by a sampling time window with its bandwidth smaller than the highest frequency of the signal. The highest frequency that can be detected by DFT from a discrete and real time series is the *Nyquist frequency*, given by

$$f_{Nq} = \frac{1}{2\Delta t} \tag{6.93}$$

in which Δt is the time interval. To perform an accurate transform of a signal, the Nyquist frequency must be equal or larger than the highest frequency, f_{max}, of the signal. From Eq. (6.93) we can readily deduce that the sample time interval must be within

$$\Delta t \leq \frac{1}{2 f_{max}} \tag{6.94}$$

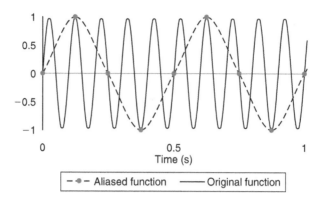

Figure 6.2　Aliasing of a signal.

which is called the *Nyquist condition*. Violation of this condition results to a distortion in the reconstructed time signal. In practice, signals have finite durations and their frequency content has no upper bound. Hence, some amount of aliasing always occurs when such functions are sampled. If the function is sampled at a high enough rate, determined by the bandwidth $[f_{min}, f_{max}]$ and Eq. (6.94), the original function can in theory be perfectly reconstructed from an infinite set of samples. Otherwise, higher frequency components merge into the frequency spectrum, and as a result, errors in the Fourier coefficients appear, leading inevitably to errors in the spectral analysis of the system response. Figure 6.2 shows a typical aliasing phenomenon in a sinusoidal function when using a sampling time interval larger than that specified by Eq. (6.94), and a Nyquist frequency, Eq. (6.93), less than the maximum frequency, $f_{Nq} < f_{max}$. This figure illustrates that, due to aliasing, a high frequency signal appears like a low frequency signal in the reconstruction of the transformed signal.

6.8.2 *Leakage*

This phenomenon occurs when the sampling time window does not meet the periodicity of the signal. Thus the time window must be an integer multiple of the period P of the signal, otherwise side lobs appear in the frequency domain. The basic cause of this phenomenon is that DFT treats the signal as a periodic function, so if the time window matches the period of the signal exactly, the Fourier transform will be accurate. If this condition is not fulfilled, side lobes occur and the transformed signal is not completely accurate. Figure 6.3 shows schematically this effect (see also Lee 2009 and Doyle 1997). The leakage can be minimized by changing the shape of the time window by, for example, multiplying the signal by a window function that tapers the amplitudes at the ends of the signal. The leakage can also be minimized by using a triangular window, also known as *Bartlett* window, or a *bell-shaped time* window. The latter has smooth discontinuities and thus produces little or no side lobes.

　　Since DFT represents a finite sample of an infinite signal on a finite period, then, and in order to reduce the probability of leakage, a scheme for increasing the apparent period is necessary. An important and simple way to doing this is by *padding* the signal with zeros (Doyle 1997). This technique makes the signal flat at its ends and prevents the occurrence of a wraparound error. Furthermore, padding in FFT makes the sample of the signal an integer power of 2, i.e. $N = 2^p$.

6.9 NUMERICAL EXAMPLES

In this section, we solve heat equations in finite, infinite and semi-infinite media subjected to different initial and boundary conditions. We use different Fourier methods for devising the

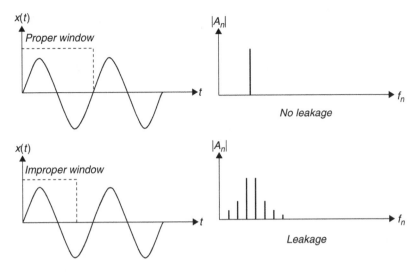

Figure 6.3 Leakage of a signal.

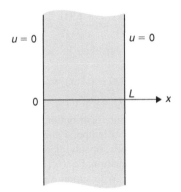

Figure 6.4 Finite domain.

solution. The objective of this exercise is to establish a methodology based on Fourier transforms for solving heat equations that will be encountered when dealing with shallow geothermal systems. This will be treated in Chapter 9 and Chapter 10.

6.9.1 *Example 1: Solution of heat equation in a finite domain*

Let $u(x, t)$ represents the temperature of a finite domain defined over the interval $0 \le x \le L$, with both ends kept at zero degrees, Figure 6.4. The initial and boundary value problem of this domain can be described as

$$\frac{\partial u(x,t)}{\partial t} = \alpha \frac{\partial^2 u(x,t)}{\partial x^2} \quad 0 < x < L, \quad t > 0 \tag{6.95}$$

$$\begin{aligned} u(x,0) &= f(x) & 0 < x < L \\ u(0,t) &= u(L,t) = 0 & t > 0 \end{aligned} \tag{6.96}$$

where $f(x)$ is a given initial temperature of the medium.

Such kind of problems can be solved elegantly using the method of *separation of variables* together with Fourier series. The method of separation of variables assumes that the equation can be split off or separated into a product of functions of the independent variables. Accordingly, $u(x, t)$ can be written as

$$u(x,t) = X(x)T(t) \tag{6.97}$$

where $X(x)$ is a function of x only, and $T(t)$ is a function of t only. The corresponding derivatives of Eq. (6.97) are

$$u_{xx} = X''(x)T(t), \quad u_t = X(x)T'(t) \tag{6.98}$$

Substituting these derivatives into Eq. (6.95), yields

$$X''(x)T(t) = \frac{1}{\alpha} X(x)T'(t) \tag{6.99}$$

This equation can also be written as

$$\frac{X''(x)}{X(x)} = \frac{1}{\alpha} \frac{T'(t)}{T(t)} \tag{6.100}$$

We note that the left-hand side of Eq. (6.100), which is solely a function of x, is equal to the right-hand side, which is solely a function of t. Since x and t are independent variables, this can only be true if both sides of the equations equal the same constant, known as the separation constant, i.e.

$$\frac{X''(x)}{X(x)} = \frac{1}{\alpha} \frac{T'(t)}{T(t)} = c \tag{6.101}$$

The resulting equation, Eq. (6.101), can then be separated into two ordinary differential equations, of the form

$$X''(x) - cX(x) = 0 \tag{6.102}$$
$$T'(t) - c\alpha T(t) = 0 \tag{6.103}$$

where Eq. (6.102) represents the temperature distribution in the spatial domain, and Eq. (6.103) represents the temperature distribution in the temporal domain. Both equations are relatively easy to handle using any of the standard methods for solving ordinary differential equations.

We first solve the spatial differential equation, Eq. (6.102). The nature of the solution depends on whether c is positive, zero or negative. With simple substitution, it can readily be shown that for $c = 0$ and $c > 0$ the solution is trivial, while for $c < 0$ the solution is non-trivial. For $c < 0$, it is convenient to write $c = -k^2$, $k > 0$, and Eq. (6.102) can then be written as

$$X''(x) + k^2 X(x) = 0 \tag{6.104}$$

Solving this equation gives

$$X(x) = A \sin kx + B \cos kx \tag{6.105}$$

where A and B are constants, determined from the boundary conditions, Eq. (6.96). Applying the first boundary condition, gives $B = 0$. Applying $X(L) = 0$, yields

$$A \sin kL = 0 \tag{6.106}$$

This is an eigenvalue problem with its eigenfunctions: $e_n(x) = \sin k_n x$. Non-trivial solution of this eigenvalue problem can be obtained only if we let $\sin kL = 0$. This condition is satisfied at the infinitely many roots of the sinusoidal function, namely at $kL = n\pi, n = 1, 2, 3, \ldots$, giving

$$k_n = \frac{n\pi}{L} \tag{6.107}$$

which represents the eigenvalues of the system. The solution of the spatial differential equation is therefore,

$$X_n(x) = A_n \sin \frac{n\pi}{L} x \tag{6.108}$$

This equation represents the nth eigenfunction of the system, and corresponds solely to eigenvalue $n\pi/L$.

We now turn our attention to the temporal equation, Eq. (6.103). Letting $c = -k^2$, Eq. (6.103) can be written as

$$T'(t) + k^2\alpha T(t) = 0 \tag{6.109}$$

This type of ordinary differential equations can readily be solved as

$$T(t) = Ce^{-k^2\alpha t} \tag{6.110}$$

where C is an arbitrary constant. Now, substituting $X(x)$ from Eq. (6.108), and $T(t)$ from Eq. (6.110) into Eq. (6.97) gives

$$
\begin{aligned}
u_n(x,t) &= Ce^{-k_n^2\alpha t} A_n \sin k_n x \\
&= D_n \sin \frac{n\pi x}{L} e^{-(n\pi/L)^2 \alpha t}
\end{aligned}
\tag{6.111}
$$

This solution is valid for the nth eigenfunction. Apparently, the general solution of the heat equation, Eq. (6.95), subjected to the boundary conditions, Eq. (6.96), must include the contribution of an infinite number of eigenfunctions, such that

$$u(x,t) = \sum_{n=1}^{\infty} D_n \sin k_n x e^{-k_n^2\alpha t} \tag{6.112}$$

To determine D_n, we apply the initial condition, Eq. (6.96), to Eq. (6.112), giving

$$u(x,0) = f(x) = \sum_{n=1}^{\infty} D_n \sin k_n x \tag{6.113}$$

This function is a typical Fourier sine series expansion of $f(x)$. The Fourier series coefficient, D_n, can then be determined using Eq. (6.24), which gives

$$D_n = \frac{2}{L} \int_0^L f(x) \sin \frac{n\pi x}{L} dx \tag{6.114}$$

The general solution can then be expressed as

$$u(x,t) = \frac{2}{L} \sum_{n=1}^{\infty} \left[\int_0^L f(x) \sin \frac{n\pi x}{L} dx \right] \cdot \sin k_n x e^{-k_n^2\alpha t} \tag{6.115}$$

If, for instance, the initial temperature is constant, $f(x) = T_0$, the general solution, after evaluating the integral in Eq. (6.115), yields

$$u(x,t) = \sum_{n=1}^{\infty} \frac{4T_0}{n\pi} \sin k_n x e^{-k_n^2\alpha t}, \quad n = 1, 3, 5, \ldots \tag{6.116}$$

This equation solves the heat equation, Eq. (6.95), and satisfies the initial and boundary conditions, Eq. (6.96) with $f(x) = T_0$.

6.9.2 Example 2: Solution of heat equation in an infinite domain

In Example 1 we discussed the solution of the heat equation in a finite domain. We saw that an elegant solution was obtained using a combination between the method of separation of variable and Fourier series. In this example, we extend the problem to an infinite domain. The fundamental

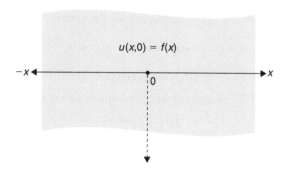

Figure 6.5   Infinite domain.

difference in solving the two problems is that the Fourier series must be replaced by the Fourier integral. We seek to solve the heat equation

$$\frac{\partial u(x,t)}{\partial t} = \alpha \frac{\partial^2 u(x,t)}{\partial x^2} \quad -\infty < x < \infty, \quad t > 0 \tag{6.117}$$

in a laterally insulated domain that extends to infinity on both sides, Figure 6.5. This entails that we do not have boundary conditions, only an initial condition of the form

$$u(x,0) = f(x) \quad -\infty < x < \infty \tag{6.118}$$

where $f(x)$ is any given initial temperature of the domain.

To solve this problem, we start as in the previous example and apply the method of separation of variables to the unknown variable, i.e. $u(x,t) = X(x)T(t)$, then substituting it into Eq. (6.117) to give

$$X''(x) - cX(x) = 0 \tag{6.119}$$

$$T'(t) - c\alpha T(t) = 0 \tag{6.120}$$

As was shown in Example 1, the solution of these two equations is

$$X(x) = A \cos kx + B \sin kx \tag{6.121}$$

$$T(t) = Ce^{-k^2 \alpha t} \tag{6.122}$$

respectively, in which we used $c = -k^2$. Hence, the solution of the heat equation, Eq. (6.117), letting $C = 1$, is

$$u(x,t) = (A \cos kx + B \sin kx)e^{-k^2 \alpha t} \tag{6.123}$$

Solution of this equation depends on whether the function is periodic or non-periodic. Since $f(x)$ in Eq. (6.118) is not periodic, we have to use the Fourier integral to solve the problem.

A and B in Eq. (6.123) are arbitrary, and we may regard them as functions of k, such that $A = A(k)$ and $B = B(k)$. For an infinite domain, and using the procedure outlined in Eqs. (6.48)-(6.56), Eq. (6.123) takes the form

$$u(x,t) = \int_0^\infty (A(k)\cos kx + B(k)\sin kx)e^{-k^2 \alpha t} \, dk \tag{6.124}$$

Applying the initial condition, Eq. (6.118), to Eq. (6.124), gives

$$u(x,0) = f(x) = \int_0^\infty (A(k)\cos kx + B(k)\sin kx) \, dk \tag{6.125}$$

which is a typical Fourier representation of an infinite medium, see Eq. (6.56). Denoting the variable of integration by v, and using Eq. (6.57), the Fourier coefficients of Eq. (6.125) can be expressed as

$$A(k) = \frac{1}{\pi} \int_{-\infty}^{\infty} f(v) \cos kv \, dv$$

$$B(k) = \frac{1}{\pi} \int_{-\infty}^{\infty} f(v) \sin kv \, dv$$

$$(6.126)$$

Substituting these coefficients into Eq. (6.124) and making use of the addition formula of sines and cosines, i.e.

$$\cos kx \cdot \cos kv + \sin kx \cdot \sin kv = \cos k(x - v) \tag{6.127}$$

yields

$$u(x,t) = \frac{1}{\pi} \int_0^{\infty} \left[\int_{-\infty}^{\infty} f(v) \cos k(x - v) e^{-k^2 \alpha t} dv \right] dk \tag{6.128}$$

This equation illustrates that the system is described by two integrations, one representing the eigenvalues (eigenmodes), k, and another representing the spatial domain, v. Note that if we haven't changed the integration variable of the initial condition from x to v, we would have the problem of having $\cos(x - x)$ in Eq. (6.128).

Inverting the order of integration in Eq. (6.128), we obtain

$$u(x,t) = \frac{1}{\pi} \int_{-\infty}^{\infty} f(v) \, dv \int_0^{\infty} \cos k(x - v) e^{-k^2 \alpha t} \, dk \tag{6.129}$$

To solve these integrations, we choose a more convenient form by introducing new independent variables, i.e.

$$b = \frac{x - v}{\sqrt{4\alpha t}}, \quad s = k\sqrt{\alpha t}, \quad \therefore dk = \frac{ds}{\sqrt{\alpha t}} \tag{6.130}$$

Inserting Eq. (6.130) into Eq. (6.129) yields

$$u(x,t) = \frac{1}{\pi} \int_{-\infty}^{\infty} f(v) \, dv \int_0^{\infty} e^{-s^2} \cos 2bs \, \frac{ds}{\sqrt{\alpha t}} \tag{6.131}$$

From literature, (also can be obtained from mathematical software, such as MAPLE (see MAPLE 13), we know

$$\int_0^{\infty} e^{-s^2} \cos 2bs \, ds = \frac{\sqrt{\pi}}{2} e^{-b^2} \tag{6.132}$$

Inserting this result into Eq. (6.131) and re-evaluating the parameters in terms of Eq. (6.130), gives

$$u(x,t) = \frac{1}{\sqrt{4\pi\alpha t}} \int_{-\infty}^{\infty} f(v) e^{-(x-v)^2/4\alpha t} \, dv \tag{6.133}$$

This equation is the general solution of the initial value problem Eqs. (6.117)–(6.118).

Next we study the solution of the system if $f(x)$ in Eq. (6.118) is bounded. Assume that we want to calculate the temperature in an infinite domain with the following initial temperature

$$f(x) = \begin{cases} U_0 & a < x < b \\ 0 & \text{else} \end{cases} \tag{6.134}$$

Substituting this condition for $f(v)$ in Eq. (6.133) yields

$$u(x,t) = \frac{U_0}{\sqrt{4\pi\alpha t}} \int_a^b e^{-(x-v)^2/4\alpha t} \, dv \tag{6.135}$$

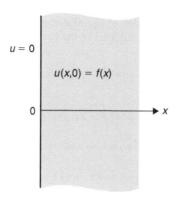

Figure 6.6   Semi-infinite domain.

For convenience, we choose a new independent variable

$$s = \frac{x - v}{\sqrt{4\alpha t}}, \quad \therefore dv = \sqrt{4\alpha t}\, ds \tag{6.136}$$

Then Eq. (6.135) becomes

$$u(x,t) = \frac{U_0}{\sqrt{\pi}} \int_{(a-x)/\sqrt{4\alpha t}}^{(b-x)/\sqrt{4\alpha t}} e^{-s^2}\, ds \tag{6.137}$$

This integral is not an elementary function, but can be expressed in terms of the *error function*, defined as

$$\operatorname{erf} x = \frac{2}{\sqrt{\pi}} \int_0^x e^{-w^2}\, dw \tag{6.138}$$

If we modify Eq. (6.137) to

$$u(x,t) = \frac{U_0}{2}\left[\frac{2}{\sqrt{\pi}}\int_0^{(b-x)/\sqrt{4\alpha t}} e^{-s^2}\, ds - \frac{2}{\sqrt{\pi}}\int_0^{(a-x)/\sqrt{4\alpha t}} e^{-s^2}\, ds\right] \tag{6.139}$$

a solution of the form

$$u(x,t) = \frac{U_0}{2}\left[\operatorname{erf}\left(\frac{b-x}{\sqrt{4\alpha t}}\right) - \operatorname{erf}\left(\frac{a-x}{\sqrt{4\alpha t}}\right)\right] \tag{6.140}$$

can be obtained. The error function is important in many engineering fields and its values have been tabulated, see, for example Kreyszig (1993).

6.9.3   Example 3: Solution of heat equation in a semi-infinite domain

In an another example, suppose that we want to calculate the temperature in a semi-infinite domain, Figure 6.6, with the following initial and boundary value problem:

$$\frac{\partial u(x,t)}{\partial t} = \alpha \frac{\partial^2 u(x,t)}{\partial x^2} \quad 0 < x < \infty, \quad t > 0 \tag{6.141}$$

$$u(x,0) = f(x) \quad x > 0 \tag{6.142}$$

$$u(0,t) = 0 \quad t > 0 \tag{6.143}$$

As for the previous cases, utilization of the method of separation of variables yields

$$X(x) = A \cos kx + B \sin kx \tag{6.144}$$

$$T(t) = Ce^{-k^2 \alpha t} \tag{6.145}$$

Applying the boundary condition, Eq. (6.143), to the spatial equation, Eq. (6.144), gives $A = 0$. The solution of the heat equation, Eq. (6.141), letting $C = 1$, can then be expressed as

$$u(x,t) = B \sin kx e^{-k^2 \alpha t} \tag{6.146}$$

Since $f(x)$ is not periodic, and the heat equation is linear and homogeneous, the solution can be expressed as

$$u(x,t) = \int_0^\infty B(k) \sin kx e^{-k^2 \alpha t} dk \tag{6.147}$$

Applying the initial condition, Eq. (6.142), to the above equation, gives

$$u(x,0) = f(x) = \int_0^\infty B(k) \sin kx \, dk \tag{6.148}$$

which is a typical Fourier expansion of a semi-infinite domain. Denoting the variable of integration by v and using Eq. (6.59), the Fourier coefficient can be expressed as

$$B(k) = \frac{2}{\pi} \int_0^\infty f(v) \sin kv \, dv \tag{6.149}$$

Substituting Eq. (6.149) into Eq. (6.147) yields

$$u(x,t) = \frac{2}{\pi} \int_0^\infty \left[\int_0^\infty f(v) \sin kv \sin kx e^{-k^2 \alpha t} dv \right] dk \tag{6.150}$$

From trigonometric relationships we have

$$2 \sin kv \cdot \sin kx = \cos k(x - v) - \cos k(x + v) \tag{6.151}$$

Applying this relationship to Eq. (6.150), and knowing that

$$\int_0^\infty e^{-k^2 \alpha t} \cdot \cos k(x - v) \, dk = \sqrt{\frac{\pi}{4\alpha t}} \cdot e^{-(x-v)^2/4\alpha t}$$

$$\int_0^\infty e^{-k^2 \alpha t} \cdot \cos k(x + v) \, dk = \sqrt{\frac{\pi}{4\alpha t}} \cdot e^{-(x+v)^2/4\alpha t} \tag{6.152}$$

we readily obtain

$$u(x,t) = \frac{1}{\sqrt{4\pi \alpha t}} \int_0^\infty f(v) \cdot (e^{-(x-v)^2/4\alpha t} - e^{-(x+v)^2/4\alpha t}) \, dv \tag{6.153}$$

This equation is the general solution of the heat equation, Eq. (6.141), of a semi-infinite domain subjected to a homogeneous boundary condition.

Let us now apply this solution to a constant initial condition of the form

$$f(x) = U_0, \quad x > 0 \tag{6.154}$$

Substituting this condition for $f(v)$ into Eq. (6.153), gives

$$u(x,t) = \frac{U_0}{\sqrt{4\pi \alpha t}} \int_0^\infty (e^{-(x-v)^2/4\alpha t} - e^{-(x+v)^2/4\alpha t}) \, dv \tag{6.155}$$

For convenience, we choose new independent variables,

$$a = -\frac{x - v}{\sqrt{4\alpha t}}, \quad \therefore dv = \sqrt{4\alpha t}\, da \quad \text{for the first integrand} \tag{6.156}$$

$$b = \frac{x + v}{\sqrt{4\alpha t}}, \quad \therefore dv = \sqrt{4\alpha t}\, db \quad \text{for the second integrand} \tag{6.157}$$

Then Eq. (6.155) becomes

$$u(x,t) = \frac{U_0}{\sqrt{\pi}} \left(\int_{-x/\sqrt{4\alpha t}}^{\infty} e^{-a^2}\, da - \int_{x/\sqrt{4\alpha t}}^{\infty} e^{-b^2}\, db \right) \tag{6.158}$$

We know that

$$\frac{1}{2}\mathrm{erfc}(x) = \frac{1}{\sqrt{\pi}} \int_{x}^{\infty} e^{-w^2}\, dw \tag{6.159}$$

in which $\mathrm{erfc} = 1 - \mathrm{erf}$ is the *complementary error function*. Applying Eq. (6.159) to Eq. (6.158) we obtain

$$u(x,t) = \frac{U_0}{2}\mathrm{erfc}\left(\frac{-x}{\sqrt{4\alpha t}}\right) - \frac{U_0}{2}\mathrm{erfc}\left(\frac{x}{\sqrt{4\alpha t}}\right) \tag{6.160}$$

6.9.4   *Example 4: Solution of heat equation in an infinite domain using Fourier transform*

In Example 2 we solved the heat equation in an infinite domain using the integral transform. In this example we solve this problem using the Fourier transform. As it is known and will be apparent here, the two methods are related to each other. We use the derivation presented by Kreyszig (1993) to relate the Fourier integral to the Fourier transform. We seek to solve the initial value problem:

$$\frac{\partial u(x,t)}{\partial t} = \alpha\frac{\partial^2 u(x,t)}{\partial x^2} \quad -\infty < x < \infty, \quad t > 0$$
$$u(x,0) = f(x) \quad -\infty < x < \infty \tag{6.161}$$

where $f(x)$ is any given initial temperature.

In Example 2, we have seen that the solution of this initial value problem can be expressed in terms of Fourier integral as

$$f(x) = \frac{1}{\pi} \int_{0}^{\infty} \left[\int_{-\infty}^{\infty} f(v) \cos k(x - v)\, dv \right] dk \tag{6.162}$$

The integral in brackets is an even function of k since $\cos(kx - kv)$ is an even function of k. The function $f(v)$ depends only on v and we integrate with respect to v (not k). Call this integral $F(k)$. Hence the integral of $F(k)$ from $k = 0 \to \infty$ is $1/2$ times the integral of $F(k)$ from $k = -\infty \to \infty$, thus

$$f(x) = \frac{1}{2\pi} \int_{-\infty}^{\infty} \left[\int_{-\infty}^{\infty} f(v) \cos k(x - v)\, dv \right] dk \tag{6.163}$$

For convenience, we express the cosine function in an exponential form using the Euler formula:

$$e^{ik(x-v)} = \cos k(x - v) + i \sin k(x - v) \tag{6.164}$$

However, in Eq. (6.163), the sinus term is missing. Hence, we must introduce a sine form of the integral in Eq. (6.163), i.e.

$$\frac{1}{2\pi} \int_{-\infty}^{\infty} \left[\int_{-\infty}^{\infty} f(v) \sin k(x - v)\, dv \right] dk \tag{6.165}$$

This integral is zero since $\sin(kx - kv)$ is an odd function of k, which makes the integral in brackets an odd function of k, and hence, the integral for $k = -\infty \rightarrow \infty$ is zero. Putting Eq. (6.163) and Eq. (6.165) together, and using the Euler formula, Eq. (6.164), yields

$$f(x) = \frac{1}{2\pi} \int_{-\infty}^{\infty} \int_{-\infty}^{\infty} f(v) e^{ik(x-v)} \, dv \, dk \qquad (6.166)$$

This is called the *complex Fourier integral*. This equation can be written as a product of two exponential functions, such that

$$f(x) = \frac{1}{\sqrt{2\pi}} \int_{-\infty}^{\infty} \left[\frac{1}{\sqrt{2\pi}} \int_{-\infty}^{\infty} f(v) e^{-ikv} \, dv \right] e^{ikx} dk \qquad (6.167)$$

The expression in brackets is a function of k, which can be denoted as $\mathcal{F}[f(x)] = \hat{F}(k)$. Bringing back x instead of v, this expression can be written as

$$\hat{F}(k) = \frac{1}{\sqrt{2\pi}} \int_{-\infty}^{\infty} f(x) e^{-ikx} \, dx \qquad (6.168)$$

With this, Eq. (6.167) becomes

$$\mathcal{F}^{-1}\left[\hat{F}(k)\right] = f(x) = \frac{1}{\sqrt{2\pi}} \int_{-\infty}^{\infty} \hat{F}(k) e^{ikx} \, dk \qquad (6.169)$$

This is the *inverse Fourier transform* of $\hat{F}(k)$. By this simple derivation we have seen how the Fourier integral and the Fourier transform are related.

Now we can go back to solve the heat conduction problem, Eq. (6.161). Our strategy is to perform the Fourier transform in space (with respect to x) and then solve the resulting ordinary differential equation in time, t. The Fourier transform of u is denoted by $\hat{u} = \mathcal{F}(u)$. Using Eq. (6.65) to transform Eq. (6.161) gives

$$\frac{1}{\alpha} \mathcal{F}\left(\frac{\partial u}{\partial t}\right) = \int_{-\infty}^{\infty} e^{-ikx} \frac{\partial^2 u}{\partial x^2} \, dx \qquad (6.170)$$

where

$$\begin{aligned}
\mathcal{F}\left[\frac{\partial u}{\partial t}\right] &= \frac{1}{\sqrt{2\pi}} \int_{-\infty}^{\infty} \frac{\partial u}{\partial t} e^{-ikx} dx \\
&= \frac{1}{\sqrt{2\pi}} \frac{\partial}{\partial t} \int_{-\infty}^{\infty} u \, e^{-ikx} dx \\
&= \frac{\partial \hat{u}}{\partial t}
\end{aligned} \qquad (6.171)$$

Thus, Eq. (6.170), using Eq. (6.77), can be expressed as

$$\frac{d\hat{u}}{dt} + \alpha k^2 \hat{u} = 0 \qquad (6.172)$$

This equation involves derivative with respect to t only, and thus it is a first order ordinary differential equation, with a solution of the form

$$\hat{u}(k, t) = A(k) e^{-\alpha k^2 t} \qquad (6.173)$$

in which $A(k)$ is an arbitrary constant, function of k. Applying the initial condition to Eq. (6.173), we obtain

$$\hat{u}(k, 0) = A(k) = \hat{F}(k) \qquad (6.174)$$

Our intermediate result is therefore

$$\hat{u}(k,t) = \hat{F}(k)e^{-\alpha k^2 t} \tag{6.175}$$

Applying the inverse Fourier transform gives

$$u(x,t) = \frac{1}{\sqrt{2\pi}} \int_{-\infty}^{\infty} \hat{F}(k)e^{-\alpha k^2 t}e^{ikx}\,dk \tag{6.176}$$

Using Eq. (6.168) and letting $x = v$, the Fourier coefficient is

$$\hat{F}(k) = \frac{1}{\sqrt{2\pi}} \int_{-\infty}^{\infty} f(v)e^{-ikv}\,dv \tag{6.177}$$

Inserting this coefficient into Eq. (6.176) and inverting the order of integration, we obtain

$$u(x,t) = \frac{1}{2\pi} \int_{-\infty}^{\infty} f(v) \left[\int_{-\infty}^{\infty} e^{-\alpha k^2 t}e^{ik(x-v)}\,dk \right] dv \tag{6.178}$$

Using Euler formula, the inner integral can be represented by sines and cosines, as

$$\int_{-\infty}^{\infty} e^{-\alpha k^2 t}e^{ik(x-v)}\,dk = \int_{-\infty}^{\infty} e^{-\alpha k^2 t}\cos k(x-v)\,dk + i\int_{-\infty}^{\infty} e^{-\alpha k^2 t}\sin k(x-v)\,dk \tag{6.179}$$

As indicated in Eq. (6.165), the imaginary part is an odd function of k, so that the integral is zero. The real part is even, so that the integral from $-\infty \rightarrow \infty$ is twice the integral from $0 \rightarrow \infty$, leading to

$$u(x,t) = \frac{1}{\pi} \int_{-\infty}^{\infty} f(v) \left[\int_{0}^{\infty} e^{-\alpha k^2 t}\cos k(x-v)\,dk \right] dv \tag{6.180}$$

By this, we obtained the same results as that given in Eq. (6.129). As a particular case, we solve the heat equation in an infinite domain subjected to a bounded initial condition of the form

$$u(x,0) = \begin{cases} U_0(x) & -L < x < L \\ 0 & \text{else} \end{cases} \tag{6.181}$$

The Fourier transform of this initial condition is

$$\begin{aligned} \hat{u}(k,0) &= \int_{-\infty}^{\infty} u(x,0)e^{-ikx}\,dx \\ &= \int_{-L}^{L} U_0(x)e^{-ikx}\,dx \end{aligned} \tag{6.182}$$

The intermediate result is hence,

$$\hat{u}(k,t) = \int_{-L}^{L} U_0(x)e^{-ikx}\,dx \cdot e^{-\alpha k^2 t} \tag{6.183}$$

Inverting $\hat{u}(k,t)$ by the use of the inversion formula, Eq. (6.66), gives

$$u(x,t) = \frac{1}{2\pi} \int_{-\infty}^{\infty} e^{ikx} \cdot e^{-\alpha k^2 t}\,dk \cdot \int_{-L}^{L} U_0(v)e^{ikv}\,dv \tag{6.184}$$

Changing the order of integration, yields

$$u(x,t) = \frac{1}{2\pi} \int_{-L}^{L} U_0(v) \left[\int_{-\infty}^{\infty} e^{-\alpha k^2 t}e^{ik(x-v)}\,dk \right] dv \tag{6.185}$$

Using Eq. (6.179), we immediately obtain

$$u(x,t) = \frac{1}{\pi} \int_{-L}^{L} U_0(v) \left[\int_0^{\infty} e^{-\alpha k^2 t} \cos k(x-v) \, dk \right] dv \qquad (6.186)$$

If we assume that $U_0(x) = U_0 = $ constant, and upon utilizing the procedure in Examples 2 and 3, we obtain

$$u(x,t) = \frac{U_0}{2} \left[\text{erf}\left(\frac{L-x}{\sqrt{4\alpha t}} \right) + \text{erf}\left(\frac{L+x}{\sqrt{4\alpha t}} \right) \right] \qquad (6.187)$$

Note that we have chosen in this example and the previous ones to examine Fourier transforms in the space-eigenvalue domain, rather than the time-frequency domain. In the latter, the Fourier transforms fail to treat the initial condition. This constitutes one of the shortcomings of the Fourier transforms, as compared, for example, to the Laplace transforms. However, in many situations, the initial condition can be added at the end of the solution, using the principle of superposition. In Chapter 9 and Chapter 10, we shall solve the boundary value problem of shallow geothermal systems in the time-frequency domain and consider the initial condition in the post processing.

CHAPTER 7

Laplace transforms

In this chapter, we present the basic principles of the Laplace transform and its inversion theorem, with emphasis on their utilization for solving transient heat conduction applications. We first give a brief introduction to commonly utilized methods for obtaining the forward Laplace transform and its inversion. Tables of transforms, analytical transform methods and numerical procedures are outlined. Next, we discuss two application examples describing heat conduction in a finite region and in an infinite region. The aim of these examples is to show how the Laplace transform provides simple tools for solving partial differential equations, but in the meanwhile, we show how the inversion might lead to rather complicated inverse transform integrals, which require rigorous computational tools to tackle.

7.1 INTRODUCTION

Classical problems of heat conduction in solids and fluids can be solved using any of the well-known mathematical methods for solving ordinary and partial differential equations. The most powerful and commonly used methods are the Laplace transforms, the Fourier transforms, the Duhamel integral, the Green's function and the Goodman's approximate integral method. For heat flow in shallow geothermal systems, however, the Laplace transform has been utilized intensively. Therefore, in this chapter, we cover the method of Laplace transforms with some details. This will help in understanding the mathematical formulation and solution of the commonly utilized models for shallow geothermal systems, described in Chapter 8. In Chapter 9 and Chapter 10 we cover the application of the Fourier transform for solving heat flow in shallow geothermal systems.

Utilization of Laplace transforms to solve partial differential equations is conducted in two steps: the forward transform, and the inverse transform. In the forward transform, solution of a partial differential equation is conducted by reducing it by at least one dimension. For instance, solving a one-dimensional transient heat conduction problem can be conducted by converting the involved partial differential equation to an ordinary differential equation by eliminating the time derivative. For two and three dimensional problems, however, the partial differential equation stays partial differential equation but reduced by one dimension. By applying multiple transforms, it is possible to obtain an ordinary differential equation of a multi-dimensional problem. This usually entails utilization of transform tables. Having obtained the reduced form of the governing equations, the solution can be conducted using any of the available mathematical techniques. After solving for the forward transform, the inverse transform must be conducted to re-construct the solution to its original domain. This usually requires utilization of tables, contour integrations or numerical procedures. An important feature of the Laplace transform is that the initial and boundary conditions are incorporated in the solution from the early stage, so that the the solution can be considerably shortened.

7.2 FORWARD LAPLACE TRANSFORM

Forward Laplace transform of a function $F(t)$ is commonly denoted by $\mathcal{L}\{F(t)\}$, defined by the integral

$$\mathcal{L}\{F(t)\} = \hat{F}(s) = \int_0^\infty F(t)e^{-st}dt \qquad (7.1)$$

where the hat ($\hat{\ }$) indicates a transformed function. The constant parameter s is the Laplace parameter and is assumed to be positive and, for most cases, large enough to ensure that the product $F(t)e^{-st}$ converges to zero as $t \to \infty$. Laplace transforms of a rather large number of functions are available in tables. They are obtained by making use of standard methods of integration and manipulation of algebraic functions. For example, Laplace transform of

$$F(t) = \sin at \tag{7.2}$$

can be worked out in a straightforward procedure by first substituting Eq. (7.2) into Eq. (7.1), giving

$$\mathcal{L}\{\sin at\} = \hat{F}(s) = \int_0^\infty \sin at \, e^{-st} dt \tag{7.3}$$

Then, using Euler's formula

$$e^{iat} = \cos at + i \sin at \tag{7.4}$$

it can readily be noticed that $\sin at$ is the imaginary part of e^{iat}, which might be written as $\text{Im}(e^{iat})$. Accordingly, Eq. (7.3) can be expressed as

$$\mathcal{L}\{\sin at\} = \hat{F}(s) = \mathcal{L}\left\{\text{Im}(e^{iat})\right\} = \text{Im}\int_0^\infty e^{iat} e^{-st} \, dt$$

$$= \text{Im}\left\{\lim_{t \to \infty} \frac{e^{t(ia-s)} - 1}{ia - s}\right\} \tag{7.5}$$

$$= \text{Im}\left\{\frac{1}{s - ia}\right\}$$

To rationalize the result, we multiply the top and bottom of Eq. (7.5) by $s + ia$, yielding

$$\mathcal{L}\{\sin at\} = \frac{a}{s^2 + a^2} \tag{7.6}$$

In the same way, Laplace transforms of many other functions can be obtained. Normally, the transformed functions are listed in tables, which can readily be used for solving large number of mathematical problems. The Laplace transforms of some elementary functions are listed in Table 7.1 (Spiegel 1965).

Table 7.1   Laplace transform of some elementary functions.

$F(t)$	$\mathcal{L}\{F(t)\} = \hat{F}(s)$
1. 1	$\dfrac{1}{s}$ $s > 0$
2. t	$\dfrac{1}{s^2}$ $s > 0$
3. t^n $n = 0, 1, 2, \ldots$	$\dfrac{n!}{s^{n+1}}$ $s > 0$
4. e^{at}	$\dfrac{1}{s - a}$ $s > a$
5. $\sin at$	$\dfrac{a}{s^2 + a^2}$ $s > 0$
6. $\cos at$	$\dfrac{s}{s^2 + a^2}$ $s > 0$
7. $\sinh at$	$\dfrac{a}{s^2 - a^2}$ $s > |a|$
8. $\cosh at$	$\dfrac{s}{s^2 - a^2}$ $s > |a|$

7.2.1 Properties of Laplace transform

Laplace transform has many important properties, but here, we present some of its properties, which are useful for solving heat transfer problems.

1. Linearity property: if $\hat{F}(s)$ and $\hat{G}(s)$ are transformed functions of $F(t)$ and $G(t)$ respectively, then for any constant c_1 and/or c_2, we get

$$L\{c_1 F(t)\} + L\{c_2 G(t)\} = c_1 \hat{F}(s) + c_2 \hat{G}(s) \tag{7.7}$$

2. Shifting property: for any constant a

$$L\{e^{at} F(t)\} = \hat{F}(s - a) \tag{7.8}$$

3. Change of scale property: if $L\{F(t)\} = \hat{F}(s)$, then

$$L\{F(at)\} = \frac{1}{a} \hat{F}\left(\frac{s}{a}\right) \tag{7.9}$$

4. Laplace transform of derivatives: for derivatives we have

$$L\{F'(t)\} = s\hat{F}(s) - F(0) \tag{7.10}$$

   This property is particularly important for solving initial value problems, as the initial condition, $F(0)$, enters into the solution at an early stage.
5. Laplace transform of integrals: if $L\{F(t)\} = \hat{F}(s)$, then

$$L\left\{ \int_0^t F(u)\,du \right\} = \frac{1}{s}\hat{F}(s) \tag{7.11}$$

6. Multiplication by t^n: for any positive integer n:

$$L\{t^n F(t)\} = (-1)^n \frac{d^n}{ds^n} \hat{F}(s) \tag{7.12}$$

7. Division by t: if $L\{F(t)\} = \hat{F}(s)$, then

$$L\left\{ \frac{1}{t} F(t) \right\} = \int_s^{\infty} \hat{F}(u)\,du \tag{7.13}$$

8. Periodic functions: if $F(t)$ is periodic with period P, so that $F(t + P) = F(t)$, then

$$L\{F(t)\} = \frac{1}{1 - e^{-sP}} \int_0^P e^{-st} F(t)\,dt \tag{7.14}$$

7.2.2 Methods of finding Laplace transform

In mathematics, several methods have been devised for the determination of the Laplace transform of functions. The most straight forward method is the direct use of the integral in Eq. (7.1), as was shown in our introductory example, Eqs. (7.5)–(7.6). The results of this direct integration process has been tabulated in what is known as the *tables of Laplace transforms*, such as Table 7.1. These tables are usually utilized directly for obtaining the Laplace transform of any function included in the list.

However, in many practical cases, the use of tables of transforms cannot be made directly and an algebraic manipulation is necessary to bring the function to a form that has a corresponding transform in the tables. One of the commonly used methods for this purpose is the *series method*. This method transforms a relatively complicated functions, such as the Bessel functions $J_v(t)$, into their series components, and then applies the Laplace transform to each individual term in

Table 7.2   Laplace transform of some special functions.

$F(t)$	$\mathcal{L}\{F(t)\} = \hat{F}(s)$
1. $J_0(at)$	$\dfrac{1}{\sqrt{s^2 + a^2}}$
2. $J_v(at)$	$\dfrac{1}{a^v \sqrt{s^2 + a^2}} (\sqrt{s^2 + a^2} - s)^v$
3. $\sin \sqrt{t}$	$\dfrac{\sqrt{\pi}}{2s^{3/2}} e^{-1/4s}$
4. $\mathrm{erf}(t)$	$\dfrac{1}{s} e^{s^2/4} \mathrm{erfc}(s/2)$
5. $\mathrm{erf}(\sqrt{t})$	$\dfrac{1}{s\sqrt{s+1}}$
6. $\mathrm{erf}\left(\dfrac{r}{2\sqrt{at}}\right)$	$\dfrac{1}{s} e^{-r\sqrt{s/a}}$
7. $\mathrm{Ei}(t) = \dfrac{1}{s} \ln(s+1)$	$\dfrac{s}{s^2 + a^2} \quad s > 0$

the series. For example, consider the Laplace transform of the Bessel function of order zero, $J_0(t)$. In terms of series, this function can be written as

$$J_0(t) = 1 - \frac{t^2}{2^2} + \frac{t^4}{2^2 4^2} - \frac{t^6}{2^2 4^2 6^2} + \cdots \tag{7.15}$$

Then, using the transforms given in 1 and 3 of Table 7.1 , and by use of the *binomial theorem*, we obtain

$$\mathcal{L}\{J_0(t)\} = \frac{1}{s} - \frac{1}{2^2}\frac{2!}{s^3} + \frac{1}{2^2 4^2}\frac{4!}{s^5} - \frac{1}{2^2 4^2 6^2}\frac{6!}{s^7} + \cdots$$

$$= \frac{1}{\sqrt{s^2 + 1}} \tag{7.16}$$

Of course there are other methods, which can be used for reducing complicated functions into functions that have counterparts in the transform tables. Among others, are the *method of differential equations*, and the *method of differentiation with respect to a parameter*. Table 7.2 lists the Laplace transform of some special functions. For more complete tables see for example Spiegel (1965).

We have seen so far that the forward Laplace transform is a neat method and relatively easy to handle. Nevertheless, in solving initial and boundary value problems, there is little importance to the forward transform, unless we carry out the inverse transform. This issue is treated next.

7.3   INVERSE LAPLACE TRANSFORM

The inverse Laplace transform of a function $\hat{F}(s)$ is commonly denoted by:

$$\mathcal{L}^{-1}\{\hat{F}(s)\} = F(t) \tag{7.17}$$

where here we have a given function in the Laplace domain and needs to be reconstructed to the original function of time, $F(t)$. The properties of the inverse Laplace transform is similar to those listed in Sub-section 7.2.1 for the forward transforms, but here, the logic must be reversed. Same goes for the inverse transform procedures. For example, in our previous example, Eq. (7.6), we found that the Laplace transform of $\sin at$ is $a/(s^2 + a^2)$. Using this information, and if the solution of a given function leads to such a function in the Laplace domain, it can then be reconstructed to the time domain, by noting that:

$$\mathcal{L}^{-1}\left\{\frac{a}{s^2 + a^2}\right\} = \sin at \qquad (7.18)$$

This is quite easy to handle since it is possible to use readily available tables of transforms to reconstruct the time domain. However, the solution of real-world physical applications is quite involved and tables might not be available. In general, the inverse Laplace transform can be carried out in various ways: direct use of tables; Bromwich integral and the direct use of calculus of residue; and numerical methods.

7.3.1 *Direct use of tables*

As mentioned above, transform tables for a quite large number of functions are available in mathematical and applied physics text books. They provide analytical and easy to handle inverse functions. However, it is reasonable to state that these tables provide solutions, which are probably significant in solving textbook exercises than solving real-world physical applications. Even in specialized heat conduction monographs, the tabulated inversion transform functions are limited to special cases. Carslaw and Jaeger (1959); Ozisik (1968); and Yener and Kakac (2008) provide useful tables of Laplace transforms for a rather large heat conduction applications.

In many practical cases, the use of table of transforms cannot be made directly and an algebraic manipulation is necessary to bring the function to a form that has an appropriate transform in the tables. The method of *partial fractions* is a suitable tool for this purpose. This method transforms a relatively complicated polynomials of the form $P(s)/Q(s)$, with the degree of $P(s)$ less than that of $Q(s)$, into the sum of other fractions such that the denominator of each new fraction is either a first degree or a quadratic polynomial raised to some power. For example, consider the inverse Laplace transform of (Stroud 1990)

$$\mathcal{L}^{-1}\left\{\frac{3s + 1}{s^2 - s - 6}\right\} \qquad (7.19)$$

Using the partial fractions method, we obtain

$$\mathcal{L}^{-1}\left\{\frac{3s + 1}{s^2 - s - 6}\right\} = \mathcal{L}^{-1}\left\{\frac{1}{s - 2} + \frac{2}{s - 3}\right\} \qquad (7.20)$$
$$= e^{2t} + 2e^{3t}$$

where we used Table 7.1 to get the transformed functions. Functions which can be computed using algebraic techniques are usually denoted as *rational functions*. For more details on the method of partial fractions, see for example Stroud (1990). There are other methods, which can be utilized for reducing relatively complicated polynomials to forms which are solvable by the transform tables. Such methods are the method of differential equations and the method of differentiation with respect to a parameter. See Speigel (1965) for more details.

7.3.2 *Bromwich integral and the calculus of residues*

For a wide class of functions, which are not rational, the inverse Laplace transform cannot be computed using algebraic techniques. In such cases, the computation of the inverse transform requires techniques based on the *complex variable theory*. For many integral transform problems, the function $F(t)$ is analytic and have a Laplace transform $\hat{F}(s)$, defined over all points in the

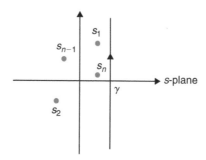

Figure 7.1   Bromwich Contour.

complex plane, except at a finite number of singular points, $s_1, s_2, \ldots, s_n$ where $\hat{F}(s)$ has *poles*. In this case, the inverse Laplace transform is defined by the contour (line) integral as

$$\mathcal{L}^{-1}\{\hat{F}(s)\} = F(t) = \begin{cases} \dfrac{1}{2\pi i} \displaystyle\int_{\gamma-i\infty}^{\gamma+i\infty} e^{st}\,\hat{F}(s)\,ds & t > 0 \\[4mm] 0 & t < 0 \end{cases} \tag{7.21}$$

This integral is called the *Bromwich integral*, also known as the *Fourier-Mellin integral*. (It must be noted that all non-rational functions listed in tables of transforms are obtained from this inversion theorem.) The basic property of this integral is that the integration contour $(\gamma - i\infty, \gamma + i\infty)$ represents an infinite vertical line in the complex plane, and the constant γ is chosen so that all singularities of $\hat{F}(s)$ are made on the left-hand side of the line, as shown in Figure 7.1.

In many cases, the Bromwich integral may be evaluated by the regular methods of contour integration and the residue theorem. The formation of the contour depends on whether the function is a *single-valued* function, which is a function that each point has a unique value; or a *multiple-valued* function, which is a "function" that a point in its range may have multiple values. In other words, it depends on the presence and location of the *branch points*. Branch points are points where various curves or surfaces of a multiple-valued function intersect. For example, the function $w = \sqrt{z}$ has two branches: one where the square root comes in with a plus sign, and the other with a minus sign. A branch cut is a curve in the complex plane such that it is possible to define a single branch point of a multiple-valued function. The branch cut allows one to transform a multiple-valued function to many single-valued functions brought together. Accordingly, the solution of the inverse integral might be obtained by use of one of the following three standard methods (Spiegel 1965):

1. If $\hat{F}(s)$ is a single-valued function with only singularities are poles, all of which lie to the left of the contour line $s = \gamma + i\tau$. In this case, the poles of $\hat{F}(s)$ are encircled inside a closed contour consisting of the portion of the Bromwich contour $C = \{s = \gamma + i\tau : -R \le \tau \le R; \tau = \sqrt{R^2 - \gamma^2}\}$ and a semicircle Γ of radius R, Figure 7.2. That is the poles lie inside the contour $C + \Gamma$. Then by taking the limit as $R \to \infty$, the integral around Γ becomes zero, such that

$$\frac{1}{2\pi i} \lim_{R \to \infty} \int_{\gamma-i\tau}^{\gamma+i\tau} e^{st}\hat{F}(s)ds = 0 \tag{7.22}$$

By the *Cauchy's theorem* we get,

$$F(t) = \sum \mathrm{Res}\, e^{st}\,\hat{F}(s), \quad \text{at poles of } \hat{F}(s) \tag{7.23}$$

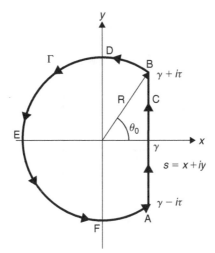

Figure 7.2   Single-valued function with poles.

The validity of the results of Eq. (7.23) is achieved if the radius, R, is suitably chosen such that all singularities are enclosed within Γ. This can be done under the condition that

$$|\hat{F}(s)| < \frac{M}{R^k} \tag{7.24}$$

where $s = R\,e^{i\theta}$, $\theta_0 \le \theta \le 2\pi - \theta_0$, and the constants $M > 0$ and $k > 0$. This condition always holds if $\hat{F}(s) = P(s)/Q(s)$ where $P(s)$ and $Q(s)$ are polynomials and the degree of $P(s)$ is less than that of $Q(s)$. This inversion solution is usually arises in heat conduction problems of finite regions.

2. If $\hat{F}(s)$ is a multiple-valued function and has branch points, the above solution technique can be utilized provided that the Bromwich contour is suitably modified. For heat conduction problems of semi-infinite regions or infinite with a source at the center, $\hat{F}(s)$ usually has a branch point at $s = 0$. In such a case we use the contour shown in Figure 7.3 with a cut along the negative real axis such that $\hat{F}(s)$ is a single-valued function of s within and on the contour. The integration is conducted by taking the limit as the radius of the large circle, R, tends to infinity, and the radius of the small circle ε, tends to zero. As described in the first case, the first integral is similar to Eq. (7.22) and thus vanishes. Then we are left with evaluating the second integral, i.e. for the small circle. This will be apparent in Example 2 below.
3. If $\hat{F}(s)$ has infinitely many singularities, the above two methods can be utilized. In such a case, the contour Γ_m is chosen with a radius R_m such that it encloses only a finite number of singularities, m, and not to pass through any singularity, Figure 7.4. The inverse Laplace transform can then be obtained by taking the upper limit of the residue summation to infinity, i.e. $m \to \infty$. This will be apparent in Example 1 below.

The essence of the Bromwich integral is that it provides analytical solutions to a large number of mathematical applications, including those related to heat conduction of relatively simple geometry and boundary conditions. Despite its apparent complexity, step-by-step procedures for tackling quite lengthy and complicated problems are available in literature. However, in many other cases, especially those related to real-world applications, the Bromwich integral leads to highly complicated integrals of oscillatory functions consisting of convolved sinusoidal or Bessel functions. In such problems, it is usually difficult to determine the location of all poles and/or branch points of the involved functions. As a result, in solving real-world geothermal systems, it is most likely inevitable to employ the numerical methods for solving the involved mathematics.

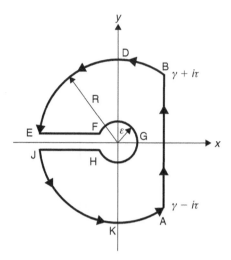

Figure 7.3   Multiple-valued function with a branch point at the center.

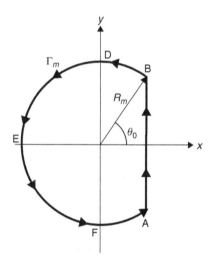

Figure 7.4   Infinitely many singularities function.

7.3.3 *Numerical inversion*

In solving physical problems, the forward Laplace transform is commonly straight forward and highly stable. That is a large change in $F(t)$ in Eq. (7.1) results to a very small and often insignificant change in $\hat{F}(s)$. In contrast to the forward transform, the inverse transform is highly unstable. A tiny change in $\hat{F}(s)$ in Eq. (7.17) may result in a wild variation of $F(t)$ (Arfken, and Weber 1995). As a consequence, in many cases, the inverse Laplace transform is difficult to conduct. It is commonly known that in using integral transform methods for solving physical applications, most of the efforts are directed towards solving the inverse integrals. Therefore, in most cases it is necessary to resort to numerical methods to solve the problem.

There is a quite large number of numerical algorithms for solving inverse Laplace transforms. One of the classical algorithms is the one introduced by Dubner and Abate (1968). They proposed an approximate solution to the Bromwich integral, Eq. (7.21), along the contour line

$\left(\gamma - i\dfrac{\pi}{T}, \gamma + i\dfrac{\pi}{T}\right)$, of the form

$$F(t) = \tilde{F}(t) - E \tag{7.25}$$

in which $\tilde{F}(t)$ is an approximation function, defined as

$$\tilde{F}(t) = \frac{2e^{\gamma t}}{T}\left[\frac{f(\gamma)}{2} + \sum_{k=1}^{\infty}\mathrm{Re}\,\hat{F}\left(\gamma + \frac{ik\pi}{T}\right)\cos\frac{k\pi t}{T}\right] \tag{7.26}$$

and E is an error function, defined by

$$E = \sum_{n=1}^{\infty}\exp(-2n\gamma t)[F(2nT+t) + \exp(2\gamma t)\,F(2nT-t)] \tag{7.27}$$

This solution has been intensively examined and found that the series in Eq. (7.25) converges slowly, and often summation of hundreds of terms is necessary to obtain a reasonable accuracy.

Other numerical procedures have been introduced with different degrees of summation efficiency and convergence rates. A well-known approach for numerical Laplace transform inversion is based on the sequence of functional developed by Gaver in 1966 (reported in Abate and Valko, 2004). Gaver discretized the solution as

$$F(t) \sim f_k(t) = \frac{k\ln 2}{t}\binom{2k}{k}\sum_{j=0}^{k}(-1)^j\binom{k}{j}\hat{F}\left[\frac{(k+j)\ln 2}{t}\right] \tag{7.28}$$

Abate and Valko argued that this approximation is very poor and requires many summations. In order to accelerate the approximation, a convergence acceleration algorithm is needed. This has been implemented by Stehfest (1970) who introduced the well-known Gaver-Stehfest algorithm. This algorithm was utilized by Bandyopadhyay *et al.* (2008) and Philippe *et al.* (2009) for solving geothermal problems. In its general form, this algorithm is defined as

$$F(t) \sim f_k(t) = \frac{\ln 2}{t}\sum_{j=1}^{k}D(j,k)\hat{F}\left(\frac{j\ln 2}{t}\right) \tag{7.29}$$

with

$$D(j,k) = (-1)^{j+M}\sum_{n=\mathrm{int}(j+1/2)}^{\min(j,M)}\frac{n^M(2n)!}{(M-n)!\,n!\,(n-1)!\,(j-n)!\,(2n-j)!} \tag{7.30}$$

and $M = k/2$, with k an even integer, chosen based on the computer precision. There are other sequence accelerators that can be used as well. Abate and Valko showed that the Wynn ρ-accelerator is one of the best for the Gaver functional. The recursive algorithm of this scheme is

$$f(t, M) = \rho_M^{(0)} \tag{7.31}$$

with

$$\rho_{-1}^{(n)} = 0, \quad \rho_0^{(n)} = f_n(t); \quad n \geq 0$$

$$\rho_k^{(n)} = \rho_{k-2}^{(n+1)} + \frac{k}{\rho_{k-1}^{(n+1)} - \rho_{k-1}^{(n)}}; \quad k \geq 1 \tag{7.32}$$

and $M = k/2$, with k an even integer. Many researchers systematically investigated the stability and accuracy of Eq. (7.29) and Eq. (7.31), and found that these algorithms are prone to round-off error propagation, and hence unstable. As M increases, the accuracy also increases but only to a certain degree, after which the accuracy decreases rapidly. To overcome this problem, Abate and Valko (2004) proposed increasing the computing precision with the increase of M. This can be achieved using a multi-precision computational environment.

Another important numerical integration scheme is proposed by Talbot in 1979, who introduced the concept of deforming the standard contour in the Bromwich integral, Eq. (7.21). The convergence of this integral is greatly improved if s takes on values with a large, negative and real component (Abate and Valko 2004). We can thus deform the contour s into any open path that wraps around the negative real axis provided no singularity of $\hat{f}(s)$ is crossed, allowing thus the use of the Cauchy's theorem. Talbot suggested the contour path of the form

$$s(\theta) = r\theta(\cot\theta + i), \quad -\pi < \theta < \pi \tag{7.33}$$

where r is a parameter. Substituting Eq. (7.33) into Eq. (7.21) yields

$$f(t) = \frac{1}{2\pi i} \int_{-\pi}^{\pi} e^{t\,s(\theta)} \hat{F}[s(\theta)]s'(\theta)\,d\theta \tag{7.34}$$

where

$$s'(\theta) = ir[1 + i\sigma(\theta)]; \quad \sigma(\theta) = \theta + (\theta\cot\theta - 1)\cot\theta \tag{7.35}$$

Abate and Valko showed that by using the trapezoidal rule with a step size π/M and $\theta_k = k\pi/M$ the integral in Eq. (7.34) can be discretized as

$$f(t, M) = \frac{r}{M}\left\{\frac{1}{2}\hat{F}(r)e^{rt} + \sum_{k=1}^{M-1}\mathrm{Re}\left[e^{t\,s(\theta_k)}\hat{F}(s(\theta_k))(1 + i\,\sigma(\theta_k))\right]\right\} \tag{7.36}$$

with $r = 2M/5t$. To control the round-off error propagation, they also utilized a multi-precision algorithm. In spite of the advantages of this algorithm it is prone to instability in computing certain transforms. One of the problems is locating the branch cut of the contours (Murli and Rizzardi 1990).

In summary, mathematicians and researchers in the field of numerical solution of inverse Laplace transform constantly endeavor to provide accurate and stable algorithms that are independent of the function under consideration. However, it seems that this is not yet possible. Even for a known class of transforms, the numerical algorithms often undergo numerical nuisance and convergence problems. To overcome this problem, we might need to resort to some innovative manipulations of our models in order to avoid the complexities of the inversion of the transform integrals. This is what we will be focusing on in Chapter 9 and Chapter 10.

7.4   NUMERICAL EXAMPLES

In this section we utilize Laplace transform to solve some academic applications of heat conduction in finite and infinite regions.

7.4.1   *Example 1: Solution of heat equation in a finite domain*

Let $u(x,t)$ represents the temperature of a one-dimensional domain, one unit in length and zero initial temperature. At $x = 0$, no heat flow is allowed, and at $x = 1$, a constant temperature T_0 is imposed at $t > 0$. That is we solve for the initial and boundary value problem:

$$\frac{\partial^2 u}{\partial x^2} - \frac{1}{\alpha}\frac{\partial u}{\partial t} = 0, \quad 0 \le x \le 1 \tag{7.37}$$

$$u(x, 0) = 0, \quad 0 \le x \le 1 \tag{7.38}$$

$$\frac{\partial u}{\partial x} = 0, \quad x = 0$$

$$u = T_0, \quad x = 1 \tag{7.39}$$

in which α is the thermal diffusivity. We employ the Laplace transform to solve this problem. Multiply Eq. (7.37) by e^{-st} and integrate with respect to t from 0 to infinity, we obtain

$$\int_0^\infty e^{-st}\frac{\partial^2 u}{\partial x^2}\,dt - \frac{1}{\alpha}\int_0^\infty e^{-st}\frac{\partial u}{\partial t}\,dt = 0 \qquad (7.40)$$

Using the Laplace properties, given in Sub-section 7.2.1, Eq. (7.40) gives

$$\frac{d^2\hat{u}}{dx^2} - \frac{s}{\alpha}\hat{u} = 0, \quad 0 \le x \le 1 \qquad (7.41)$$

where the hat $(\hat{\ })$ denotes that the temperature is in the Laplace domain. Note that the Laplace transform has reduced the partial differential equation, Eq. (7.37), to an ordinary differential equation, Eq. (7.41). The transformed equation is usually referred to as the *subsidiary equation*. In the same way, the Laplace transform of the boundary conditions in Eq. (7.39), gives

$$\frac{d\hat{u}}{dx} = 0, \quad x = 0$$

$$\hat{u} = \frac{T_0}{s}, \quad x = 1 \qquad (7.42)$$

The solution of this boundary value problem, Eqs. (7.41)-(7.42), can readily be shown to be

$$\hat{u}(s) = T_0\frac{\cosh x\sqrt{s/\alpha}}{s\cosh\sqrt{s/\alpha}} \qquad (7.43)$$

Having solved the subsidiary equation, Eq. (7.43) needs to be reconstructed back to the time domain. Applying the inversion theorem, gives

$$u(t) = T_0\mathcal{L}^{-1}\left\{\frac{\cosh x\sqrt{2/\alpha}}{s\cosh\sqrt{2/\alpha}}\right\}$$

$$= \frac{T_0}{2\pi i}\int_{\gamma-i\infty}^{\gamma+i\infty} e^{st}\frac{\cosh x\sqrt{s}}{s\cosh\sqrt{s}}\,ds \qquad (7.44)$$

where for simplicity of notation, we replaced $\sqrt{s/\alpha} \to \sqrt{s}$, assuming at the moment that $\alpha = 1$. To solve this inversion equation we use the residue theorem, following closely the procedure outlined by Spiegel (1965). First determine the singularities of

$$\hat{u}(s) = \frac{\cosh x\sqrt{s}}{s\cosh\sqrt{s}} \qquad (7.45)$$

Because of the presence of $\sqrt{s}$, it appears, but not definite, that there is a branch point at $s = 0$. To investigate this, we apply the expansion technique, which gives

$$\frac{\cosh x\sqrt{s}}{s\cosh\sqrt{s}} = \frac{1}{s}\left\{\frac{1+(x\sqrt{s})^2/2!+(x\sqrt{s})^4/4!+\cdots}{1+(\sqrt{s})^2/2!+(\sqrt{s})^4/4!+\cdots}\right\} \qquad (7.46)$$

Apparently, there is no branch point at $s = 0$. Rather, there is a simple pole at this point. Furthermore, $\hat{u}(s)$ has infinitely many simple poles given by the roots of

$$\cosh\sqrt{s} = \cos i\sqrt{s} = 0 \qquad (7.47)$$

These occurs where $i\sqrt{s} = (n+1/2)\pi$, $n = 0, \pm 1, \pm 2, \ldots$. Thus $\hat{u}(s)$ has simple poles at

$$s = 0$$

$$s = s_n = -\left(n - \frac{1}{2}\right)^2\pi^2, \quad n = 1, 2, 3, \ldots \qquad (7.48)$$

Now, the solution can be proceeded using the Bromwich contour given in Figure 7.4. Line AB is chosen so that all the poles, Eq. (7.48), must lie at its left-hand side. The curved portion Γ_m is chosen with center at the origin and radius $R_m = m^2\pi^2$, where m is a positive integer. This choice insures that the contour does not pass through any of the poles. We begin by finding the residue of

$$e^{st}\frac{\cosh x\sqrt{s}}{s\cosh\sqrt{s}} \tag{7.49}$$

at the poles. At $s = 0$, the residue is:

$$\lim_{s\to 0}(s-0)\left\{e^{st}\frac{\cosh x\sqrt{s}}{s\cosh\sqrt{s}}\right\} = 1 \tag{7.50}$$

At $s = s_n = -(n-\frac{1}{2})^2\pi^2, n = 1, 2, 3, \ldots$, the residue is:

$$\lim_{s\to s_n}(s-s_n)\left\{e^{st}\frac{\cosh x\sqrt{s}}{s\cosh\sqrt{s}}\right\} = \lim_{s\to s_n}\left\{\frac{s-s_n}{\cosh\sqrt{s}}\right\}\lim_{s\to s_n}\left\{e^{st}\frac{\cosh x\sqrt{s}}{s}\right\} \tag{7.51}$$

Using *L'Hospital's rule*, which states that for given functions $P(x)$ and $Q(x)$, the limit

$$\lim_{x\to a}\frac{P(x)}{Q(x)} = \lim_{x\to a}\frac{P'(x)}{Q'(x)} \tag{7.52}$$

applies. Introducing this rule to the first limit on the right-hand side of Eq. (7.51), and solving, yields

$$\lim_{s\to s_n}(s-s_n)\left\{e^{st}\frac{\cosh x\sqrt{s}}{s\cosh\sqrt{s}}\right\} = \lim_{s\to s_n}\left\{\frac{1}{(\sinh\sqrt{s})(1/2\sqrt{s})}\right\}\lim_{s\to s_n}\left\{e^{st}\frac{\cosh x\sqrt{s}}{s}\right\}$$

$$= \frac{4(-1)^n}{\pi(2n-1)}e^{-(n-1/2)^2\pi^2 t}\cos(n-1/2)\pi x \tag{7.53}$$

Then the Bromwich integral can be evaluated using Eq. (7.23) as

$$\frac{1}{2\pi i}\oint_{C_m}e^{st}\frac{\cosh x\sqrt{s}}{s\cosh\sqrt{s}}ds = 1 + \frac{4}{\pi}\sum_{n=1}^{m}\frac{(-1)^n}{(2n-1)}e^{-(n-1/2)^2\pi^2 t}\cos(n-1/2)\pi x \tag{7.54}$$

Taking the limit as $m \to \infty$, and noting that the integral around Γ_m approaches zero as the radius $R_m \to \infty$ (see Eq. (7.22)), the final solution can then be expressed as

$$u(t) = T_0 + \frac{4T_0}{\pi}\sum_{n=1}^{\infty}\frac{(-1)^n}{(2n-1)}e^{-\alpha(2n-1)^2\pi^2 t/4}\cos\frac{(2n-1)\pi x}{2} \tag{7.55}$$

where we reversed $\sqrt{s} \to \sqrt{s/\alpha}$.

If the domain spans between $x = 0$ and $x = L$, the solution becomes

$$u(t) = T_0 + \frac{4T_0}{\pi}\sum_{n=1}^{\infty}\frac{(-1)^n}{(2n-1)}e^{-\alpha(2n-1)^2\pi^2 t/4L^2}\cos\frac{(2n-1)\pi x}{2L} \tag{7.56}$$

If the temperature at $x = L$ is a function of time, $T(t)$, we have to solve for

$$\hat{u}(s) = \hat{T}(s)\frac{\cosh x\sqrt{s/\alpha}}{\cosh L\sqrt{s/\alpha}} \tag{7.57}$$

In this case we are entitled to utilize the convolution theory. This theory states that if $\mathcal{L}^{-1}\{\hat{f}(s)\} = F(t)$ and $\mathcal{L}^{-1}\{\hat{g}(s)\} = G(t)$ then the Laplace of the convolved functions is given by

$$\mathcal{L}^{-1}\{\hat{f}(s)\hat{g}(s)\} = \int_0^t F(\tau) G(t-\tau) d\tau = F * G \tag{7.58}$$

Following this procedure, Carslaw and Jaeger (1959) have shown that the solution of this problem can be expressed as

$$u(t) = \frac{\alpha\pi}{L^2} \sum_{n=0}^{\infty} (-1)^n (2n+1) e^{-\alpha(2n+1)^2 \pi^2 t/4L^2} \cos\frac{(2n+1)\pi x}{2L} \cdot$$
$$\int_0^t e^{\alpha(2n+1)^2\pi^2\tau/4L^2} T(\tau) d\tau \tag{7.59}$$

7.4.2 *Example 2: Solution of heat equation in an infinite domain*

Consider a heat conduction case of an infinite domain, $-\infty \le r \le \infty$ ($r = \sqrt{x^2 + y^2 + z^2}$) subjected to a point heat source at $r = 0$. Assuming constant thermo-physical properties and that the body is initially at a uniform temperature, $T_0 = 0$, the governing partial differential equation in the spherical coordinate system, assuming no variations along the φ and ψ coordinates, can be expressed as

$$\frac{1}{\alpha}\frac{\partial u}{\partial t} = \frac{\partial^2 u}{\partial r^2} + \frac{2}{r}\frac{\partial u}{\partial r} \tag{7.60}$$

If a point heat source of a constant rate q_0 is subjected at $r = 0$, the initial and boundary conditions can be described as

$$u(r, 0) = 0 \tag{7.61}$$

$$\lim_{r \to 0}\left[-4\pi r^2 \lambda \frac{\partial u}{\partial r}\right] = q_0, \quad u(\infty, t) = 0 \tag{7.62}$$

in which λ is the heat conductivity. Applying Laplace transform to Eqs. (7.60)–(7.62) yields

$$\frac{d^2\hat{u}}{dr^2} + \frac{2}{r}\frac{d\hat{u}}{dr} - \frac{s}{\alpha}\hat{u}(r, s) = 0 \tag{7.63}$$

$$\lim_{r \to 0}\left[r^2\frac{d\hat{u}}{dr}\right] = -\frac{q_0}{4\pi\lambda s}, \quad \hat{u}(\infty, s) = 0 \tag{7.64}$$

in which $\hat{u}(r, s)$ is the transformed temperature. Eq. (7.63) is the subsidiary equation. The solution of this equation, satisfying the boundary conditions, Eq. (7.64), can readily be obtained as (Yener and Kakac 2008)

$$\hat{u}(r, s) = \frac{q_0}{4\pi r\lambda s} \exp(-r\sqrt{s/\alpha}) \tag{7.65}$$

Having solved the problem in the Laplace domain, we now need to reconstruct the function back into the time domain. That is we solve for

$$f(t) = \mathcal{L}^{-1}\left\{\frac{e^{-r\sqrt{s}}}{s}\right\} \tag{7.66}$$

where, as for the previous example, we set, for simplicity, $\sqrt{s/\alpha} \to \sqrt{s}$, assuming at the moment $\alpha = 1$. To solve this problem we again follow closely the procedure presented by Spiegel (1965). By the complex inversion theorem, the Bromwich integral of Eq. (7.66) is expressed as

$$f(t) = \frac{1}{2\pi i}\int_{\gamma-i\infty}^{\gamma+i\infty} \frac{e^{st-r\sqrt{s}}}{s} ds \tag{7.67}$$

This function is not a single-valued function due to the presence of $\sqrt{s}$, and hence it has a branch point at $s = 0$. For this we choose the integration contour, C, shown in Figure 7.3. C consists of the line AB at $s = \gamma$, the arcs BDE and JKA of a circle of radius R and center at origin O; and the arc FGH of a circle of radius ε and center at O. The lines EF and HJ are actually coincident with the x-axis but are shown separated for visual purposes.

Since the only singularity $s = 0$ is not inside or along C, the integral of Eq. (7.67) must be evaluated counterclockwise along the contour line AB, BDE, EF, FGH, HJ and JKA. That is

$$\frac{1}{2\pi i} \oint_C \frac{e^{st-r\sqrt{s}}}{s} \, ds = \frac{1}{2\pi i} \int_{AB} \frac{e^{st-r\sqrt{s}}}{s} \, ds + \frac{1}{2\pi i} \int_{BDE} \frac{e^{st-r\sqrt{s}}}{s} \, ds$$

$$+ \frac{1}{2\pi i} \int_{EF} \frac{e^{st-r\sqrt{s}}}{s} \, ds + \frac{1}{2\pi i} \int_{FGH} \frac{e^{st-r\sqrt{s}}}{s} \, ds \qquad (7.68)$$

$$+ \frac{1}{2\pi i} \int_{HJ} \frac{e^{st-r\sqrt{s}}}{s} \, ds + \frac{1}{2\pi i} \int_{JKA} \frac{e^{st-r\sqrt{s}}}{s} \, ds$$

The integral of Eq. (7.68) is analytic within the region bounded by and on C. Then, by the Cauchy's theorem, the integral on the left is zero. Furthermore, the radius of the arcs BDE and JKA must satisfy the condition of Eqs. (7.22) and (7.24), and hence in the limit as $R \to \infty$ the integrals around these arcs approach zero. It follows that

$$f(t) = \frac{1}{2\pi i} \int_{\gamma - i\infty}^{\gamma + i\infty} \frac{e^{st-r\sqrt{s}}}{s} \, ds$$

$$= \lim_{\substack{R \to \infty \\ \varepsilon \to 0}} \frac{1}{2\pi i} \int_{AB} \frac{e^{st-r\sqrt{s}}}{s} \, ds \qquad (7.69)$$

$$= -\lim_{\substack{R \to \infty \\ \varepsilon \to 0}} \frac{1}{2\pi i} \left\{ \int_{EF} \frac{e^{st-r\sqrt{s}}}{s} \, ds + \int_{FGH} \frac{e^{st-r\sqrt{s}}}{s} \, ds + \int_{HJ} \frac{e^{st-r\sqrt{s}}}{s} \, ds \right\}$$

In the complex plane, we have $s = R e^{i\theta}$, $\theta_0 \le \theta \le 2\pi - \theta_0$ with $\theta_0 = \tan^{-1}[(\gamma + i\tau)/\gamma]$, see Figure 7.3. Hence, along EF, $s = xe^{\pi i}$, $\sqrt{s} = \sqrt{x}\, e^{\pi i/2} = i\sqrt{x}$ (using Euler's formula), and as s goes from $-R$ to $-\varepsilon$, x goes from R to ε. Hence,

$$\int_{EF} \frac{e^{st-r\sqrt{s}}}{s} \, ds = \int_{-R}^{-\varepsilon} \frac{e^{st-r\sqrt{s}}}{s} \, ds = \int_R^{\varepsilon} \frac{e^{-xt-ri\sqrt{x}}}{x} \, dx \qquad (7.70)$$

Along FGH, $s = \varepsilon e^{i\theta}$, $\sqrt{s} = \sqrt{\varepsilon} e^{i\theta/2}$. Thus

$$\int_{FGH} \frac{e^{st-r\sqrt{s}}}{s} \, ds = \int_\pi^{-\pi} \frac{e^{\varepsilon e^{i\theta} t - r\sqrt{\varepsilon} e^{i\theta/2}}}{\varepsilon e^{i\theta}} \, i\varepsilon e^{i\theta} \, d\theta$$

$$= i \int_\pi^{-\pi} e^{\varepsilon e^{i\theta} t - r\sqrt{\varepsilon} e^{i\theta/2}} \, d\theta \qquad (7.71)$$

Along HJ, $s = xe^{-\pi i}$, $\sqrt{s} = \sqrt{x}\, e^{-\pi i/2} = -i\sqrt{x}$ and as s goes from $-\varepsilon$ to $-R$, x goes from ε to R. Hence

$$\int_{HJ} \frac{e^{st-r\sqrt{s}}}{s} \, ds = \int_{-\varepsilon}^{-R} \frac{e^{st-r\sqrt{s}}}{s} \, ds = \int_\varepsilon^R \frac{e^{-xt+ri\sqrt{x}}}{x} \, dx \qquad (7.72)$$

Substituting Eqs. (7.70), (7.71), and (7.72) into Eq. (7.69) yields

$$f(t) = -\lim_{\substack{R \to \infty \\ \varepsilon \to 0}} \frac{1}{2\pi i} \left\{ \int_R^\varepsilon \frac{e^{-xt - ri\sqrt{x}}}{x} dx + i \int_\pi^{-\pi} e^{\varepsilon e^{i\theta} t - r\sqrt{\varepsilon} e^{i\theta/2}} d\theta + \int_\varepsilon^R \frac{e^{-xt + ri\sqrt{x}}}{x} dx \right\}$$

$$= -\lim_{\substack{R \to \infty \\ \varepsilon \to 0}} \frac{1}{2\pi i} \left\{ \int_R^\varepsilon \frac{e^{-xt}(e^{ri\sqrt{x}} - e^{-ri\sqrt{x}})}{x} dx + i \int_\pi^{-\pi} e^{\varepsilon e^{i\theta} t - r\sqrt{\varepsilon} e^{i\theta/2}} d\theta \right\} \qquad (7.73)$$

$$= -\lim_{\substack{R \to \infty \\ \varepsilon \to 0}} \frac{1}{2\pi i} \left\{ 2i \int_R^\varepsilon \frac{e^{-xt} \sin r\sqrt{x}}{x} dx + i \int_\pi^{-\pi} e^{\varepsilon e^{i\theta} t - r\sqrt{\varepsilon} e^{i\theta/2}} d\theta \right\}$$

Since

$$\lim_{\varepsilon \to 0} \frac{1}{2\pi i} \left\{ i \int_\pi^{-\pi} e^{\varepsilon e^{i\theta} t - r\sqrt{\varepsilon} e^{i\theta/2}} d\theta \right\} = \frac{1}{2\pi} \left\{ \int_\pi^{-\pi} \lim_{\varepsilon \to 0} (e^{\varepsilon e^{i\theta} t - r\sqrt{\varepsilon} e^{i\theta/2}}) d\theta \right\}$$

$$= \frac{1}{2\pi} \left\{ \int_\pi^{-\pi} 1 \, d\theta \right\} = -1 \qquad (7.74)$$

the solution in the time domain can then be expressed as

$$f(t) = 1 - \frac{1}{\pi} \int_0^\infty \frac{e^{-xt} \sin r\sqrt{x}}{x} dx \qquad (7.75)$$

This can be written, reversing $\sqrt{s} \to \sqrt{s/\alpha}$, as (see Spiegel 1965)

$$f(t) = 1 - \mathrm{erf}(r/2\sqrt{\alpha t}) = \mathrm{erfc}(r/2\sqrt{\alpha t}) \qquad (7.76)$$

Then the solution of the initial and boundary value problem, Eqs. (7.60)–(7.62), at any distance r from the origin, at time t, is given by

$$u(r, t) = \frac{q_0}{4\pi r \lambda} \mathrm{erfc}(r/2\sqrt{\alpha t}) \qquad (7.77)$$

Note that this result can be directly obtained from the table of Laplace transforms, Table 7.2 transform number 6 (see also Yener and Kakas 2008). Letting $t \to \infty$, and knowing that $\mathrm{erfc}(0) = 1$, the steady-state temperature distribution gives

$$u_{\mathrm{st}} = \frac{q_0}{4\pi\lambda r} \qquad (7.78)$$

The application in this example constitutes the basis for most of the computational models currently utilized in shallow geothermal systems. This will be covered in the next chapter.

CHAPTER 8

Commonly used analytical models for ground-source heat pumps

In this chapter, we present the basic mathematical formulations and computational models currently available and intensively utilized for the simulation of heat flow in ground-source heat pumps. Emphasis will be placed on the utilization of the line and cylindrical heat source models. The aim of this chapter is to present modeling ideas which have been developed on the basis of the point heat source model of Carslaw and Jaeger in 1947. In the meanwhile, we show how simple modeling procedures might lead to rather complicated inverse transform integrals, which require rigorous computational tools to tackle.

8.1   INTRODUCTION

A borehole heat exchanger (BHE) constitutes an important technology for heating and cooling of buildings and other facilities, and is widely used in the renewable energy ground-source heat pump systems. It consists of one or more U-tubes inserted in a borehole that is filled with grout, see Figure 1.2.

The borehole heat exchanger is a very slender heat pipe with dimensions of the order of 30 mm in diameter for the U-tube, and 150 mm in diameter and 100 m in length for the borehole. The U-tubes carry a working fluid that collects heat from the surrounding soil and interacts with other BHE components. These two factors, slenderness and thermal interactions, exert significant computational difficulties, and as a consequence, considerable efforts are placed on their modeling. Several theoretical and computational assumptions and approximations were introduced in order to circumvent these difficulties and obtain feasible solutions. Most of the currently utilized analytical and semi-analytical models are based on the work of Carslaw and Jaeger in 1947 (see Carslaw and Jaeger 1959), who seem to be the first to introduce a comprehensive treatment of heat conduction in solids subjected to different combinations of initial and boundary conditions. Heat flow in finite, semi-infinite and infinite domains subjected to point, line, plane, cylindrical and spherical heat sources were studied in their works between 1947 and 1959. In the meanwhile, and on the basis of Carslaw and Jaeger works, Ingersoll *et al.* (1948 and 1954) made a significant contribution to the field of heat conduction in solids and provided a practical framework for modeling geothermal systems.

The idea of the point source of heat in an infinite solid was due to Kelvin systematic use of Fourier's theory of heat conduction and the study of transmission of electric signals through submarine cables, around the year 1880. The *instantaneous point heat source*, corresponding to an instantaneous release of a finite quantity of heat at a given point and time in an infinite domain, has proven to be a useful tool for the theory of heat conduction. One great advantage is that by integrating with time, we obtain the solution for the *continuous point heat source*, which corresponds to a prescribed rate of heat subjected at a given point in space. Upon integrating the solution of a point source with regards to an appropriate space variable, we obtain solutions for continuous line, plane, spherical surface, and cylindrical surface sources. Using this concept, the solutions of a great many problems can be constructed in the form of definite integrals.

This simple mathematical interpretation of the problem has encouraged the utilization of this theory for modeling heat transfer in shallow geothermal systems. The borehole heat exchanger is represented as a line or a cylindrical heat source, with constant or time dependent heat flow rates. The soil mass is represented by an infinite or a semi-infinite medium subjected to a line or a cylindrical heat source.

Current analytical and semi-analytical models for shallow geothermal systems are covered in many journal papers. The governing equations of these models are usually solved based on the Laplace transforms. In this chapter, we make no pretense to cover all available models in this field, but tempt to give an overview on some of the commonly utilized models and their solution procedures. The purpose is to illustrate the advantages and disadvantages of the current modeling procedures, and to serve as the basis for tackling some of the computational problems that will be dealt with in the subsequent chapters.

Most of the available mathematical formulations for shallow geothermal systems treat the borehole heat exchanger as a constant heat source embedded in a soil mass. They either regard the borehole heat exchanger as a point heat source in a thin layer system, an infinite line heat source, a finite line heat source, or a cylindrical surface heat source. In another type of models, the borehole heat exchanger is explicitly modeled and the temperatures in some or all of the BHE components are calculated. This type of models describes the soil-BHE thermal interaction in more details. In what follows we give a brief account on the current soil and borehole heat exchanger models.

8.2 MODELING SOIL MASS

In the literature, there are several analytical models, which can be classified in three distinct categories: the infinite line source model (ILS), the finite line source model (FLS), and the infinite cylindrical source model (ICS). These models calculate heat conduction in an axially symmetric soil mass, embedded in which a vertical borehole heat exchanger with a constant heat flow rate per unit length. The particular attributes and physical interpretation of these models are best described by Philippe *et al.* (2009), from which we base our description of these models.

8.2.1 *Infinite line source model*

The infinite line source model (ILS) is a direct utilization of the point and infinite line source solutions provided by Carslaw and Jaeger (1959). Ingersoll (1954) provided practical applications of these solutions and established the framework for more elaborate modeling of shallow geothermal systems. ILS model describes heat flow in an infinite soil mass subjected to a constant heat flux from a borehole heat exchanger. The borehole heat exchanger is represented by an infinite line embedded along the vertical axis. This modeling set-up entails that heat flow in the soil mass occurs only in the radial direction, and furthermore, the contact between the soil and the BHE goes along the centerline of the borehole, not along its surface area.

The governing partial differential equation of the ILS model is described in the cylindrical coordinate system in terms of the temperature difference, $u(r,t) = T(r,t) - T_0$, with T_0 represents the initial temperature of the soil, as

$$\frac{1}{\alpha}\frac{\partial u}{\partial t} = \frac{\partial^2 u}{\partial r^2} + \frac{1}{r}\frac{\partial u}{\partial r}$$

$$u(r,0) = 0$$

$$u(r \to \infty, t) = 0 \tag{8.1}$$

$$-\lambda\frac{\partial u}{\partial r}\cdot 2\pi r\Big|_{r\to 0} = q_0$$

in which q_0 is the BHE heat flux. Solving this initial and boundary value problem is usually conducted using the Laplace transforms. Accordingly, the temperature in the Laplace domain can be described as

$$\hat{u}(r,s) = \mathcal{L}[u(r,t)] = \int_0^\infty e^{-st}u(r,t)\,dt \tag{8.2}$$

where the transform is conducted on the time domain. Applying Eq. (8.2) to the time derivative in Eq. (8.1) gives

$$\mathcal{L}\left[\frac{\partial u}{\partial t}\right] = \int_0^\infty e^{-st}\frac{\partial u}{\partial t}\,dt = \lim_{P\to\infty}\int_0^P e^{-st}\frac{\partial u}{\partial t}\,dt \tag{8.3}$$

Integrating by parts, yields

$$\mathcal{L}\left[\frac{\partial u}{\partial t}\right] = \lim_{P\to\infty}\left\{e^{-st}u(r,t)\Big|_0^P + s\int_0^P e^{-st}u(r,t)\,dt\right\}$$

$$= -u(r,0) + s\int_0^\infty e^{-st}u(r,t)\,dt \tag{8.4}$$

$$= s\hat{u}(r,s) - u(r,0)$$

Note that we could get the same result if we directly applied property number 4 in Sub-section 7.2.1, Chapter 7. Also note that, as shown in the last term of Eq. (8.4), we could include the initial condition at this early stage. In the same way, applying Eq. (8.2) to the first and second order spatial derivatives in Eq. (8.1), gives

$$\mathcal{L}\left[\frac{\partial u}{\partial r}\right] = \int_0^\infty e^{-st}\frac{\partial u}{\partial r}\,dt = \frac{d}{dr}\left\{\lim_{P\to\infty}\int_0^P e^{-st}u\,dt\right\} = \frac{d\hat{u}}{dr}$$

$$\mathcal{L}\left[\frac{\partial^2 u}{\partial r^2}\right] = \int_0^\infty e^{-st}\frac{\partial^2 u}{\partial r^2}\,dt = \frac{d^2}{dr^2}\left\{\lim_{P\to\infty}\int_0^P e^{-st}u\,dt\right\} = \frac{d^2\hat{u}}{dr^2} \tag{8.5}$$

Substituting Eq. (8.4) and Eq. (8.5) into Eq. (8.1), and considering the initial condition in Eq. (8.1), yields

$$\frac{d^2\hat{u}}{dr^2} + \frac{1}{r}\frac{d\hat{u}}{dr} - \frac{s}{\alpha}\hat{u} = 0$$

$$\hat{u}(r\to\infty,s) = 0 \tag{8.6}$$

$$-\lambda\frac{d\hat{u}}{dr}\cdot 2\pi r\bigg|_{r\to 0} = \frac{q_0}{s}$$

The solution of this ordinary differential equation can be expressed as

$$\hat{u}(r,s) = AK_0(r\sqrt{s/\alpha}) + BI_0(r\sqrt{s/\alpha}) \tag{8.7}$$

in which I_0 and K_0 are the modified Bessel functions of zero order of the first and second kind, respectively. A and B are arbitrary constants which need to be determined from the boundary conditions. Applying the first boundary condition in Eq. (8.6), and as I_0 is unbounded at infinity, yields $B=0$, and Eq. (8.7) becomes

$$\hat{u}(r,s) = AK_0(r\sqrt{s/\alpha}) \tag{8.8}$$

Substituting Eq. (8.8) into the second boundary condition of Eq. (8.6), and knowing that

$$\lim_{r\to 0} K_1(r) \simeq \frac{1}{r} \tag{8.9}$$

where K_1 is the modified Bessel functions of the first order, we obtain

$$A = \frac{q_0}{2\pi\lambda s} \tag{8.10}$$

The solution of $\hat{u}(r,s)$ in the Laplace domain is thus

$$\hat{u}(r,s) = \frac{q_0}{2\pi\lambda s}K_0(r\sqrt{s/\alpha}) \tag{8.11}$$

Having solved the heat equation in the Laplace domain, we now have to reconstruct the temperature back to the time domain, using the inverse Laplace transform. This entails solving for

$$u(r,t) = \mathcal{L}^{-1}[\hat{u}(r,s)] = \frac{q_0}{2\pi\lambda} \mathcal{L}^{-1}\left[\frac{1}{s} K_0(r\sqrt{s/\alpha})\right] \tag{8.12}$$

Making use of the Laplace transform of integrals (see property number 5 in Sub-section 7.2.1, Chapter 7):

$$\frac{1}{s}\hat{f}(s) = \mathcal{L}\left[\int_0^t f(\tau)d\tau\right] \tag{8.13}$$

Eq. (8.12) can be written as

$$u(r,t) = \frac{q_0}{2\pi\lambda} \int_0^t \mathcal{L}^{-1}\left[K_0(r\sqrt{s/\alpha})\right] d\tau \tag{8.14}$$

Using the inverse Laplace transform tables provided by Carslaw and Jaeger (1959), we obtain

$$\mathcal{L}^{-1}\left[K_0(r\sqrt{s/\alpha})\right] = \frac{1}{2t} \exp(-r^2/4\alpha t) \tag{8.15}$$

Substituting Eq. (8.15) into Eq. (8.14) yields the temperature in the time domain, as

$$u(r,t) = \frac{q_0}{4\pi\lambda} \int_0^t \frac{1}{\tau} \exp\left(-r^2/4\alpha\tau\right) d\tau \tag{8.16}$$

This concludes the solution of the infinite line source model. For convenience, the integrand can be written using an independent variable, such that

$$v = \frac{r^2}{4\alpha\tau}; \quad d\tau = -\frac{r^2}{4\alpha v^2} dv \tag{8.17}$$

Substituting Eq. (8.17) into Eq. (8.16), gives

$$u(r,t) = \frac{q_0}{4\pi\lambda} \int_{r^2/4\alpha t}^{\infty} \frac{e^{-v}}{v} dv \tag{8.18}$$

In the literature, Eq. (8.18) is usually expressed as

$$
\begin{aligned}
T(r,t) - T_0 &= \frac{q_0}{2\pi\lambda} \int_{r/2\sqrt{\alpha t}}^{\infty} \frac{e^{-v^2}}{v} dv \\
&= \frac{q_0}{4\pi\lambda} \mathrm{Ei}\left(\frac{r^2}{4\alpha t}\right) \\
&= \frac{q_0}{2\pi\lambda} \mathrm{Ei}\left(\frac{r}{2\sqrt{\alpha t}}\right)
\end{aligned}
\tag{8.19}
$$

where Ei(x) is the *exponential integral function*, defined as

$$\mathrm{Ei}(x) = \int_x^{\infty} \frac{e^{-v}}{v} dv \tag{8.20}$$

Note that the last two formulations in Eq. (8.19) are used in practice and are equal since,

$$-\mathrm{Ei}(r^2/4\alpha t) = -\frac{1}{2}\mathrm{Ei}(r/2\sqrt{\alpha t}) \tag{8.21}$$

Ingersoll and Plass (1948) provided tabulated values of the exponential integral function. For $r/2\sqrt{\alpha t} < 0.2$, the integral in Eq. (8.19) can be approximated by

$$I(v) = \ln\frac{1}{v} + \frac{v^2}{2} - \frac{v^4}{8} - 0.2886 \tag{8.22}$$

Ingersoll and Plass recommended using the ILS model only for applications with Fourier's number, $\alpha t/r_b^2 > 20$. For smaller values, the solution gets distorted in the shorter time scale because the effect of the actual finite length of the BHE becomes significant.

Hart and Couvillion (1986) introduced an algebraic approximation to Eq. (8.19) of the form

$$T(r,t) - T_0 = \frac{q_0}{2\pi\lambda} \left[\ln\frac{r_\infty}{r} - 0.9818 + \frac{2r^2}{r_\infty^2} - \frac{1}{8}\left(\frac{4r^2}{r_\infty^2}\right)^2 + O\left(\frac{4r^2}{r_\infty^2}\right) \right] \tag{8.23}$$

in which r_∞ is a hypothetical far-field radius where the effect of the line source vanishes. They proposed aradius of the form

$$r_\infty = 4\sqrt{\alpha t} \tag{8.24}$$

Hart and Couvillion asserted that this solution provides computationally efficient calculations for $r_\infty > 15r_p$, with r_p the pipe diameter. Lower than this value, the calculations become more involved. Various other algebraic approximations to Eq. (8.19) can be found in the mathematical functions handbook by Abramowitz and Stegun (1972).

From Eq. (8.19) we note that as $t \to \infty$, $\mathrm{Ei}(0) = \infty$, and thus there exists no steady-state solution to this model. However, for a small value of v, i.e., large values of t, Carslaw and Jaeger have shown that the temperature can be approximated as

$$T(r,t) \simeq \frac{q_0}{4\pi\alpha} \ln\frac{4\alpha t}{r^2} - \frac{q_0}{4\pi\alpha}\gamma \tag{8.25}$$

in which $\gamma = 0.5772\ldots$, is *Euler's constant*. Furthermore, for the same reason, in this model the temperature at the center, where $r = 0$, cannot be calculated. Despite of these two disadvantages, the ILS model is usually used in practice to get a quick estimate of the soil temperature.

For an instantaneous line heat source, where a heat flux is released suddenly at $t = 0$ with a strength q_i (J/m), Yener and Kakac (2008) have shown that the solution of the heat equation leads to

$$u(r,t) = T(r,t) - T_0 = \frac{q_i}{4\pi\lambda}\frac{1}{t}\exp(-r^2/4\alpha t) \tag{8.26}$$

8.2.2 *Infinite cylindrical source model*

Similar to the ILS model, the infinite cylindrical source model (ICS) simulates heat conduction in a soil mass subjected to a constant heat flow rate. The difference, however, is that the contact area between the borehole heat exchanger and the soil is along the surface area of the borehole, i.e. at $r = r_b$, the borehole radius. The governing partial differential equation of the ICS model is described in the cylindrical coordinate system in terms of $u(r,t) = T(r,t) - T_0$, with T_0 represents the initial temperature of the soil, as

$$\frac{1}{\alpha}\frac{\partial u}{\partial t} = \frac{\partial^2 u}{\partial r^2} + \frac{1}{r}\frac{\partial u}{\partial r}, \quad r > r_b$$

$$u(r,0) = 0$$

$$u(r \to \infty, t) = 0 \tag{8.27}$$

$$-\lambda\frac{\partial u}{\partial r} \cdot 2\pi r \bigg|_{r=r_b} = q_0$$

Using Laplace transform, the subsidiary equation can be expressed as

$$\frac{d^2\hat{u}}{dr^2} + \frac{1}{r}\frac{d\hat{u}}{dr} - \frac{s}{\alpha}\hat{u} = 0$$

$$\hat{u}(r \to \infty, s) = 0 \tag{8.28}$$

$$-\lambda \frac{d\hat{u}}{dr} \cdot 2\pi r \bigg|_{r=r_b} = \frac{q_0}{s}$$

The solution of this boundary value problem can readily be obtained as

$$\hat{u}(r,t) = \frac{q_0}{2\pi r_b \lambda s} \frac{K_0(r\sqrt{s\alpha})}{K_1(r_b\sqrt{s/\alpha})} \tag{8.29}$$

in which K_0 and K_1 are modified Bessel functions of the second kind of order 0 and 1, respectively. Then, using the Bromwich integral (see Chapter 7), the inverse Laplace transform can be expressed as

$$u(r,t) = \frac{q_0}{4\pi^2 r_b \lambda_i} \int_{\gamma - i\infty}^{\gamma + i\infty} e^{st} \frac{K_0(r\sqrt{s/\alpha})}{sK_1(r_b\sqrt{s/\alpha})} ds \tag{8.30}$$

The integrand in Eq. (8.30) has a branch point at $s = 0$, and hence the integral can be solved using the contour line of Figure 8.1. Carslaw and Jaeger (1959) provided the solution of this integrand, of the form

$$T(r,t) - T_0 = -\frac{2Q_0}{\pi\lambda} \int_0^\infty (1 - e^{-\alpha v^2 t}) \frac{J_0(vr) Y_1(vr_b) - Y_0(vr) J_1(vr_b)}{v^2 [J_1^2(vr_b) + Y_1^2(vr_b)]} dv \tag{8.31}$$

in which $Q_0 = q_0/2\pi r_b$, at $r = r_b$.

Solving the integral in Eq. (8.31) is quite difficult. Nevertheless, Carslaw and Jaeger provided approximate solutions for short and long time scales. For a short time scale, i.e. small values of Fourier's number $\alpha t/r_b^2$, they showed that Eq. (8.31) can be represented as

$$T(r,t) = \frac{2Q_0}{\lambda} \sqrt{\frac{\alpha r_b t}{r}} \left[\text{ierfc}\frac{r - r_b}{2\sqrt{\alpha t}} - \frac{(3r + r_b)\sqrt{\alpha t}}{4rr_b} i^2 \text{erfc}\frac{r - r_b}{2\sqrt{\alpha t}} + \cdots \right] \tag{8.32}$$

where i^nerfc is the iterated integrals of the complementary error function, defined as

$$i^n \text{erf}(z) = \int_z^\infty i^{n-1} \text{erf}(\zeta) d\zeta \tag{8.33}$$

For a long time scale, the evaluation of the integral in Eq. (8.31) becomes more involved and difficult to solve analytically. Therefore it is important to begin from the subsidiary equation and solve the problem by proper evaluation of the Bromwich integral, Eq. (8.30). As described in Chapter 7, and as the radius of the circle tends to infinity, the integrals over the arcs AKJ and EDB tend to zeros; and the line $(\gamma - i\infty, \gamma + i\infty)$ can be transformed into the contour $JHGFE$, Figure 8.1. Hence the integral will be performed along the contour which begins from $-\infty$ in the lower half plane, passes around the branch point at the origin in the positive direction, and ends at $-\infty$ in the upper plane. This contour is denoted as $(-\infty, 0^+)$. Hence, the counter integration in Eq. (8.30) can be approximated as

$$u(r,t) = \frac{Q_0}{2\pi r_b i\lambda} \int_{-\infty}^{0^+} e^{v^2 \alpha t} \frac{2K_0(rv)}{vK_1(r_b v)} dv \tag{8.34}$$

in which $v = \sqrt{s/\alpha}$. That is the contour integral along the complex plane $(\gamma - i\infty, \gamma + i\infty)$ is reduced to a real integral along $(-\infty, 0^+)$. This makes the integration somewhat easier. To facilitate

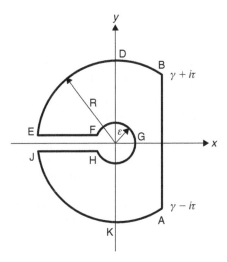

Figure 8.1   Contour path of the ICS model.

the integrand in Eq. (8.34), the modified Bessel function is approximated using:

$$K_n(z) = (-1)^{n+1}\{\ln(z/2) + \gamma\}I_n(z)$$

$$+ \frac{1}{2}(-1)^n \sum_{x=0}^{\infty} \frac{(z/2)^{n+2x}}{x!\,(n+x)!} \left[\sum_{m=1}^{n+x} m^{-1} + \sum_{m=1}^{x} m^{-1}\right] \tag{8.35}$$

$$+ \frac{1}{2} \sum_{x=0}^{n-1} (-1)^x (z/2)^{-n+2x} \frac{(n-x-1)!}{x!}$$

where n is any positive integer, and I_n is the modified Bessel function of the first kind of order n, and $\gamma = \ln C = 0.57722\ldots$, is Euler's constant. Now, using some other typical integrals given in Carslaw and Jaeger (1959), together with Eq. (8.35), Carslaw and Jaeger showed that the solution for the long time scale can be approximated by

$$u(r,t) = \frac{Q_0 r_b}{2\lambda} \left\{\ln\frac{4\alpha t}{Cr^2} + \frac{r_b^2}{2\alpha t}\ln\frac{4\alpha t}{Cr^2} + \frac{1}{4\alpha t}\left[r_b^2 + r^2 - 2r_b^2\ln\frac{r_b}{r}\right] + \cdots\right\} \tag{8.36}$$

Note that Eq. (8.36) is equivalent to Eq. (8.25) for a line source emitting $2\pi r_b q_0$ units of heat per unit of time per unit of length. In this context, Ingersoll (reported in Yavuzturk 1999) asserted that for a heat source of less than 50 mm in diameter, the infinite line source model can still produce accurate results. Beyond that, error occurs.

Ingersoll *et al.* (1954) provided tabulated values of the integral in Eq. (8.31) for different values of Fourier's number, but limited to only four values of r/r_b. Philippe *et al.* (2009) presented and utilized the solution given by Baudoin (1988) (reported in Philippe *et al.* 2009), who solved Eq. (8.29) using the Gaver-Stehfest numerical Laplace transform inversion algorithm, which yields

$$u(r,t) = -\frac{q_0}{2\pi\lambda r_b} \sum_{j=1}^{10} \left[\frac{D_j}{j\mu_j} \frac{K_0(\mu_j r)}{K_1(\mu_j r_b)}\right] \tag{8.37}$$

with

$$\mu_j = \sqrt{j\ln 2/\alpha t}$$

$$D_j = \sum_{k=\text{Int}((j+1)/2)}^{\min(j,5)} \frac{(-1)^{j-5}k^5(2k)!}{(5-k)!\,(k-1)!\,k!\,(j-k)!\,(2k-j)!} \tag{8.38}$$

Philippe *et al.* (2009) compared the numerical results obtained from eq. (8.37) with the finite different solution of the ICS model given by Lee and Lam (2008), and showed that there is a quite good matching between the two results.

It can be noticed that modeling heat conduction in infinite and semi-infinite regions, and even by applying simplified physical assumptions such as small or large Fourier's numbers, or using numerical procedures for solving the semi-infinite integral in Eq. (8.31), it is still a formidable task. On the other hand, for a finite region, and as we have seen in the numerical examples of Chapter 6 and 7, the solution can been obtained in terms of series summation, which converge relatively rapidly. In cylindrical coordinate system, for example, Yener and Kakac (2008) have shown that the solution of the heat equation of a finite domain in the range $r_b \leq r \leq R$, subjected to a heat flux $q'(t)$ per unit length, can be expressed as

$$T(r,t) - T_0 = \frac{1}{\pi R^2} \sum_{n=1}^{\infty} e^{-\alpha\beta_n^2 t} \frac{J_0(\beta_n r)J_0(\beta_n r_b)}{J_1^2(\beta_n R)} \int_0^t e^{\alpha\beta_n^2 \tau}q'(\tau)\,d\tau \tag{8.39}$$

where β_n is the positive roots of $J_0(\beta R) = 0$. In practice, in many of the currently utilized models, heat flux coming from the borehole heat exchanger is considered constant, i.e. $q'(t) = q'$. In this case, and by solving the integral of Eq. (8.39), the temperature distribution in the medium can be expressed as

$$T(r,t) - T_0 = \frac{q'}{\pi\lambda R^2} \sum_{n=1}^{\infty} \frac{1-e^{-\alpha\beta_n^2 t}}{\beta_n^2} \frac{J_0(\beta_n r)J_0(\beta_n r_b)}{J_1^2(\beta_n R)} \tag{8.40}$$

Obviously, Eq. (8.40) is much more elegant and relatively easy to handle. Therefore, and in order to circumvent the hassle of solving infinite integrals involved in infinite regions, it is recommended, when possible, to terminate the far field distance at, say $r = R$, where it is known analytically or intuitively that heat flux generated by the source vanishes. That is, despite dealing with an infinite region, our region of interest is finite. In Chapter 9 we shall pursue the concept of the "*region-of-interest*" in order to facilitate our solution.

8.2.3 *Finite line source model*

The finite line source model (FLS) approximates the borehole heat exchanger by a finite line constituting a series of point sources. The temperature in the soil is calculated by integrating the solution of the continuous point source case over the length of the borehole.

We have seen in Chapter 7 that the temperature at point r in an infinite domain subjected to a constant heat flux, q_0, is described as

$$T(r,t) = \frac{q_0}{4\pi r\lambda}\text{erfc}(r/2\sqrt{\alpha t}) \tag{8.41}$$

In an infinitely thin layer, $d\zeta$, the temperature variation due to total heat flux $q_0 d\zeta$, can be described as

$$dT(r,t) = \frac{q_0}{4\pi r\lambda\zeta}\text{erfc}(r/2\sqrt{\alpha t})\,d\zeta \tag{8.42}$$

Imagine that we have many thin layers laid on top of each other and inserted in an infinite domain. Each layer is subjected to a point source at its origin, as shown in Figure 8.2. Points in the domain are subjected to an accumulated heat flow coming from all involved point heat sources.

In this case, the temperature at a point, $P(r, z)$ for instance, can be described by integrating Eq. (8.42) over the length of the point heat sources as

$$T(r, z, t) = \frac{q_0}{4\pi\lambda} \int_0^L \frac{1}{r_1} \mathrm{erfc}(r_1/2\sqrt{\alpha t}) \, d\zeta \tag{8.43}$$

in which $r_1 = \sqrt{r^2 + (z - \zeta)^2}$, with ζ is any point along the line source (borehole).

Physically, the soil mass is semi-infinite and the surface temperature variation has significant influence on the soil temperature. To model the soil mass as a semi-infinite region in the z-direction, and impose a prescribed temperature at the ground surface, the *method of images* can be utilized (Carslaw and Jaeger 1959). This method was first introduced in the mathematical theory of electricity, and elegantly adopted to the solution of heat conduction in semi-infinite medium, where a constant temperature is imposed at the surface. The image has identical values as those in the original domain but opposite in magnitude. In this case, we have a source at point ζ and a sink at point $-\zeta$, Figure 8.2. Based on this technique, Eskilson (1987) introduced the FLS model, expressed as

$$T(r, z, t) - T_0 = \frac{q_0}{4\pi\lambda} \int_0^L \left[\frac{1}{r_1} \mathrm{erfc}\left(\frac{r_1}{2\sqrt{\alpha t}}\right) - \frac{1}{r_2} \mathrm{erfc}\left(\frac{r_2}{2\sqrt{\alpha t}}\right) \right] d\zeta \tag{8.44}$$

in which T_0 is the initial temperature, and

$$r_1 = \sqrt{r^2 + (z - \zeta)^2}, \quad r_2 = \sqrt{r^2 + (z + \zeta)^2} \tag{8.45}$$

This model is valid for

$$5r_b^2/\alpha < t < t_s/10 \tag{8.46}$$

where t_s is the time when the steady-state condition has been reached. This time ranges between few hours to few years.

On the basis of this model, Claesson and Eskilson (1987) introduced the concept of what is known as the *g-function*. This function is a dimensionless function representing the temperature at the borehole wall, described as

$$T_b - T_0 = \frac{q_0}{2\pi\lambda} g\left(\frac{t}{t_s}, \frac{r_b}{L}\right), \quad t_s = \frac{L^2}{9\alpha} \tag{8.47}$$

in which T_b is the borehole surface temperature, and t_s is a steady-state time scale. In the last two decades, and since the introduction of the Eskilson model in 1987, this model has been utilized intensively and seems to dominate the research works for ground source heat pumps, especially those dealing with analytical and semi-analytical procedures.

Claesson and Eskilson treated Eq. (8.47) numerically and analytically. Numerically, the temperature distribution in the soil mass due to a unit step heat pulse is calculated using the finite difference method. The response to multiple heat sources coming from different boreholes is calculated as a superposition of a series of step functions, such that

$$T_b - T_0 = \sum_i \frac{\Delta q_i}{2\pi\lambda} g\left(\frac{t - t_i}{t_s}, \frac{r_b}{L}\right) \tag{8.48}$$

These functions are calculated for various borehole geometry and configurations, and stored in the database of commonly utilized design and analysis software. The limitation of the numerical g-function is that it is only valid for a time greater than $L^2/9\alpha$. Analytically, Claesson and Eskilson gave two asymptotic approximations to the g-function, such that

$$g\left(\frac{t}{t_s}, \frac{r_b}{L}\right) = \begin{cases} \ln\left(\frac{L}{2r_b}\right) + \frac{1}{2}\ln\left(\frac{t}{t_s}\right) & 5r_b^2/\alpha < t \leq t_s \\ \\ \ln\left(\frac{L}{2r_b}\right) & t \geq L^2/9\alpha \end{cases} \tag{8.49}$$

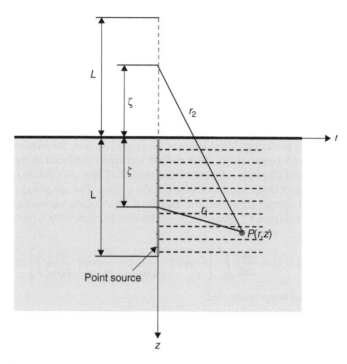

Figure 8.2   Finite line source model.

They seemed to adopt the numerical model on their calculations, and accordingly provided tabulated numerical results for many different borehole layout configurations. Following Lamarche and Beauchamp (2007a), the reason for this choice is that the analytical g-function, Eq. (8.49), consistently overestimates the temperature.

The concept of the g-function has received considerable interest from researchers in the field and soon after its introduction, several models with different complexity and computational efficiency have been introduced. Zeng *et al.* (2002 and 2003) suggested somewhat similar, non-dimensional g-function, defined as

$$g(t, L') = \int_0^1 \left[\frac{1}{\tilde{r}_1} \mathrm{erfc}\left(\frac{\tilde{r}_1}{2\sqrt{\mathrm{Fo}}} \right) - \frac{1}{\tilde{r}_2} \mathrm{erfc}\left(\frac{\tilde{r}_2}{2\sqrt{\mathrm{Fo}}} \right) \right] dL' \tag{8.50}$$

where

$$\tilde{r}_1 = \sqrt{R^2 + (Z - L')^2}, \quad \tilde{r}_2 = \sqrt{R^2 + (Z + L')^2}, \quad \mathrm{Fo} = \alpha t / L^2$$
$$Z = z/L, \quad L' = \zeta/L, \quad R = r/L \tag{8.51}$$

However, this model does not really produce significant improvements over Eq. (8.49). Lamarche and Beauchamp derived a computationally more efficient g-function based on the continuous infinite line source model (ILS). After an elaborate mathematical manipulation, they arrived at a g-function of the form

$$g(t^*, \beta) = \int_\beta^{\sqrt{\beta^2 + 1}} \frac{\mathrm{erfc}(\gamma z)}{\sqrt{z^2 - \beta^2}} dz - D_A - \int_{\sqrt{\beta^2 + 1}}^{\sqrt{\beta^2 + 4}} \frac{\mathrm{erfc}(\gamma z)}{\sqrt{z^2 - \beta^2}} dz - D_B \tag{8.52}$$

in which

$$D_A = \int_{\beta}^{\sqrt{\beta^2+1}} \mathrm{erfc}(\gamma z)\, dz$$

$$= \sqrt{\beta^2 + 1}\, \mathrm{erfc}(\gamma\sqrt{\beta^2 + 1}) - \beta\, \mathrm{erfc}(\gamma\beta) - \frac{1}{\gamma\sqrt{\pi}}\left[e^{-\gamma^2(\beta^2+1)} - e^{-\gamma^2\beta^2}\right]$$

(8.53)

and

$$D_B = \frac{1}{2}\left[\int_{\beta}^{\sqrt{\beta^2+1}} \mathrm{erfc}(\gamma z)\, dz + \int_{\sqrt{\beta^2+1}}^{\sqrt{\beta^4+1}} \mathrm{erfc}(\gamma z)\, dz\right]$$

$$= \sqrt{\beta^2 + 1}\, \mathrm{erfc}(\gamma\sqrt{\beta^2 + 1}) - \frac{1}{2}\beta\, \mathrm{erfc}(\gamma\beta) + \sqrt{\beta^2 + 4}\, \mathrm{erfc}(\gamma\sqrt{\beta^2 + 4})$$

$$- \frac{1}{\gamma\sqrt{\pi}}\left[e^{-\gamma^2(\beta^2+1)} - \frac{1}{2}\left(e^{-\gamma^2\beta^2} + e^{-\gamma^2(\beta^2+4)}\right)\right]$$

(8.54)

where $t^* = t/t_s$ with t_s is as shown in Eq. (8.47), and $\gamma = 3/2\sqrt{t^*}$. For the detailed solution of the integrals in Eqs. (8.53) and (8.54) the reader is referred to Lamarche and Beauchamp (2007b). Note that the first integral in Eq. (8.52) is *convergent improper* and need special numerical tools for solving it. Lamarche and Beauchamp provided information on the possible utilization of some available numerical tools, which are suitable for solving Eq. (8.52). They conducted numerical examples for two β values and showed that, despite the apparent intricate formulation, the CPU time needed for solving their g-function was ten of orders less than those of Eskilson and Zeng. Marcotte and Pasquier (2008), solved this model in the frequency domain using the Fast Fourier Transform algorithm (FFT). They also extended the model to account for borehole inclination and borehole head located below the ground surface (Marcotte and Pasquier 2009).

Philippe *et al.* (2009) used the recursive adaptive Simpson quadrature for solving the integral in Eq. (8.44) numerically. They compared the numerical results with the finite difference solution of the FLS model introduced by Lee and Lam (2008), and showed that there is a quite good matching between the two results.

In contrast to ISL, FSL model allows for the calculation of the steady-state temperature, which can be described as

$$T(r,z) - T_0 = -\frac{q_0}{4\pi\lambda}\int_0^L \left(\frac{1}{r_1} - \frac{1}{r_2}\right) d\zeta$$

(8.55)

Solving Eq. (8.55) gives (Philippe *et al.* 2009)

$$T(r,z) - T_0 = -\frac{q_0}{4\pi\lambda}\ln\left(\frac{\sqrt{r^2 + (z-L)^2} - (z-L)}{\sqrt{r^2 + (z+L)^2} + (z+L)} \cdot \frac{\sqrt{r^2 + z^2} + z}{\sqrt{r^2 + z^2} - z}\right)$$

(8.56)

8.2.4   *Short-time transient response*

Models presented so far, among many others, are formulated for analyzing relatively long-term performance of the geothermal systems. The short-term transient response is ignored in order to make the models computationally feasible for utilization in engineering practice. These models constitute the basic tools, which are commonly utilized for designing borehole heat exchangers. However, it was found that the short-term response has significant impact on the performance of the heat pump and the heat flow build up stages in both heating and cooling modes. Studies have shown that optimization of the electric demands for system operation requires accurate knowledge of the short-term behavior.

In the literature, there are many numerical and analytical models capable of simulating the short-term response of shallow geothermal systems. Yavuzturk and Spitler (1999) seemed to be

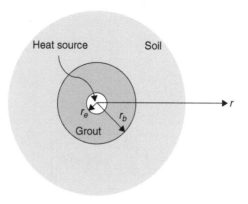

Figure 8.3   Lamarche and Beauchamp short-term model.

the first to extend Eskilson's concept of the *g*-function to include the short-term behavior, using the implicit finite volume method. However, Yavuzturk and Spitler model is limited to simulating heat flow in the radial direction only. The model is used to calculate the average fluid temperature in the borehole. This temperature is adjusted by the borehole thermal resistance to determine the average temperature at the borehole wall. Then it is normalized to formulate a *g*-function for short-time steps. They showed that the resulting short time-step *g*-function curve matches very well the lower boundary of the long time-step *g*-function developed by Eskilsonand Claesson. Since then, most of the subsequent analytical and design procedures are either direct application of these *g*-functions, or enhanced by some modifications, but mostly along the same line. In an another development, Al-Khoury and Bonnier (2006) introduced fully transient three-dimensional model using the Petrov-Galerkin finite element method. This model will be treated in Chapter 12. Lamarche and Beauchamp (2007b) introduced a comprehensive analytical model for short-term analysis based on Laplace transform. Recently, Al-Khoury (2010) introduced a fully transient analytical model based on the spectral analysis method. This model will be treated in Chapter 9. Here we briefly present Lamarche and Beauchamp model.

Lamarche and Beauchamp short-term model assumes that the shallow geothermal system is described by two concentric domains: an inner, representing the borehole, and an outer, representing the soil mass, Figure 8.3. The model describes heat flow in the radial direction. The U-tube (pipe-in and pipe-out) is not modeled, but its effect is included as a heat source on the grout inner surface area. The proposed governing equations are:

$$
\frac{1}{\alpha_g}\frac{\partial T_g}{\partial t} = \frac{\partial^2 T_g}{\partial r^2} + \frac{1}{r}\frac{\partial T_g}{\partial r}
$$

$$
\frac{1}{\alpha_s}\frac{\partial T_s}{\partial t} = \frac{\partial^2 T_s}{\partial r^2} + \frac{1}{r}\frac{\partial T_s}{\partial r}
$$

(8.57)

where the subscript *g* refers to the grout and the subscript *s* refers to the soil. Two different boundary conditions are applied. One prescribes heat flux at the grout, and the other applies a convection boundary condition. They are described as

Heat flux:

$$
T_g(r,0) = T_s(r,0) = T_0
$$

$$
T_g(r_b,t) = T_s(r_b,t)
$$

$$
-\lambda_g\frac{\partial T_g}{\partial r} = q_b''(t)
$$

(8.58)

$$
-\lambda_g\frac{\partial T_g}{\partial r}\bigg|_{r=r_b} = -\lambda_s\frac{\partial T_s}{\partial r}\bigg|_{r=r_b}
$$

Convection:

$$T_g(r,0) = T_s(r,0) = T_0$$

$$T_g(r_b,t) = T_s(r_b,t)$$

$$-\lambda_g \frac{\partial T_g}{\partial r}\bigg|_{r=r_e} = \bar{h}[T_f - T_g(r_e,t)] \tag{8.59}$$

$$-\lambda_g \frac{\partial T_g}{\partial r}\bigg|_{r=r_b} = -\lambda_s \frac{\partial T_s}{\partial r}\bigg|_{r=r_b}$$

in which $\bar{h}$ is the average convective heat transfer coefficient, λ is the thermal conductivity, α is the thermal diffusivity, and the subscript f refers to the circulating fluid in the U-tube.

Lamarche and Beauchamp utilized Laplace transform to solve this set of partial differential equations. Following this, it can readily be shown that the forward Laplace transform yields

$$\hat{T}_g = c_1 K_o(q_1 r) + c_2 I_o(q_1 r)$$
$$\hat{T}_s = c_3 K_o(q_2 r) + c_4 I_o(q_2 r) \tag{8.60}$$

in which I_o and K_o are the modified Bessel functions of the first and second kind respectively, $q_1 = s/\alpha_g$ and $q_2 = s/\alpha_s$ with s being the Laplace parameter. To keep the temperature finite at infinity, $c_4 = 0$. The other constants must be determined from the boundary conditions. Applying a step heat flux, and substituting it into the transformed boundary conditions of Eq. (8.58), the solution in the Laplace domain can then be expressed as

$$\hat{T}_g(r,s) = \frac{q_b''}{q_1 \lambda_g s \Delta}[\{K_0(q_2 r_b)I_1(q_1 r_b) + \tilde{k}\gamma K_1(q_2 r_b)I_0(q_1 r_b)\}K_0(q_1 r)$$
$$+ \{K_0(q_2 r_b)K_1(q_1 r_b) - \tilde{k}\gamma K_1(q_2 r_b)K_0(q_1 r_b)\}I_0(q_1 r)] \tag{8.61}$$

in which

$$\Delta = \{K_0(q_2 r_b)I_1(q_1 r_b) + \tilde{k}\gamma K_1(q_2 r_b)I_0(q_1 r_b)\}K_1(q_1 r_e)$$
$$- \{K_0(q_2 r_b)K_1(q_1 r_b) - \tilde{k}\gamma K_1(q_2 r_b)K_0(q_1 r_b)\}I_1(q_1 r_e) \tag{8.62}$$

and $\gamma = \sqrt{\alpha_g/\alpha_s}, \tilde{k} = \lambda_s/\lambda_g$.

Having solved the forward transform, the time domain must be re-constructed using the inverse Laplace transform, given by the Bromwich integral:

$$T_g(r,t) - T_0 = \frac{1}{2\pi i}\int_{a-i\infty}^{a+i\infty} e^{st}\hat{T}_g(r,s)\,ds \tag{8.63}$$

where a is a vertical contour on the complex plane, chosen such that all singularities of $\hat{T}_g(r,s)$ lie to the left of it. Lamarche and Beauchamp solved Eq. (8.63) in the complex plane analytically and gave this integral formulation:

$$T_g(\tilde{r},\tilde{t}) - T_0 = \frac{q_b''}{\lambda_g}\frac{4\tilde{k}}{\pi^4 \delta^2}\int_0^\infty \frac{\chi(1 - e^{-\beta^2 \tilde{t}})}{\beta^4(\varphi^2 + \psi^2)}\,d\beta \tag{8.64}$$

in which

$$\chi = [Y_0(\beta\tilde{r})J_1(\beta) - J_0(\beta\tilde{r})Y_1(\beta)]$$

$$\varphi = Y_1(\beta)[Y_0(\beta\delta\gamma)J_1(\beta\delta) - Y_1(\beta\delta\gamma)J_0(\beta\delta)\tilde{k}\gamma] - J_1(\beta)[Y_0(\beta\delta\gamma)Y_1(\beta\delta) - Y_1(\beta\delta\gamma)Y_0(\beta\delta)\tilde{k}\gamma]$$

$$\psi = J_1(\beta)[J_0(\beta\delta\gamma)Y_1(\beta\delta) - J_1(\beta\delta\gamma)Y_0(\beta\delta)\tilde{k}\gamma] - Y_1(\beta)[J_0(\beta\delta\gamma)J_1(\beta\delta) - J_1(\beta\delta\gamma)J_0(\beta\delta)\tilde{k}\gamma]$$

$$\tag{8.65}$$

and $\tilde{r} = r/r_e, \tilde{t} = \alpha_g t/r_e^2, \delta = r_b/r_e$.

They denote:

$$g_{stq}(\tilde{r},\tilde{t}) = \frac{4\tilde{k}}{\pi^4\delta^2} \int_0^\infty \frac{\chi(1 - e^{-\beta^2\tilde{t}})}{\beta^4(\varphi^2 + \psi^2)} \, d\beta \tag{8.66}$$

as the g-function for the short-term for the heat flux case. By applying a similar procedure, they derived the solution for the convective boundary condition, Eq. (8.59), as

$$T_g(\tilde{r},\tilde{t}) - T_f = (T_0 - T_f)\frac{8\tilde{k}\tilde{H}}{\pi^3\delta^2} \int_0^\infty \frac{\chi e^{-\beta^2\tilde{t}}}{\beta^3(\varphi^2 + \psi^2)} \, d\beta \tag{8.67}$$

in which

$$\begin{aligned}
\chi &= \beta[Y_0(\beta\tilde{r})J_1(\beta) - J_0(\beta\tilde{r})Y_1(\beta)] \\
&\quad + \tilde{k}\gamma\tilde{H}[Y_0(\beta\tilde{r})J_0(\beta) - J_0(\beta\tilde{r})Y_0(\beta)] \\
\phi &= [\tilde{H}J_0(\beta)+\beta J_1(\beta)][J_0(\beta\delta\gamma)Y_1(\beta\delta) - J_1(\beta\delta\gamma)Y_0(\beta\delta)\tilde{k}\gamma] \\
&\quad - [\tilde{H}Y_0(\beta)+\beta Y_1(\beta)][J_0(\beta\delta\gamma)J_1(\beta\delta) - J_1(\beta\delta\gamma)J_0(\beta\delta)\tilde{k}\gamma] \\
\psi &= [\tilde{H}J_0(\beta) + \beta J_1(\beta)][Y_0(\beta\delta\gamma)Y_1(\beta\delta) - Y_1(\beta\delta\gamma)Y_0(\beta\delta)\tilde{k}\gamma] \\
&\quad - [\tilde{H}Y_0(\beta) + \beta Y_1(\beta)][Y_0(\beta\delta\gamma)J_1(\beta\delta) - Y_1(\beta\delta\gamma)J_0(\beta\delta)\tilde{k}\gamma]
\end{aligned} \tag{8.68}$$

and $\tilde{H} = \bar{h} r_e/\lambda_g$. As for the heat flux case, they denote

$$g_{stc}(\tilde{r},\tilde{t}) = \frac{8\tilde{k}\tilde{H}}{\pi^3\delta^2} \int_0^\infty \frac{\chi e^{-\beta^2\tilde{t}}}{\beta^3(\phi^2 + \psi^2)}) \, d\beta \tag{8.69}$$

as the g-function for the short-term for the convection case. Solving the semi-infinite integrations in Eq. (8.66) and Eq. (8.69) analytically is, if possible, a formidable task. It is therefore feasible to resort to numerical procedures to solve these integrals.

Bandyopadhyay *et al.* (2008) used somewhat similar model to the one presented above, but included the temperature of the circulating fluid. They also utilized the Laplace transform to solve the problem. However, they conducted the Gaver-Stehfest algorithm for the numerical inversion of the Laplace domain equations. As shown in Chapter 7, this algorithm is defined as

$$F(t) \sim f_k(t) = \frac{\ln 2}{t} \sum_{j=1}^k D(j,k)\hat{F}\left(\frac{j\ln 2}{t}\right) \tag{8.70}$$

with

$$D(j,k) = (-1)^{j+M} \sum_{n=\mathrm{int}(j+1/2)}^{\min(j,M)} \frac{n^M(2n)!}{(M - n)!\,n!\,(n - 1)!\,(j - n)!\,(2n - j)!} \tag{8.71}$$

and $M = k/2$, with k is an even integer, chosen based on the computer precision. For a single precision computer of 16 decimal digits, they found that $k = 10 - 14$ gave stable and smooth results. Increasing k beyond 16 gave unstable and oscillatory results.

8.3 MODELING BOREHOLE HEAT EXCHANGER

Most of the current analytical models for heat flow in a borehole heat exchanger follow the Eskilson and Claesson (1988) model. They approximated the BHE by two interacting channels conveying a circulating fluid. The original model was formulated to describe a local steady-state convective heat transfer in the system. Later, modifications were applied by several researchers to take different accounts on the involved processes, including the transient effect and the pipe configurations.

Eskilson and Claesson proposed local steady-state heat balance equations for a typical borehole heat exchanger, consisting of pipe-in and pipe-out, of the form

$$\rho c Q \frac{dT_i}{dz} = \frac{T_b - T_i}{R_i} - \frac{T_i - T_o}{R_{io}}$$

$$-\rho c Q \frac{dT_o}{dz} = \frac{T_b - T_o}{R_o} + \frac{T_i - T_o}{R_{io}}$$

$$(8.72)$$

in which Q is the pumping rate or the flow rate in the pipes, T_i is pipe-in temperature, T_o is pipe-out temperature, T_b is the temperature at the borehole wall, R_i is the thermal resistance in pipe-in, R_o is the thermal resistance in pipe-out, and $R_{io} = R_{oi}$ is the thermal resistance between pipe-in and pipe-out. The relevant boundary conditions are

$$T_i(0,t) = T_{in}(t)$$
$$T_i(L,t) = T_o(L,t)$$

$$(8.73)$$

in which T_{in} is the incoming temperature at the inlet of pipe-in, and L is the length of the pipe. Eskilson and Claesson utilized Laplace transforms to solve this boundary value problem in the region $0 \leq z \leq L$, and provided this result:

$$T_i(z,t) = T_i(0,t) f_1(z) + T_o(0,t) f_2(z) + \int_0^z T_b(\zeta,t) f_4(z - \zeta) d\zeta$$

$$T_o(z,t) = -T_i(0,t) f_2(z) + T_o(0,t) f_3(z) - \int_0^z T_b(\zeta,t) f_5(z-\zeta) d\zeta$$

$$(8.74)$$

where

$$f_1(z) = e^{\beta z}[\cosh \gamma z - \delta \sinh \gamma z]$$

$$f_2(z) = e^{\beta z} \frac{\beta_{12}}{\gamma} \sinh \gamma z$$

$$f_3(z) = e^{\beta z}[\cosh \gamma z + \delta \sinh \gamma z]$$

$$f_4(z) = e^{\beta z}[\beta_1 \cosh \gamma z - \left(\delta \beta_1 + \frac{\beta_2 \beta_{12}}{\gamma}\right) \sinh \gamma z]$$

$$f_5(z) = e^{\beta z}[\beta_2 \cosh \gamma z + \left(\delta \beta_2 + \frac{\beta_1 \beta_{12}}{\gamma}\right) \sinh \gamma z]$$

$$(8.75)$$

in which

$$\beta_1 = \frac{1}{R_i \rho c Q}, \quad \beta_2 = \frac{1}{R_o \rho c Q}, \quad \beta_{12} = \frac{1}{R_{io} \rho c Q}$$

$$\beta = \frac{\beta_2 - \beta_1}{2}, \quad \gamma = \sqrt{\frac{(\beta_1 + \beta_2)^2}{4} + \beta_{12}(\beta_1 + \beta_2)}$$

$$\delta = \frac{1}{\gamma}\left(\beta_{12} + \frac{\beta_1 + \beta_2}{2}\right)$$

$$(8.76)$$

At the outlet, and by the use of Eq. (8.73), they showed that the temperature is expressed as

$$T_{out}(t) = \frac{f_1(L) + f_2(L)}{f_3(L) - f_2(L)} T_{in}(t) + \int_0^L \frac{T_b(\zeta,t)[f_4(L - \zeta) + f_5(L - \zeta)]}{f_3(L) - f_2(L)} d\zeta$$

$$(8.77)$$

The soil-pipe thermal interaction is postulated to be of the form

$$2\pi r_b \lambda_s \left.\frac{dT_s}{dr}\right|_{r=r_b} = \frac{T_b - T_i}{R_i} + \frac{T_b - T_o}{R_o}, \quad 0 \leq z \leq L$$

$$(8.78)$$

in which T_s is the soil temperature, λ_s is the soil thermal conductivity, and r_b is the radius of the borehole. They utilized the Delta-configuration shown in Figure 5.8 to account for the thermal interactions between pipe-in and pipe-out, and between the grout and the soil. They used explicit forward difference method to solve the soil temperature in two-dimensional domain. The heat capacitance of the individual borehole components such as the pipe wall and the grout are neglected.

As mentioned earlier, this model describes heat flow in the system as pseudo transient, referred in the literature as a local steady-state. Heat flow in the soil is transient, whereas, heat flow in the borehole heat exchanger is steady-state. As a result of this combination, the size of the time step must be larger than the time of validity of the soil model, i.e. t in Eq. (8.49), and the time at which the fluid completes at least a whole cycle, going through the whole length of pipe-in and pipe-out, such that

$$t \geq \frac{5r_b^2}{\alpha} + \pi r_p^2 \frac{2L}{Q} \tag{8.79}$$

in which α is the thermal diffusivity of the soil, Q the fluid pumping rate, and r_p is the inner radius of pipe-in. In practice, only the first term on the right-hand side of Eq. (8.79) is considered. Note that, due to the relatively small values of α, the first term in Eq. (8.79) might be very large, and such a model might require, in many cases, time steps of the order of hours to months to be valid. For a typical borehole, Yavuzturk and Spitler (1999) have shown that this condition might imply times of the order of 3 to 6 hours. However, for a saturated soil mass and a 76 m deep borehole, they showed that the g-function of a single borehole is only applicable for times in excess of 60 days.

Zeng et al. (2003) provided different solution to the heat flow in U-tube borehole heat exchangers using heat equations similar to Eq. (8.72). They described heat flow in a single U-tube BHE in a dimensionless form, as

$$\left. \begin{aligned} -\frac{d\Theta_d}{dZ} &= \frac{\Theta_d}{S_1} + \frac{\Theta_d - \Theta_u}{S_{12}} \\ \frac{d\Theta_u}{dZ} &= \frac{\Theta_u - \Theta_d}{S_{12}} + \frac{\Theta_u}{S_1} \end{aligned} \right\} \qquad 0 \leq Z \leq 1 \tag{8.80}$$

The boundary conditions are:

$$\begin{aligned} \Theta_d(1) &= \Theta_u(1) \\ \Theta_d(0) &= 1 \end{aligned} \tag{8.81}$$

The subscripts d and u represent the downward and upward fluid flow, and the dimensionless parameters are

$$Z = \frac{z}{L}$$

$$\Theta_d = \frac{T_d(z) - T_b}{T_f' - T_b}, \qquad \Theta_u = \frac{T_u(z) - T_b}{T_f' - T_b}$$

$$S_1 = \frac{Mc}{L}(R_{11} + R_{13}) \tag{8.82}$$

$$S_{12} = \frac{Mc}{L} \frac{R_{11}^2 - R_{13}^2}{R_{13}}$$

in which T_f' is the inlet fluid temperature, T_b is the borehole wall temperature, M is the mass flow rate of the circulating fluid, c is the fluid heat capacity, L is the pipe length, and R_{11} and R_{13} are the thermal resistances defined as

$$R_{11} = \frac{1}{2\pi\lambda_b} \left[\ln\left(\frac{r_b}{r_p}\right) - \frac{\lambda_b - \lambda_s}{\lambda_b + \lambda_s} \ln\left(\frac{r_b^2 - D^2}{r_b^2}\right) \right] + R_p$$

$$R_{13} = \frac{1}{2\pi\lambda_b} \left[\ln\left(\frac{r_b}{2D}\right) - \frac{\lambda_b - \lambda_s}{\lambda_b + \lambda_s} \ln\left(\frac{r_b^2 + D^2}{r_b^2}\right) \right]$$

(8.83)

where λ_s denotes the thermal conductivity of the soil surrounding the borehole, λ_b is the thermal conductivity of the grout, and R_p is the thermal resistance of the pipe material. Using Laplace transform, Zeng *et al.* introduced the solution to the boundary value problem, Eqs. (8.80)–(8.81), as

$$\Theta_d(Z) = \cosh(\beta Z) - \frac{1}{\beta S_{12}} \left[\left(\frac{S_{12}}{S_1} + 1\right) - \frac{\beta S_1 \cosh(\beta) - 1}{\beta S_1 \cosh(\beta) + 1} \right] \sinh(\beta Z)$$

$$\Theta_u(Z) = \frac{\beta S_1 \cosh(\beta) - 1}{\beta S_1 \cosh(\beta) + 1} \cosh(\beta Z) - \frac{1}{\beta S_{12}} \left[1 - \left(\frac{S_{12}}{S_1} + 1\right) \frac{\beta S_1 \cosh(\beta) - 1}{\beta S_1 \cosh(\beta) + 1} \right] \sinh(\beta Z)$$

(8.84)

where

$$\beta = \sqrt{\frac{1}{S_1^2} + \frac{2}{S_1 S_{12}}}$$

(8.85)

Following this formulation, the outlet fluid temperature can be calculated to give

$$\Theta_o = \frac{\beta S_1 \cosh(\beta) - 1}{\beta S_1 \cosh(\beta) + 1}$$

(8.86)

Note that both formulations, Eq. (8.74) and Eq. (8.84), are steady-state. Solution of the transient case will be presented in Chapter 9 and Chapter 12.

CHAPTER 9

Spectral analysis of shallow geothermal systems

In this chapter, we solve the initial and boundary value problem of a typical shallow geothermal system using the spectral analysis method. We introduce a framework for devising a spectral model capable of simulating fully transient conductive-convective heat transfer processes in a shallow geothermal system consisting of a borehole heat exchanger embedded in a soil mass. The model combines the exactness of the analytical methods to an important extent of the generality of the numerical methods in describing initial and boundary conditions. It calculates temperature distribution in all involved borehole heat exchanger components and the surrounding soil mass. We use the fast Fourier transform algorithm for modeling the time domain, and Fourier series and Fourier-Bessel series for modeling the spatial domain.

9.1 INTRODUCTION

Methods for solving transient problems can be classified into two major groups. The first group solves the problems in the time domain. The rigor of the solution of this group depends on the type of the mathematical method used to solve the dependency on time. The numerical methods, such as the finite difference and the finite element methods, simulate the time function as a discrete number of time steps. As such, complicated time functions can be treated. The analytical methods, on the other hand, such as the Laplace transform, the integral transform or the Green's functions, typically simulate the dependency on time using relatively simple functions, such as sinusoidal, time step or instantaneous (Dirac) functions.

The second group solves the transient problems in the frequency domain. The spectral analysis method (SAM) is one of the most commonly utilized procedures for solving transient problems by transforming functions in the time domain to functions in the frequency domain and vice versa. SAM is basically associated with continuous Fourier transforms. The application of the continuous Fourier transforms is possible if the function to be transformed is analytical. However, in practice, only few cases can be described analytically, and most practical problems involve digital signals, usually obtained from measured data. In this case, the solution is approximated using the discrete Fourier transform (DFT), where the signal is represented by a finite number of discrete frequencies. The discrete Fourier transform is associated with the well-known fast Fourier transform algorithm (FFT), see Chapter 6. This transform possesses all the advantages of the continuous transform methods, adding to that, it is computationally very efficient in the forward and inverse transforms. Due to this efficiency, the use of the FFT algorithm makes the spectral analysis very attractive for utilization to solve engineering problems involving complicated time functions.

Dealing with the spatial domain, in Chapter 6 we presented several applications of Fourier series, Fourier integrals and Fourier transforms for solving heat equations of different domains subjected to different initial and boundary conditions. We saw that the solution of heat equation of a finite domain was obtained using Fourier series, leading to simple summation over algebraic terms. On the other hand, solution of heat equations of infinite and semi-infinite domains was obtained using Fourier integrals, or equivalently Fourier transforms. These two transforms led to semi-infinite integrals of transcendental functions. Solving such integrals, especially for real physical cases, is cumbersome since it requires dealing with complex functions, which are oscillatory in nature. Comparing the different techniques, it is evident that the computational requirement necessary to conduct summation over algebraic terms of the Fourier series is very much less than conducting semi-infinite integrations over complex transcendental functions. Therefore, in order to obtain a feasible solution, it is recommended to use, whenever possible, the Fourier series, or equivalently the Fourier-Bessel series, for discretizing the spatial domain.

A shallow geothermal system involves borehole heat exchangers of finite lengths, embedded in a soil mass of a semi-infinite dimension. Therefore, in principle, Fourier integrals or transforms must be utilized to solve the heat flow problem in the soil. To take advantage of the Fourier-Bessel series algebraic summation, we adjust the semi-infinite boundary of the soil mass to a pseudo finite boundary, by introducing what we denote as the region-of-interest. This region is chosen such that, at some distance far from the borehole, where it is known, intuitively or otherwise, that heat flux coming from the borehole vanishes. That is, we impose a fictitious homogeneous boundary condition at the far field boundary of the region-of-interest. The advantage of this technique will be apparent in what follows.

In this chapter, we use the spectral analysis method to solve the initial and boundary value problem of a typical shallow geothermal system. The solution is conducted by first transforming the governing equations from the time domain to the frequency domain, using the forward FFT algorithm. Then, for each frequency, we solve the resulting system of equations analytically, using the Fourier or the Fourier-Bessel series. After that, we reconstruct the solution back to the time domain, using the inverse FFT algorithm.

9.2   MODELING SHALLOW GEOTHERMAL SYSTEM

A shallow geothermal system, particularly a geothermal heat pump, consists basically of two thermally interacting components: the borehole heat exchanger and the soil mass. For a system consisting of one borehole heat exchanger, the geometry of the shallow geothermal system can be described effectively using axial-symmetric coordinate system, where the axis of symmetry coincides with the centerline of the borehole heat exchanger.

Before operating the geothermal heat pump, the soil temperature, T_{soil}, arises as a result of an existing steady-state condition, and a transient condition. The steady-state temperature, T_{st}, arises mainly due to temperature gradient between the soil surface temperature, T_0, and some deeper layer temperature, T_b. The transient temperature, T_t, develops as a result of the short term temperature changes in the air or in the soil surface, $T_a(t)$. In both cases the temperature gradient is in the vertical direction, as in the radial direction, within the region-of-interest, there is no temperature gradient. Thus, heat flow in the soil mass before operating the system is mainly vertical, and the soil temperature can be described as

$$T_z(t, z) = T_{st}(z) + T_t(t, z) \tag{9.1}$$

During system operation, the temperature in the soil mass arises as a result of the interaction with the borehole heat exchanger. In this case, the soil is subjected to temperature variations in at least three sides: the upper side, the bottom side, and the side where the BHE interacts with the soil mass. This system represents a typical non-homogeneous *Dirichlet problem*. Solving such a problem directly is somewhat involved. However, and as the system is linear, it is possible to utilize the superposition principle. Accordingly, the system can be decomposed into two sub-systems, each of which has homogeneous boundary conditions on its parallel boundaries, as shown in Figure 9.1. The first sub-system represents a one-dimensional steady-state and transient heat flow generated by the temperature gradient between the surface and the bottom boundaries. The second sub-system represents an axial-symmetric transient heat flow generated by the interaction with the borehole heat exchanger. The resulting temperature in the soil will be propagating in the z-direction and diffusing to infinity in the radial direction. Superposing these temperatures, the total temperature in the soil can then be described as

$$\begin{aligned} T_{soil}(r, z, t) &= T_z(t, z) + T_{rz}(r, z, t) \\ &= T_{st}(z) + T_t(z, t) + T_{rz}(r, z, t) \end{aligned} \tag{9.2}$$

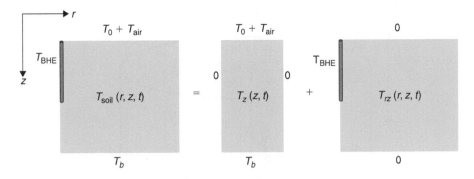

Figure 9.1   A schematic representation of the soil temperature.

in which $T_{rz}(r, z, t)$ is the soil temperature generated by the thermal interaction between the soil mass and the borehole heat exchanger.

The borehole heat exchanger, on the other hand, is subjected to an inlet temperature coming from a heat pump, $T_{in}(t)$, a steady-state soil temperature, $T_{st}(z)$, and a transient soil temperature, $T_t(z, t)$, as shown schematically in Figure 9.2.

Apparently, the two sub-systems are thermally interacting. Solving the governing heat equations of such a system analytically, using classical methods such as Laplace transforms or Green's functions, results to rather complicated functions. We saw in Chapter 7 and Chapter 8 that even with overly simplified assumptions, formidable integrals were obtained. In this chapter, we try to circumvent some of the complications of existing solutions and present a feasible and more comprehensive solution such that temperature distribution in all involved components of the system can be calculated.

In what follows, we solve the initial and boundary value problem of a fully transient conduction-convection heat transfer in a shallow geothermal system consisting of a single U-tube borehole heat exchanger embedded in a soil mass. The problem is divided into two sub-systems describing the borehole heat exchanger and the soil mass.

9.2.1   *Sub-system 1: borehole heat exchanger*

In this section, we first establish the governing equations and the relevant initial and boundary conditions describing heat transfer in the involved BHE components and their thermal interactions. Then, we solve the problem using the spectral analysis method.

In Chapter 4, we established heat equations of different types of borehole heat exchangers. We show that the heat equation of a single U-tube BHE consisting of pipe-in, pipe-out and grout can be described as

$$\rho c \frac{\partial T_i}{\partial t} dV_i - \lambda \frac{\partial^2 T_i}{\partial z^2} dV_i + \rho c u \frac{\partial T_i}{\partial z} dV_i - b_{ig}(T_i - T_g)dS_{ig} = 0$$

$$\rho c \frac{\partial T_o}{\partial t} dV_o - \lambda \frac{\partial^2 T_o}{\partial z^2} dV_o - \rho c u \frac{\partial T_o}{\partial z} dV_o - b_{og}(T_o - T_g)dS_{og} = 0 \qquad (9.3)$$

$$\rho c_g \frac{\partial T_g}{\partial t} dV_g - \lambda_g \frac{\partial^2 T_g}{\partial z^2} dV_g - b_{ig}(T_g - T_i)dS_{ig} - b_{og}(T_g - T_o)dS_{og} = 0$$

in which:

T_i : pipe-in temperature (K)
T_o : pipe-out temperature (K)
T_g : grout temperature (K)

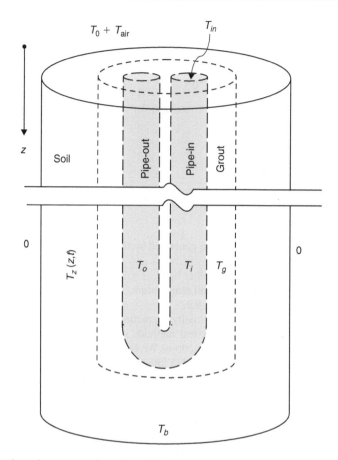

Figure 9.2   A schematic representation of the BHE temperature.

b_{ig} : thermal interaction coefficient between pipe-in and grout (W/m$^2 \cdot$ K)
b_{og} : thermal interaction coefficient between pipe-out and grout (W/m$^2 \cdot$ K)
c : specific heat capacity of the circulating fluid (J/kg.K)
ρ : mass density of the circulating fluid (kg/m^3)
ρc_g : volume heat capacity of the grout (J/m$^3 \cdot$ K)
λ : thermal conductivity of the circulating fluid (W/m $\cdot$ K)
λ_g : thermal conductivity of the grout (W/m $\cdot$ K)
dV_i : partial volume of pipe-in (m^3)
dV_o : partial volume of pipe-out (m^3)
dV_g : partial volume of grout (m^3)
dS_{ig} : partial surface area of pipe-in in contact with the grout (m^2)
dS_{og} : partial surface area of pipe-out in contact with the grout (m^2)

This formulation illustrates that the grout works as an intermediate medium. There is a direct coupling between the grout and each individual pipe, and an indirect coupling between pipe-in and pipe-out via the grout.
The involved initial and boundary conditions are

$$T_i(z,0) = T_o(z,0) = T_g(z,0) = T_{st}(z,0) \tag{9.4}$$

$$T_i(0,t) = T_{in}(t) \tag{9.5}$$

$$-\lambda_g \frac{\partial T_g(z,t)}{\partial z} dS_{gs} - b_{ig}(T_g - T_i)dS_{ig} - b_{og}(T_g - T_o)dS_{og} = b_{gs}[T_g - T_t(z,t)]dS_{gs} \qquad (9.6)$$

$$T_i(L,t) = T_o(L,t) \qquad (9.7)$$

in which T_{st} is any initial soil temperature, $T_t(z,t)$ is the one-dimensional transient temperature of the soil immediately surrounding the BHE, L denotes the length of the BHE, b_{gs} is the thermal interaction coefficient between the grout and the soil, and dS_{gs} is the surface area of the borehole in contact with the soil. It is worth mentioning that the second boundary condition, Eq. (9.6), can be described as

$$-\lambda_g \frac{\partial T_g}{\partial z} = b_{gs}[T_g - T_t(z,t)] \qquad (9.8)$$

assuming that the BHE temperature at the surface can be represented by the grout temperature. However, numerical analyses have shown that this assumption produces, especially in the short term, limited thermal interaction between the BHE and the soil. In Eq. (9.6), it is assumed that the soil temperature is in equilibrium with the temperatures in all components (pipe-in, pipe-out and grout), motivated by the fact that the grout is relatively thin, and it is reasonable to assume that a stronger interaction between the soil and the inner pipes exists. Mathematically, and physically, the two formulations are valid, but experimental work can verify the most appropriate one.

The third boundary condition, Eq. (9.7), can also be described by equating the heat flux from both pipes, that is

$$q_i(L,t) = q_o(L,t) \qquad (9.9)$$

where q_i and q_o is the heat flux going out of pipe-in and going into pipe-out respectively.

In Chapter 5 we presented different thermal resistance coefficients. Any formulation can be utilized, but as mentioned, the thermal resistance coefficient which is based on the analogy between Fourier's law and Ohm's law is reliable. Accordingly, the thermal interaction coefficients of pipe-in – grout, pipe-out – grout, and the borehole-soil are assumed of the form

$$b_{ig} = \frac{1}{R_{ig}}, \qquad b_{og} = \frac{1}{R_{og}}, \qquad b_{gs} = \frac{1}{R_b} \qquad (9.10)$$

where R_{ig}, R_{og} and R_b are shown in Eqs. (5.56)–(5.59).

9.2.1.1 Spectral analysis of the BHE initial and boundary value problem

Here we solve the governing heat equations of the borehole heat exchanger, Eqs. (9.3)–(9.7), using the spectral analysis method (Doyle 1997 and 1988b). This method has been briefly treated in Chapter 6 within the context of the discrete Fourier transform (DFT) and the fast Fourier algorithm (FFT). Using FFT, the spectral representation of a discrete temperature in time can be described as

Forward:

$$\hat{T}(z,\omega_n) = \frac{1}{N}\sum_m T(z,t_m)e^{-i\omega_n t_m} \qquad (9.11)$$

Inverse:

$$T(z,t_m) = \sum_n \hat{T}(z,\omega_n)e^{i\omega_n t_m} \qquad (9.12)$$

where $\hat{T}(z,\omega_n)$ is a spatially dependent Fourier coefficient at frequency ω_n. N is the number of the involved discrete samples, where in the fast Fourier transform it is usually made a power of 2, such that $N = 2^\gamma = 2,4,8,\dots,2048,\dots$. For a real time signal, the transform is symmetric about a middle frequency, referred to as the Nyquist frequency, see Chapter 6. This means that N real points are transformed into $N/2$ complex points.

Applying Eq. (9.12) to Eq. (9.3), the transformation from the time domain to the frequency domain yields

$$i\omega\rho c\hat{T}_i dV_i - \lambda \frac{d^2\hat{T}_i}{dz^2} dV_i + \rho\, cu \frac{d\hat{T}_i}{dz} dV_i - b_{ig}(\hat{T}_i - \hat{T}_g) dS_{ig} = 0$$

$$i\omega\rho c\hat{T}_o dV_o - \lambda \frac{d^2\hat{T}_o}{dz^2} dV_o - \rho\, cu \frac{d\hat{T}_o}{dz} dV_o - b_{og}(\hat{T}_o - \hat{T}_g) dS_{og} = 0 \qquad (9.13)$$

$$i\omega\rho c_g \hat{T}_g dV_g - \lambda_g \frac{d^2\hat{T}_g}{dz^2} dV_g - b_{ig}(\hat{T}_g - \hat{T}_i) dS_{ig} - b_{og}(\hat{T}_g - \hat{T}_o) dS_{og} = 0$$

where the transform is represented by $T \Leftrightarrow \hat{T}$. It is obvious that the spectral representation of the time derivative

$$\frac{\partial T}{\partial t} = \frac{\partial}{\partial t} \sum \hat{T}_n e^{i\omega_n t} = \sum i\omega_n \hat{T}_n e^{i\omega_n t} \qquad (9.14)$$

has been written as

$$\frac{\partial T}{\partial t} \Rightarrow i\omega\hat{T} \qquad (9.15)$$

where $\hat{T}_n = \hat{T}(z, \omega_n)$. Similarly, the spectral representation of the spatial derivatives

$$\frac{\partial^m T}{\partial z^m} = \frac{\partial^m}{\partial z^m} \sum \hat{T}_n e^{i\omega_n t} = \sum \frac{\partial^m \hat{T}_n}{\partial z^m} e^{i\omega_n t} \qquad (9.16)$$

are represented by

$$\frac{\partial^m T}{\partial z^m} \Rightarrow \frac{\partial^m \hat{T}}{\partial z^m} \qquad (9.17)$$

In contrary to the time derivatives, there is no reduction in the order of the spatial derivatives. The utilization of the spectral analysis has reduced the partial differential equations, Eq. (9.3), to ordinary differential equations, Eq. (9.13), by transforming the time derivative to an algebraic expression. Here lies the advantage of the spectral approach to solving transient partial differential equations. However, the resulting equations are frequency dependent and need to be solved for each frequency, ω_n.

The spectral functions, Eq. (9.13), are second order ordinary differential equations of Strum-Liouville type. As we have seen in Chapter 6, this type of equations can be solved elegantly using eigenfunction expansions. Since Eq. (9.13) has constant coefficients, the solution might be of the form

$$\hat{T}_i = Ae^{-ikz}$$

$$\hat{T}_g = \overline{A}e^{-ikz} \qquad (9.18)$$

$$\hat{T}_o = \overline{\overline{A}}e^{ikz}$$

in which $A, \overline{A}, \overline{\overline{A}}$ are integration constants, and k denotes the eigenvalues of the system that needs to be determined. Note that, the proposed solution, Eq. (9.18), states that heat flow in pipe-in and grout is in $z > 0$ direction, and heat flow in pipe-out is in the opposite direction, Figure 9.2. Substituting Eq. (9.18) into Eq. (9.13) leads to

$$\lambda k^2 Ae^{-ikz} dV_i - \rho cuik Ae^{-ikz} dV_i + (i\omega\rho c dV_i - b_{ig} dS_{ig}) Ae^{-ikz} + b_{ig}\,\overline{A}e^{-ikz} dS_{ig} = 0$$

$$\lambda k^2 \overline{\overline{A}}e^{ikz} dV_o - \rho cuik \overline{\overline{A}}e^{ikz} dV_o + (i\omega\rho c dV_o - b_{og} dS_{og})\overline{\overline{A}}e^{ikz} + b_{og}\overline{A}e^{-ikz} dS_{og} = 0 \qquad (9.19)$$

$$\lambda_g k^2 \overline{A}e^{-ikz} dV_g + (i\omega\rho c_g dV_g - b_{ig}\, dS_{ig} - b_{og} dS_{og})\overline{A}e^{-ikz} + b_{ig} Ae^{-ikz} dS_{ig} + b_{og}\overline{\overline{A}}e^{ikz} dS_{og} = 0$$

Dividing the first and third equations in Eq. (9.19) by e^{-ikz} and the second equation by e^{ikz} gives

$$\lambda k^2 A dV_i - \rho cuik A dV_i + (i\omega\rho c dV_i - b_{ig}dS_{ig})A + b_{ig}\overline{A}dS_{ig} = 0$$

$$\lambda k^2 \overline{\overline{A}}dV_o - \rho cuik\overline{\overline{A}}dV_o + (i\omega\rho c dV_o - b_{og}dS_{og})\overline{A} + b_{og}\overline{A}e^{-2ikz}dS_{og} = 0 \qquad (9.20)$$

$$\lambda_g k^2 \overline{A}dV_g + (i\omega\rho c_g dV_g - b_{ig}dS_{ig} - b_{og}dS_{og})\overline{A} + b_{ig}A dS_{ig} + b_{og}\overline{\overline{A}}e^{2ikz}dS_{og} = 0$$

In a matrix form, Eq. (9.20) can be written as

$$\begin{pmatrix} \lambda k^2 dV_i - \rho cuik dV_i + \\ i\omega\rho c dV_i - b_{ig}dS_{ig} & 0 & b_{ig}dS_{ig} \\ \\ b_{ig}dS_{ig} & b_{og}e^{2ikz}dS_{og} & \lambda_g k^2 dV_g + i\omega\rho c_g dV_g - \\ & & b_{ig}dS_{ig} - b_{og}dS_{og} \\ \\ 0 & \lambda k^2 dV_o - \rho cuik dV_o + \\ & i\omega\rho c dV_o - b_{og}dS_{og} & b_{og}e^{-2ikz}dS_{og} \end{pmatrix} \begin{pmatrix} A \\ \overline{\overline{A}} \\ \overline{A} \end{pmatrix} = 0 \qquad (9.21)$$

To determine the eigenvalues, k, of the system, we need to solve Eq. (9.21). This equation presents a challenge, since the determination of the constants, $A, \overline{A}, \overline{\overline{A}}$ requires solving for $\mathbf{a} = \mathbf{A}^{-1}\mathbf{0}$, where $\mathbf{a}$ denotes $[A, \overline{A}, \overline{\overline{A}}]^T$ and $\mathbf{A}^{-1}$ represents the inverse of the matrix in Eq. (9.21), if exists. This product produces $\mathbf{a} = \mathbf{0}$, which is trivial. Thus the assumption that $\mathbf{A}^{-1}$ does exist, is not possible. We know from linear algebra that a matrix does not have an inverse if its determinant $|\mathbf{A}|$ is zero. Therefore, and as it has been shown in Chapter 6, the non-trivial solution of Eq. (9.21) can only be obtained by letting the determinant equal to zero.

The determinant of Eq. (9.21) yields a complex six degree polynomial of the form

$$a_6 k^6 + a_5 k^5 + a_4 k^4 + a_3 k^3 + a_2 k^2 + a_1 k + a_0 = 0 \qquad (9.22)$$

from which k values can be obtained by solving for the roots of the equation. The zeros of the determinant may exist for certain values of $k_n, n = 1, 2, \ldots$, but not for others. Those values of k_n for which non-trivial solutions do exist are the eigenvalues of the system. The corresponding non-trivial solutions are the eigenfunctions. The exact forms of the constants, $a_0, \ldots, a_6$, of Eq. (9.22) are listed in Appendix 9.1. The convenient form of Eq. (9.22) can be obtained by any of the available analytical software, such as MAPLE (see MAPLE 13).

The determination of the eigenvalues must be done for every frequency. Six eigenvalues in three complex conjugates are obtained, representing three basic modes, one for each BHE component. Accordingly, solution of the temperature distribution in the three BHE components can be written as

$$\hat{T}_i = Ae^{-ik_1 z} + Be^{-ik_2 z} + Ce^{-ik_3 z}$$

$$\hat{T}_g = \overline{A}e^{-ik_1 z} + \overline{B}e^{-ik_2 z} + \overline{C}e^{-ik_3 z} \qquad (9.23)$$

$$\hat{T}_o = \overline{\overline{A}}e^{ik_1 z} + \overline{\overline{B}}e^{ik_2 z} + \overline{\overline{C}}e^{ik_3 z}$$

where only flow-off the heat source is taken into consideration. Since T_i, T_g, and T_o are coupled, the constants, $A, B, \ldots, \overline{\overline{C}}$, must be related to each other (Doyle 1988b). According to Eq. (9.21), the relationship between pipe-in constants and grout constants can be expressed as

$$\overline{A} = \overline{Y}A \qquad (9.24)$$

where

$$\overline{Y} = \frac{1}{b_{ig}dS_{ig}}(\lambda k^2 dV_i - \rho cuik dV_i + i\omega\rho c dV_i + b_{ig}dS_{ig}) \qquad (9.25)$$

In the same way, the relationship between pipe-out constants and grout constants can be expressed as

$$\overline{A} = \overline{\overline{Y}}\overline{A} \tag{9.26}$$

where

$$\overline{\overline{Y}} = \frac{1}{b_{og}e^{-2ikz}dS_{og}}(\lambda k^2 dV_o - \rho cuikdV_o + i\omega\rho cdV_o + b_{og}dS_{og}) \tag{9.27}$$

Note that for each k there is a corresponding $\overline{Y}$ and $\overline{\overline{Y}}$.

9.2.1.2   *Satisfying the boundary conditions*

The constants, $A, B, \ldots, \overline{\overline{C}}$, in Eq. (9.23) must be determined from boundary conditions. The spectral representation of the boundary conditions, Eqs. (9.5), (9.6) and (9.7), can be expressed as

$$\hat{T}_i(0, \omega) = \hat{T}_{in}(\omega) \tag{9.28}$$

$$-\lambda_g \frac{\partial \hat{T}_g(z, \omega)}{\partial z} dS_{gs} - b_{ig}[\hat{T}_g(z, \omega) - \hat{T}_i(z, \omega)]dS_{ig} - b_{og}[\hat{T}_g(z, \omega) - \hat{T}_o(z, \omega)]dS_{og}$$
$$= b_{gs}[\hat{T}_g(z, \omega) - \hat{T}_t(z, \omega)]dS_{gs} \tag{9.29}$$

$$\hat{T}_i(L, \omega) = \hat{T}_o(L, \omega) \tag{9.30}$$

The imposition of the third boundary condition, Eq. (9.30), indicates that heat flow in pipe-in is continuing in pipe-out, but in the opposite direction. This entails that the system is, in effect, consisting of two components, pipe-in and grout, that can be described basically by two eigenvalues (eigenmodes). The third mode, which is related to pipe-out, would eventually diminish, i.e. $C = \overline{C} = \overline{\overline{C}} = 0$, (Al-Khoury 2010). For $C = \overline{C} = \overline{\overline{C}} = 0$ to be valid implicates that the properties of both pipes, pipe-in and pipe-out, such as geometry, material and thermal resistance must be identical. In practice, this is the case. The U-tube pipe is in-effect one pipe, which is bent in a U-shape and inserted in the borehole. This makes the properties of the two pipes identical. If, however, the two pipe properties are made different, especially that the thermal resistance of pipe-in is made different from that of pipe-out, i.e. $b_{ig} \neq b_{og}$, the BHE system needs to be treated as two sub-systems, one representing pipe-in – grout, and another pipe-out – grout.

The two sub-systems are coupled at the point where pipe-in and pipe-out meet, i.e. $z = L$. The spectral equations, Eqs. (9.28), (9.29) and (9.30), can then be expanded to become:

Pipe-in – grout:

$$\hat{T}_i(0, \omega) = \hat{T}_{in}(\omega)$$
$$-\lambda_g \frac{\partial \hat{T}_{gi}(z, \omega)}{\partial z} dS_{gs} - b_{ig}[\hat{T}_{gi}(z, \omega) - \hat{T}_i(z, \omega)] dS_{ig} = b_{gs}[\hat{T}_{gi}(z, \omega) - \hat{T}_t(z, \omega)] dS_{gs} \tag{9.31}$$

Pipe-out – grout:

$$\hat{T}_o(L, \omega) = \hat{T}_i(L, \omega)$$
$$-\lambda_g \frac{\partial \hat{T}_{go}(z, \omega)}{\partial z} dS_{gs} - b_{og}[\hat{T}_{go}(z, \omega) - \hat{T}_o(z, \omega)] dS_{og} = b_{gs}[\hat{T}_{go}(z, \omega) - \hat{T}_t(z, \omega)] dS_{gs} \tag{9.32}$$

with the temperature in the grout is taken as an average between the two sub-systems, such that

$$\hat{T}_g = \frac{1}{2}(\hat{T}_{gi} + \hat{T}_{go}) \tag{9.33}$$

where $\hat{T}_{gi}$ and $\hat{T}_{go}$ represent grout temperature in contact with pipe-in and pipe-out respectively. Accordingly, Eq. (9.23) can be modified to account for the two sub-systems, as

Pipe-in – grout:

$$\hat{T}_i = Ae^{-ik_1z} + Be^{-ik_2z}$$
$$\hat{T}_{gi} = \overline{A}e^{-ik_1z} + \overline{B}e^{-ik_2z} \tag{9.34}$$

Pipe-out – grout:

$$\hat{T}_o = \overline{\overline{A}}e^{ik_1z} + \overline{\overline{B}}e^{ik_2z}$$
$$\hat{T}_{go} = \overline{A}e^{-ik_1z} + \overline{B}e^{-ik_2z} \tag{9.35}$$

Substituting Eqs. (9.34) and (9.35) into Eq. (9.31) and Eq. (9.32), and assuming that the soil temperature is constant along the BHE length, i.e. $\hat{T}_t(z, \omega) = \hat{T}_t(\omega) = \hat{T}_t$, yields

Pipe-in – grout:

$$A + B = \hat{T}_{in}$$
$$ik_1\lambda_g\overline{A}e^{-ik_1z}dS_{gs} + ik_2\lambda_g\overline{B}e^{-ik_2z}dS_{gs} - b_{ig}(\overline{A}e^{-ik_1z} + \overline{B}e^{-ik_2z} - Ae^{-ik_1z} - Be^{-ik_2z})\,dS_{ig}$$
$$= b_{gs}[\overline{A}e^{-ik_1z} + \overline{B}e^{-ik_2z} - \hat{T}_t]\,dS_{gs} \tag{9.36}$$

Pipe-out – grout:

$$\overline{\overline{A}}e^{ik_1L} + \overline{\overline{B}}e^{ik_2L} = \hat{T}_i(L, \omega) = \hat{T}_{iL}$$
$$ik_1\lambda_g\overline{A}e^{-ik_1z}\,dS_{gs} + ik_2\lambda_g\overline{B}e^{-ik_2z}\,dS_{gs} - b_{og}(\overline{A}e^{-ik_1z} + \overline{B}e^{-ik_2z} - \overline{\overline{A}}e^{ik_1z} - \overline{\overline{B}}e^{ik_2z})\,dS_{og}$$
$$= b_{gs}[\overline{A}e^{-ik_1z} + \overline{B}e^{-ik_2z} - \hat{T}_t]\,dS_{gs} \tag{9.37}$$

Solving Eq. (9.36) for A and B and using Eq. (9.24), leads to

$$A = \frac{1}{\Delta_i}[a_{11}\hat{T}_{in} - \hat{T}_t]$$
$$B = \frac{1}{\Delta_i}[a_{21}\hat{T}_{in} + \hat{T}_t] \tag{9.38}$$

in which

$$\Delta_i = a_{11} + a_{21}$$
$$a_{11} = -\frac{e^{-ik_2z}}{b_{gs}dS_{gs}}[\overline{Y}_2(ik_2\lambda_g dS_{gs} - b_{ig}dS_{ig} - b_{gs}dS_{gs}) + b_{ig}dS_{ig}] \tag{9.39}$$
$$a_{21} = \frac{e^{-ik_1z}}{b_{gs}dS_{gs}}[\overline{Y}_1(ik_1\lambda_g dS_{gs} - b_{ig}dS_{ig} - b_{gs}dS_{gs}) + b_{ig}dS_{ig}]$$

Solving Eq. (9.37) for $\overline{\overline{A}}$ and $\overline{\overline{B}}$ and using Eq. (9.26), leads to

$$\overline{\overline{A}} = \frac{1}{\Delta_o}[b_{11}\hat{T}_{iL} - e^{ik_2L}\hat{T}_t]$$
$$\overline{\overline{B}} = \frac{1}{\Delta_o}[b_{21}\hat{T}_{iL} + e^{ik_1L}\hat{T}_t] \tag{9.40}$$

in which

$$\Delta_o = b_{11}e^{ik_1L} + b_{21}e^{ik_2L}$$
$$b_{11} = -\frac{e^{-ik_2z}}{b_{gs}dS_{gs}}[\overline{\overline{Y}}_2(ik_2\lambda_g dS_{gs} - b_{og}dS_{og} - b_{gs}dS_{gs}) + e^{2ik_2z}b_{og}dS_{og}] \tag{9.41}$$
$$b_{21} = \frac{e^{-ik_1z}}{b_{gs}dS_{gs}}[\overline{\overline{Y}}_1(ik_1\lambda_g dS_{gs} - b_{og}dS_{og} - b_{gs}dS_{gs}) + e^{2ik_1z}b_{og}dS_{og}]$$

where $\overline{Y}_1$ and $\overline{\overline{Y}}_1$ are associated with eigenvalue, k_1, and $\overline{Y}_2$ and $\overline{\overline{Y}}_2$ are associated with eigenvalue, k_2.

9.2.1.3   *General solution of the BHE heat equations*

Having determined the eigenvalues and the integration constants, the general solution of the BHE system of equations can then be obtained by summing over all eigenfunctions (corresponding to k_1 and k_2) and frequencies, as

$$T_i(z,t) = \sum_n (Ae^{-ik_1 z} + Be^{-ik_2 z})e^{i\omega_n t}$$

$$T_o(z,t) = \sum_n (\overline{\overline{A}}e^{ik_1 z} + \overline{\overline{B}}e^{ik_2 z})e^{i\omega_n t}$$

$$T_g(z,t) = \frac{1}{2}\left[\sum_n [(\overline{Y}_1 A + \overline{Y}_1 \overline{\overline{A}})e^{-ik_1 z} + (\overline{Y}_2 B + \overline{Y}_2 \overline{\overline{B}})e^{-ik_2 z}]e^{i\omega_n t}\right]$$

(9.42)

The reconstruction of the time domain is obtained using the inverse FFT algorithm, see Eq. (9.12). Note that FFT algorithms are available in a large number of software, many of which can be downloaded free from the internet.

9.2.1.4   *Variable soil temperature*

In practice, soil temperature surrounding the BHE varies with depth. For any arbitrary distribution of soil temperature in space, it can be described using the complex Fourier series

$$T_t(z) = \sum_m \hat{F}_m e^{-i\xi_m z}$$

(9.43)

in which

$$\hat{F}_m = \frac{2\pi r_g}{L} \int_0^L T_t(z)e^{i\xi_m z} dz$$

(9.44)

where ξ_m represents the eigenvalues of the soil temperature distribution along the axial axis. Accordingly, the general solution of the system of equations can be obtained by summing over all significant frequencies, eigenvalues and soil temperature eigenmodes, i.e.

$$T_i(z,t) = \sum_n \sum_m (A_i e^{-ik_1 z} + B_i e^{-ik_2 z})\hat{F}_m e^{i\omega_n t}$$

$$T_o(z,t) = \sum_n \sum_m (A_o e^{ik_1 z} + B_o e^{ik_2 z})\hat{F}_m e^{i\omega_n t}$$

$$T_g(z,t) = \frac{1}{2}\left[\sum_n \sum_m ((\overline{Y}_1 A_i + \overline{\overline{Y}}_1 A_o)\hat{F}_m e^{-ik_1 z} + (\overline{Y}_2 B_i + \overline{\overline{Y}}_2 B_o)\hat{F}_m e^{-ik_2 z})e^{i\omega_n t}\right]$$

(9.45)

Summing over m can be simply made by algebraic summation, and over n can be made using the inverse FFT algorithm.

9.2.2   *Sub-system 2: soil mass*

Earlier, we have seen that heat transfer in a soil mass of a typical shallow geothermal system is characterized as a non-homogeneous Dirichlet problem in at least three boundaries. At the surface, the soil is subjected to air temperature (or surface temperature), at the bottom, the soil is subjected to a temperature arises from heat transfer between bottom of the earth and the surface, and at the axial-symmetric axis, the soil is subjected to the borehole heat exchanger temperature. To solve this problem it is convenient to decompose the system into two sub-systems, which are homogeneous in two parallel boundaries. As mentioned above, the soil temperature can be described as

$$T_{\text{soil}}(r,z,t) = T_z(z,t) + T_{rz}(r,z,t)$$

(9.46)

Along this line, in the sequel, we formulate the boundary value problem of heat flow in a soil mass, and then solve the governing equations under some relevant initial and boundary conditions.

9.2.2.1 *One-dimensional transient soil temperature, $T_z(z,t)$*

At the beginning, the soil temperature is assumed to be undisturbed and is under steady-state condition. Then, the soil surface is subjected to a change in air temperature or soil surface temperature. Along the vertical boundaries, zero temperatures are imposed, and at the bottom, at some depth h, the temperature is equal to the initial temperature, T_b, Figure 9.1. Heat flow in this problem is obviously one-dimensional.

Assuming that heat conductivity is homogeneous, $\lambda_r = \lambda_z = \lambda$, the heat conduction equation for a one-dimensional domain can be described as

$$\frac{1}{\alpha}\frac{\partial T_z}{\partial t} = \frac{\partial^2 T_z}{\partial z^2} \tag{9.47}$$

in which $\alpha = \lambda/\rho c$ is the thermal diffusivity. The relevant initial and boundary conditions are

$$T_z(0,0) = T_0; \quad T_z(h,0) = T_b; \quad T_z(0,t) = T_a(t) \tag{9.48}$$

The initial condition is employed to ensure that the initial temperature in the soil before operating the geothermal system is equal to that of the steady-state condition.

Recall Eq. (9.1)

$$T_z(z,t) = T_{st}(z) + T_t(z,t) \tag{9.49}$$

where the subscript st stands for the steady-state, and t stands for the transient condition. Substituting Eq. (9.49) into Eq. (9.47) yields

$$\frac{\partial^2}{\partial z^2}[T_{st}(z) + T_t(z,t)] = \frac{1}{\alpha}\frac{\partial}{\partial t}[T_{st}(z) + T_t(z,t)] \tag{9.50}$$

Ignoring the independent variables between the brackets, Eq. (9.50) can be written in terms of two equations

$$\frac{d^2 T_{st}}{dz^2} = 0 \tag{9.51}$$

$$\frac{\partial^2 T_t}{\partial z^2} = \frac{1}{\alpha}\frac{\partial T_t}{\partial t} \tag{9.52}$$

The initial and boundary conditions, Eq. (9.48), can then be modified to

$$T_t(z,0) = 0 \tag{9.53}$$

$$\begin{aligned}
T_{st}(0) + T_t(0,t) &= T_a(t), \quad \therefore T_t(0,t) = T_a(t) - T_0 = \Delta T \\
T_{st}(h) + T_t(h,t) &= T_b, \qquad \therefore T_t(h,t) = 0
\end{aligned} \tag{9.54}$$

Newton's law of cooling can also be applied here. The second boundary condition in Eq. (9.54) indicates that at the bottom the temperature is constant, no transient effect.

9.2.2.2 *Axial-symmetric soil temperature, $T_{rz}(r,z,t)$*

During system operation, the borehole heat exchanger exerts heat flux to the surrounding soil mass. In an axial-symmetric system, the heat flux propagates in the z direction (along the fluid flow direction), and diffuses away in the radial direction. In such a system, the transient heat conduction equation can be described, ignoring the independent variables, as

$$\frac{1}{\alpha}\frac{\partial T_{rz}}{\partial t} - \frac{\partial^2 T_{rz}}{\partial r^2} - \frac{1}{r}\frac{\partial T_{rz}}{\partial r} - \frac{\partial^2 T_{rz}}{\partial z^2} = 0 \tag{9.55}$$

The corresponding boundary conditions are assumed as

$$-\lambda_s \frac{\partial T_{rz}(0, z, t)}{\partial z} dS_{gs} = b_{gs}(T_{rz} - T_g) dS_{gs} + b_{ig}(T_g - T_i) dS_{ig} + b_{og}(T_g - T_o) dS_{og} \tag{9.56}$$

$$T_{rz}(R, z, t) = 0 \tag{9.57}$$

In the first boundary condition, Eq. (9.56), we postulate that the centerline of the soil overlaps with the centerline of the BHE, though the physical contact surface areas are taken into consideration. We also assume that the soil temperature is affected by heat flow generated by all BHE components. This choice is attributed to the fact that the diameter of the borehole is small, and heat flow in the inner pipes has a direct effect on the soil, especially in the short term. In the second boundary condition, Eq. (9.57), we utilized the concept of region-of-interest, which we discussed earlier. R represents a fictitious homogeneous boundary, faraway from the borehole heat exchanger where it is known, intuitively or analytically, that heat flux from the BHE vanishes. This choice will result to an algebraic summation over Fourier-Bessel series, alleviating thus the need to solve semi-infinite integrals of oscillatory transcendental functions.

9.2.2.3   *Spectral analysis of the soil initial and boundary value problem*
Having formulated the initial and boundary value problem, we now use the spectral analysis method to solve the resulting system of equations.

9.2.2.3.1   *Solution of steady-state soil temperature, $T_{st}(z)$*
The steady-state heat equation, Eq. (9.51), is a simple ordinary differential equation, which can be solved readily using direct integration. Integrating Eq. (9.51) twice yields

$$T_{st} + c_1 z + c_2 = 0 \tag{9.58}$$

in which c_1 and c_2 are integration constants, which need to be determined from boundary conditions. Introducing the boundary conditions in Eq. (9.48), or equivalently in Eq. (9.54), into Eq. (9.58) leads to

$$T_{st}(z) = T_0 \left(1 - \frac{z}{h}\right) + T_b \frac{z}{h} \tag{9.59}$$

with T_0 the temperature at the surface, and T_b the temperature at the bottom of the system, at some depth, h, deeper than the bottom of the BHE.

This solution assumes constant thermal conductivity. For shallow geothermal systems, where the temperature variation is relatively low, this assumption is reasonably valid. However, for completeness, it might be worth presenting the solution of the steady-state heat equation for a varying thermal conductivity with depth, $\lambda = \lambda(z)$. The heat equation, Eq. (9.51), in this case, is written as

$$\frac{d}{dz}\left[\lambda(z)\frac{dT}{dz}\right] = 0 \tag{9.60}$$

The relevant boundary conditions are

$$\begin{aligned} T = T_0, \quad z = 0 \\ T = T_b, \quad z = h \end{aligned} \tag{9.61}$$

Integrating Eq. (9.60) gives

$$\lambda(z)\frac{dT}{dz} + c_1 = 0 \tag{9.62}$$

This equation might be written as

$$dT = -c_1 \frac{dz}{\lambda(z)} \tag{9.63}$$

Integrating Eq. (9.63) again, leads to

$$T(z) = -c_1 \int_0^z \frac{dz}{\lambda(z)} + c_2 \qquad (9.64)$$

Applying the first boundary conditions, at $z = 0$, we obtain

$$T_0 = -c_1 \int_0^0 \frac{dz}{\lambda(z)} + c_2, \quad \therefore c_2 = T_0 \qquad (9.65)$$

Applying the second boundary condition, at $z = h$, we get

$$T_b = -c_1 \int_0^h \frac{dz}{\lambda(z)} + T_0, \quad \therefore c_1 = \frac{T_0 - T_b}{\int_0^h \frac{dz}{\lambda(z)}} \qquad (9.66)$$

Substituting Eqs. (9.65) and (9.66) into Eq. (9.64) gives

$$T = T_0 - \frac{T_0 - T_b}{\int_0^h \frac{dz}{\lambda(z)}} \int_0^z \frac{dz}{\lambda(z)} \qquad (9.67)$$

The thermal conductivity might also vary with temperature, $\lambda = \lambda(T)$. In this case, the heat equation, Eq. (9.51), is written as

$$\frac{d}{dz}\left[\lambda(T)\frac{dT}{dz}\right] = 0 \qquad (9.68)$$

This equation is obviously nonlinear. Yener and Kakac (2008) have shown that for a linearly varying thermal conductivity, such as

$$\lambda(T) = \lambda_0(1 + \beta T) \qquad (9.69)$$

in which λ_0 is the thermal conductivity under some reference temperature and β is a material parameter, the solution can be expressed as

$$\frac{T - T_b}{T_0 - T_b} = 1 - \frac{1 + \beta T_0}{\beta(T_0 - T_b)}\left[1 - \left(\beta\frac{T_0 - T_b}{1 + \beta T_0}\left(2 - \beta\frac{T_0 - T_b}{1 + \beta T_0}\right)\frac{z}{h}\right)^{1/2}\right] \qquad (9.70)$$

9.2.2.3.2   *Solution of the transient soil temperature, $T_t(z,t)$*
Fourier transform of Eq. (9.52) yields

$$\frac{d^2 \hat{T}_t}{dz^2} - \kappa^2 \hat{T}_t = 0 \qquad (9.71)$$

in which

$$\kappa = \left(\frac{i\omega\rho c}{\lambda}\right)^{1/2} \qquad (9.72)$$

represents the spectrum relationship of a typical one-dimensional heat flow. Due to the presence of the imaginary component in the spectrum relationship, the temperature in the soil exhibits strong attenuation. Also, since the relationship in Eq. (9.72) is nonlinear in frequency, the propagated temperature reveals changing in its shape.

Fourier transform of the corresponding boundary conditions, Eq. (9.54), are

$$\hat{T}_t(0, \omega) = \hat{T}_a(\omega) - T_0\delta(\omega) = \Delta\hat{T}(\omega)$$

$$\hat{T}_t(h, \omega) = 0 \qquad (9.73)$$

where δ is the Dirac delta function. Solution of the ordinary differential equation, Eq. (9.71), can be expressed as

$$\hat{T}_t(z, \omega) = A e^{\kappa z} + B e^{-\kappa z} \tag{9.74}$$

in which A and B are constants to be determined from boundary conditions.

At $z = 0$, Eq. (9.74) reads

$$A + B = \Delta\hat{T}(\omega) \tag{9.75}$$

At $z = h$, Eq. (9.74) reads

$$A e^{\kappa h} + B e^{-\kappa h} = 0 \tag{9.76}$$

Solving Eqs. (9.75) and (9.76) simultaneously gives

$$A = -\frac{\Delta\hat{T}(\omega)}{1 - e^{-2\kappa h}} e^{-2\kappa h}$$

$$B = \frac{\Delta\hat{T}(\omega)}{1 - e^{-2\kappa h}} \tag{9.77}$$

Hence, for a given frequency, ω_n, the solution of Eq. (9.74) becomes

$$\hat{T}_t(z, \omega_n) = -\frac{\Delta\hat{T}(\omega_n)}{1 - e^{-2\kappa_n h}} e^{-2\kappa_n h} e^{\kappa_n z} + \frac{\Delta\hat{T}(\omega_n)}{1 - e^{-2\kappa_n h}} e^{-\kappa_n z} \tag{9.78}$$

The general solution in the time domain can then be obtained by summing over all significant frequencies, using the inverse fast Fourier transform algorithm as

$$T_t(z, t) = \sum_n \left(-\frac{\Delta\hat{T}(\omega_n)}{1 - e^{-2\kappa_n h}} e^{\kappa_n(z - 2h)} + \frac{\Delta\hat{T}(\omega_n)}{1 - e^{-2\kappa_n h}} e^{-\kappa_n z} \right) e^{i\omega_n t} \tag{9.79}$$

in which $n = 0, 1, 2, \ldots, N - 1$, with N denotes the total number of samples.

9.2.2.3.3 Solution of the axial-symmetric soil temperature, $T_{rz}(r, z, t)$

Fourier transform of Eq. (9.55), gives

$$\frac{i\omega}{\alpha} \hat{T}_{rz} - \frac{\partial^2 \hat{T}_{rz}}{\partial r^2} - \frac{1}{r} \frac{\partial \hat{T}_{rz}}{\partial r} - \frac{\partial^2 \hat{T}_{rz}}{\partial z^2} = 0 \tag{9.80}$$

Fourier transform of the corresponding boundary conditions, Eqs. (9.56) and (9.57), are

$$-\lambda_s \frac{\partial \hat{T}_{rz}(0, z)}{\partial z} dS_{gs} = b_{gs}[\hat{T}_g - \hat{T}_{rz}(0, z)] dS_{gs} + b_{ig}(\hat{T}_g - \hat{T}_i) dS_{ig} + b_{og}(\hat{T}_g - \hat{T}_o) dS_{og} \tag{9.81}$$

$$\hat{T}_{rz}(R, z) = 0 \tag{9.82}$$

Solution of Eq. (9.80) can be carried out elegantly using the method of separation of variables. Assume

$$\hat{T}_{rz}(r, z) = \hat{R}(r)\hat{Z}(z) \tag{9.83}$$

in which $\hat{R}(r)$ is a function of r only and $\hat{Z}(z)$ is a function of z only. Substituting Eq. (9.83) into Eq. (9.80), dividing through by $\hat{R}(r)\hat{Z}(z)$ and equating both sides of the equation by some arbitrary constant, say $-\xi^2$, two independent ordinary differential equations for $\hat{R}(r)$ and $\hat{Z}(z)$ can be obtained, such that

$$\frac{d^2 \hat{R}}{dr^2} + \frac{1}{r} \frac{d\hat{R}}{dr} + \xi^2 \hat{R} = 0 \tag{9.84}$$

$$\frac{d^2 \hat{Z}}{dz^2} + \left(-\frac{i\omega}{\alpha} - \xi^2 \right) \hat{Z} = 0 \tag{9.85}$$

Setting $s = \xi r$, and using the chain rule, Eq. (9.84) reduces to the Bessel equation,

$$\frac{d^2 \hat{R}}{ds^2} + \frac{1}{s} \frac{d\hat{R}}{ds} + \hat{R} = 0 \tag{9.86}$$

Solution of Eq. (9.86) is the Bessel functions J_0 and Y_0 of the first and second kind of order zero. Since the temperature is finite at the origin, $r = 0$, and because Y_0 is infinite at this point, the Y_0 solution is discarded, and the solution of $\hat{R}(r)$ is left with

$$\hat{R}(r) = A_1 J_0(s) = A_1 J_0(\xi r) \tag{9.87}$$

in which A_1 is a constant, and ξ represents the eigenvalues (spatial heat modes) in the radial direction.

For $\hat{Z}(z)$, solution of Eq. (9.85) may take the form

$$\hat{Z}(z) = A_2 e^{i \zeta z} \tag{9.88}$$

in which A_2 is a constant, and

$$\zeta = \left(-\frac{i\omega}{\alpha} - \xi^2 \right)^{1/2} \tag{9.89}$$

represents the eigenvalues in the vertical direction, z. As for the vertical temperature distribution, the spectrum relationship, Eq. (9.89), reveals that, upon propagation, the temperature exhibits attenuation and changing in shape.

Collecting solutions, the solution of the axial-symmetric heat equation in the spectral domain, Eq. (9.80), can then be expressed as

$$\hat{T}_{rz}(r, z) = A e^{i\xi z} J_0(\xi r) \tag{9.90}$$

This equation exhibits propagating temperature in the vertical direction, and diffusive in the radial direction. This function is the eigenfunction of the system, and ξ and ζ are its corresponding eigenvalues, which need to be determined from the boundary conditions.

Fourier representation of the second boundary condition, Eq. (9.82), is

$$\hat{T}_{rz}(R, z) = \hat{Z}(z) \cdot \hat{R}(R) = 0 \tag{9.91}$$

The non-trivial solution of this equation is

$$\hat{R}(R) = J_0(\xi R) = 0 \tag{9.92}$$

This condition can be satisfied at the infinity many positive roots, β_m, of the Bessel function, J_0. This implies

$$\xi_m = \frac{\beta_m}{R} \tag{9.93}$$

The radial function, Eq. (9.87), should then be modified to read

$$\hat{R}_m(r) = J_0(\xi_m r) = J_0 \left(\frac{\beta_m}{R} r \right) \tag{9.94}$$

with each m corresponds to the mth eigenvalue of the system.

It can be noticed that the imposition of the homogeneous boundary condition at $r = R$ has inevitably lead to a priori determined discrete set of eigenvalues. This property renders this approach computationally very efficient, as compared to those, which involve discretization of the spatial domain assuming infinite boundaries. In the later, ξ_m has to be evaluated at the poles and the branch points of the involved contour integrals for each calculation point. Such evaluation is computationally cumbersome due to the infinite upper limit of the involved integrands, and the oscillatory nature of the Bessel function. These problems exacerbate at points relatively far from the source, causing spurious oscillations.

The general solution of the heat equation in the spatial domain can then be obtained by summing over all significant eigenvalues as

$$\hat{T}_{rz}(r, z) = \sum_m A_m e^{i\zeta_m z} J_0(\xi_m r) \tag{9.95}$$

This equation is a Fourier-Bessel series with its coefficient ($A_m e^{i\zeta_m z}$) needs to be determined from the boundary conditions. Substituting Eq. (9.95) into the boundary condition, Eq. (9.81), results to

$$\sum_m \tilde{A}_m e^{i\zeta_m z} = \tilde{T}(z) \tag{9.96}$$

in which

$$\tilde{A}_m = A_m(-i\zeta_m \lambda_s + b_{gs}) dS_{gs} \tag{9.97}$$

and

$$\tilde{T}(z) = (-b_{gs} dS_{gs} + b_{ig} dS_{ig} + b_{og} dS_{og})\hat{T}_g(z) - b_{ig} dS_{ig} \hat{T}_i(z) - b_{og} dS_{og} \hat{T}_o(z) \tag{9.98}$$

Eq. (9.96) is a typical complex Fourier series with its coefficient expressed as

$$\tilde{A}_m = \frac{1}{L} \int_0^L \tilde{T}(z) e^{-i\zeta_m z} dz \tag{9.99}$$

in which L is the BHE length.

9.2.2.4 General solution of the soil heat equation, $T_{soil}(r, z, t)$

Collecting the above solutions, and applying the inverse Fourier transform, the general solution of the soil heat equations in the time domain can then be expressed as

$$T_{soil}(r, z, t) = T_{st}(z) + T_t(z, t) + T_{rz}(r, z, t) \tag{9.100}$$

in which

$$T_{st}(z) = T_0 \left(1 - \frac{z}{h}\right) + T_b \frac{z}{h} \tag{9.101}$$

$$T_t(z, t) = \sum_n \left(-\frac{\Delta \hat{T}(\omega)}{1 - e^{-2\kappa_n h}} e^{\kappa_n(z - 2h)} + \frac{\Delta \hat{T}(\omega)}{1 - e^{-2\kappa_n h}} e^{-\kappa_n z}\right) e^{i\omega_n t} \tag{9.102}$$

and

$$T_{rz}(r, z, t) = \sum_n \sum_m A_m J_0(\xi_m r) e^{i\omega_n t} \tag{9.103}$$

with

$$A_m = \frac{1}{L(-i\zeta_m \lambda_s + b_{gs}) dS_{gs}} \int_0^L \tilde{T}(z) e^{-i\zeta_m z} dz \tag{9.104}$$

and κ_n, ζ_m, ξ_m and $\tilde{T}(z)$ are given in Eqs. (9.72), (9.89), (9.93), and (9.98), respectively.

As it is known, Fourier transform does not account for the initial condition. This is manifested by $T_{st}(z)$ in Eq. (9.101), which does not go into the spectral analysis procedure. However, its values can readily be added to the transient solution in the post processing. This temperature might have any distribution, $f(z)$.

The reader might notice that the spectral analysis offers the means for solving rather complicated initial and boundary value problems. Solution of the transient temperature distribution in all BHE components (pipe-in, pipe-out and grout) and their thermal interactions with fully transient soil temperatures is carried out elegantly using series summations. Compared to commonly utilized analytical methods, and as we have seen in Chapter 7 and Chapter 8, the use of the Laplace transform, despite its versatility, has restricted the solution to a rather simplified combination of geometry and initial and boundary conditions.

An important feature of the spectral analysis which has been employed in this chapter is that, in addition to the ease in summing over algebraic terms, the eigenvalues are determined a priori using Eq. (9.93). As a result, the calculation of the system of equations of typical geothermal systems can be performed only once for practically any number of output points. The response of the system at any point can be calculated as a post-processing, without notable extra computational time. This is not the case in integral formulations, where evaluation of the involved integrals must be done for each calculation point.

9.3   VERIFICATION OF THE BHE MODEL

The BHE analytical model, Eq. (9.42), was devised to simulate heat flow in each individual BHE component and its thermal interactions with other components. To verify the model accuracy, we compare its results with those obtained from an analytical solution of a simplified case. We use the solution introduced by van Genuchten and Alves (1982) for one-dimensional convective-dispersive solute transport. Using this model, the temperature distribution of a fluid moving in a throw-off heat pipe can be described as

$$T(z,t) = T_{in} - \frac{T_s - T_{in}}{2}\left\{e^{(u-v)z/2\alpha}\text{erfc}\left(\frac{z-vt}{2\sqrt{\alpha t}}\right) + e^{(u+v)z/2\alpha}\text{erfc}\left(\frac{z+vt}{2\sqrt{\alpha t}}\right)\right\} \tag{9.105}$$

in which erfc is the complementary error function, $\alpha = \lambda/\rho c$ is the thermal diffusivity, u is the fluid velocity, and

$$v = u\sqrt{1 + \frac{4\phi\alpha}{u^2}} \tag{9.106}$$

with $\varphi = 2b_{ig}/(r_i\rho c)$. T_s represents the temperature in the soil, and T_{in} represents the temperature of the fluid at the inlet of the pipe. The relevant initial and boundary conditions are

$$T(z,0) = T_s; \quad T(0,t) = T_{in}; \quad \partial T(\infty,t)/\partial z = 0 \tag{9.107}$$

The geometry and material parameters are as the following:

Pipe length	=	1 m
Pipe radius, r_i	=	0.013 m
Fluid ρc	=	4.1298E6 J/m³K
Fluid λ	=	0.38 W/m K
Fluid velocity, u	=	3.75E − 4 m/s
Pipe $b_{ig} = b_{gs}$	=	12 W/m² K

The external boundary conditions are: $T_s = 10°C$, and $T_{in} = 50°C$. In the spectral analysis, the external boundary conditions are described as

$$T_{in}(t) = \begin{cases} 50 & 0 < t \leq 3000\,s \\ 0 & 3000 < t < \infty\,s \end{cases}, \quad \text{and} \quad T_s(t) = 10 \quad 0 < t < \infty\,s \tag{9.108}$$

T_{in} is in effect equal to $T_s + \Delta T_{in}$, where in this case $\Delta T_{in} = 40°C$.

We begin by transforming the time history of T_{in} to the frequency domain using forward FFT algorithm. The number of samples is 2048, with a sample rate of 10 s, giving a time window of 20480 s. The calculation results of the temperature along the pipe after 1728 s, as calculated by van Genuchten model, Eq. (9.105), and the BHE spectral model, Eq. (9.42), are shown in

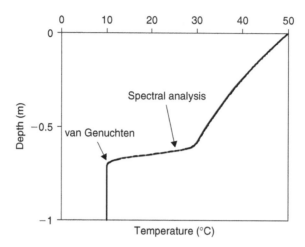

Figure 9.3   BHE spectral model vs. van Genuchten model.

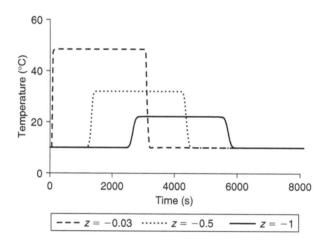

Figure 9.4   Time snapshots for temperatures at different locations.

Figure 9.3. Apparently, the two results are nearly identical. Figure 9.4 shows time reconstruction of the fluid temperature at different depths. As expected the temperature is attenuated and the signal is propagating at a constant speed.

To examine the capability of the spectral model to describe the temperature distribution along pipe-in and pipe-out of a single U-tube BHE, a parametric analysis was performed. Same geometry, material parameters and initial and boundary conditions as in the previous example were utilized. Two cases were studied. In one case, the pipe-out – grout thermal coefficient, b_{og} was made equal to b_{ig}, i.e. $b_{og} = 12\,\text{W/m}^2\,\text{K}$, and in another case, b_{og} was made very small, $b_{og} = 12\text{E} - 5\,\text{W/m}^2\,\text{K}$. Figure 9.5 shows the temperature distribution in pipe-in and pipe-out for both cases. As expected, in the first case, the fluid along pipe-out continues to interact with the surrounding medium, and hence the fluid temperature is continuously changing. In the second case, the fluid temperature at the bottom of pipe-in stayed the same throughout pipe-out. This indicates that pipe-out is totally insulated.

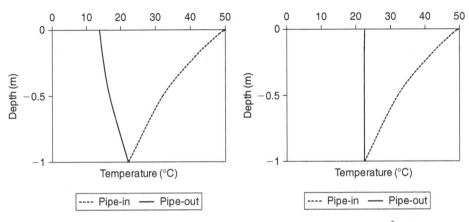

Figure 9.5 Temperature distributions along pipe-in and pipe-out. Left: $b_{og} = 12\,\text{W/m}^2\,\text{K}$, Right: $b_{og} = 12\text{E}{-}5\,\text{W/m}^2\,\text{K}$

9.4 VERIFICATION OF THE SOIL MODEL

Here, we use the infinite line-source model (ILS) for comparison with the soil spectral model, Eq. (9.100). In such a medium the heat flow is normal to the line source, resulting to a radial temperature distribution, described as

$$T = T_0 + \frac{q}{2\pi\lambda} \int_\beta^\infty \frac{e^{-\beta^2}}{\beta} d\beta \tag{9.109}$$

in which

$$\beta = \frac{r}{2\sqrt{\alpha t}}, \quad \text{and} \quad \alpha = \frac{\lambda}{\rho c} \tag{9.110}$$

with r the radial distance from the source, t the time, λ the soil heat conductivity, ρ the mass density, c the heat capacity, T_0 the initial soil temperature, and q heat extraction per meter length of the line source (pipe). Solution for this semi-infinite upper limit integral is available in exact form for $\beta < 0.2$ and in tables for larger values of β. For $\beta < 0.2$ the solution to the integrand of Eq. (9.109) is:

$$I(\beta) = \ln\frac{1}{\beta} + \frac{\beta^2}{2} - \frac{\beta^4}{8} - 0.2886 \tag{9.111}$$

The tabulated values $\beta < 3.1$ are available in Ingersoll et al. (1954). In such a medium, it is assumed that no thermal interaction exists between the heat source and the medium, i.e. there is a full contact between the heat source and the medium. The above theory is applicable to a point source centered in a thin plane sheet. To simulate such a medium, some manipulations are necessary.

1. The temperature along the pipe must be constant. This can be simulated by introducing high flow rates of the refrigerant.
2. The temperatures in each individual component of the pipe (pipe-in, pipe-out and grout) must be equal. This can be done by specifying high b_{ig} (low thermal resistance), and low b_{og} not to be affected by the soil.

Four calculations with different transient times of 5 days, 10 days, 50 days and 100 days were conducted. The material parameters of the soil mass are:

$$\lambda = 2.5\,\text{W/m K}, \quad \alpha = \lambda/\rho c = 1.14\text{E}{-}6\,\text{m}^2/\text{s}, \quad q = 30\,\text{W/m}, \quad \text{and} \quad T_0 = 10^\circ\text{C}.$$

Table 9.1   Spectral model versus ILS model.

Radial distance [m]	$\beta = r/2\sqrt{\alpha t}$	$\dfrac{q}{2\pi\lambda}\displaystyle\int_{\beta}^{\infty}\dfrac{e^{-\beta^2}}{\beta}\,d\beta$	Line source model	Spectral model
5 days				
0.075	0.0535	2.64[†]	4.96	4.87
1.5	1.07	0.085[‡]	9.84	9.12
3	2.14	0.001[‡]	10.00	9.62
10 days				
0.075	0.0379	2.99[†]	4.30	4.30
1.5	0.757	0.24[‡]	9.54	8.74
3	1.51	0.0173[‡]	9.97	9.44
50 days				
0.075	0.0169	3.79[†]	2.76	3.08
1.5	0.339	0.865[‡]	8.35	7.70
3	0.677	0.308[‡]	9.41	8.73
100 days				
0.075	0.0120	4.14[†]	2.10	2.6
1.5	0.239	1.18[‡]	7.75	7.28
3	0.479	0.555[‡]	8.94	8.37

[†]Obtained from Eq. (9.111), [‡]Obtained from tables

In the spectral model, we used: fluid velocity, $u = 50$ m/s, $b_{ig} = 100000$ W/m² K, $b_{og} = 0.001$ W/m² K, and R (the homogeneous boundary of the region-of-interest) $= 25$ m. The number of FFT samples were 2048, with a sampling rate of 10000 s. The number of the Fourier-Bessel series terms were 500.

The calculated results at the pipe surface (0.075 m from the center) up to 3 m are shown in Table 9.1 and Figure 9.6, together with those obtained from the line source solutions, Eq. (9.109). The results show that there is a fairly good matching between the two solutions. The difference in the results, though insignificant, can be attributed to the difference in geometry, and in the solution procedure.

In Figure 9.6 it must be noted that only three points were calculated. The lines connecting these points are not necessarily representing the actual temperature distribution between the points.

The CPU time for conducting any of the above analyses, with 2048 frequency samples and 500 Fourier-Bessel series samples was in the order of 1 second in an Intel PC. The computational results include temperature distributions in all BHE components: pipe-in, pipe-out, and grout; and the soil.

9.5   COMPUTER IMPLEMENTATION

Spectral analysis of the initial and boundary value problem of a shallow geothermal system involves discretization in space and time. For the time domain, the FFT algorithm is utilized, and for the spatial domain the eigenfunction expansion is utilized. Discretization of both domains requires a proper choice of the number of samples and the sampling rates. These two requirements are important for both accuracy and computational efficiency. In Chapter 6 we show that by wrong choice of the sampling rate, aliasing phenomenon occurs in the reconstructed function that distorts the accuracy of the results. The same is valid for the spatial sampling rate and windowing.

Discretization in the temporal domain using FFT algorithm is computationally very efficient for both short terms, in the order of few seconds, and long terms, in the order of many years. The fast

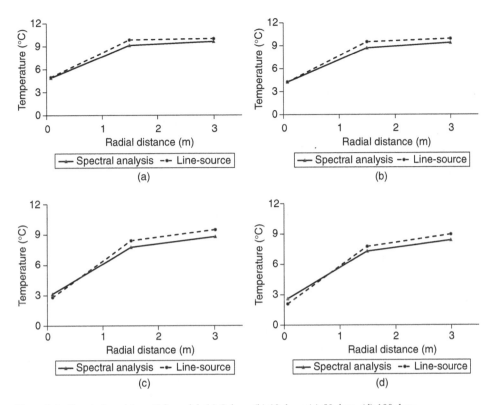

Figure 9.6 Spectral model vs. ILS model. (a) 5 days, (b) 10 days, (c) 50 days, (d) 100 days

Fourier transform procedure requires constant sampling rates. This assumption is perfectly reliable if the analysis is conducted for a consistent time window. Both short and long time windows can be sampled using constant time steps. However, for a time window spanning from short to long time, the use of a constant time step size would be less efficient. Sampling the time function considering only the short term would result to small time steps, which are not necessary for the long term. On the other hand, sampling the time function considering only the long term would neglect detailed behavior of the function in the short term. This issue can be resolved by varying the sampling rate. For this a Fourier-based transform or a Wavelet transform with adaptive sampling rates are more appropriate.

Discretization in the spatial domain requires a decision upon the number of the Bessel function roots, and the radial distance R, the far boundary of the region-of-interest where the BHE heat flow vanishes. The choice of the number of the Bessel function roots is not really a problem since such a function converges fast and the number of samples can easily be determined by trial-and-error. The choice of R needs to be studied. Apparently it is dependent on time. The longer the time of BHE operation is, the further the heat can reach. This problem might be approached either analytically, or by collecting data from field tests. For example, for a point source in an infinite region, Carslaw and Jaeger (1959) had shown that the temperature at point r from the source has its maximum value at time

$$t = \frac{r^2}{6\alpha}$$
(9.112)

This entails that the mean square distance of heat from the source at time t is $6\alpha t$. In this case, R might be chosen such that

$$R > \sqrt{6\alpha t}$$
(9.113)

Similar approach might be developed for other cases.

APPENDIX 9.1

The determinant of the eigenvalue problem, Eq. (9.21), is described in Eq. (9.22), as

$$a_6 k^6 + a_5 k^5 + a_4 k^4 + a_3 k^3 + a_2 k^2 + a_1 k + a_0 = 0 \qquad \text{(A9.1)}$$

The exact form of the involved constants are

$$a_6 = \lambda^2 \lambda_g dV_i dV_o dV_g$$

$$a_5 = -2i\lambda\lambda_g \rho cu dV_i dV_o dV_g$$

$$a_4 = b_{ig}\lambda_g \lambda dS_{ig} dV_g dV_o - \rho^2 c^2 u^2 \lambda_g dV_i dV_g dV_o + 2i\lambda\lambda_g \omega \rho c dV_i dV_g dV_o$$
$$\quad + \lambda^2 b_{og} dV_i dS_{og} dV_o + \lambda\lambda_g b_{og} dV_i dS_{og} dV_g + i\lambda^2 \omega \rho_g c_g dV_i dV_g dV_o$$
$$\quad + \lambda^2 b_{ig} dV_i dS_{ig} dV_o$$

$$a_3 = 2\rho^2 c^2 u \lambda_g \omega dV_i dV_g dV_o - i\rho cu \lambda_g b_{og} dV_i dV_g dS_{og} - ib_{ig}\lambda_g \rho cu dV_o dV_g dS_{ig}$$
$$\quad - 2i\lambda b_{og} \rho cu dV_o dV_i dS_{og} + 2\lambda \rho cu \omega \rho_g c_g dV_o dV_i dV_g - 2i\lambda b_{ig} \rho cu dV_o dV_i dS_{ig}$$

$$a_2 = \lambda b_{ig} b_{og} dV_i dS_{ig} dS_{og} - 2i\lambda b_{ig} \omega \rho c dV_i dV_o dS_{ig} - \omega^2 \rho^2 c^2 \lambda_g dV_i dV_o dV_g$$
$$\quad + ib_{ig}\omega \rho_g c_g \lambda dV_g dV_o dS_{ig} - 2\lambda \omega^2 \rho_g c_g \rho c dV_i dV_o dV_g - \rho^2 c^2 u^2 b_{ig} dV_i dV_o dS_{ig}$$
$$\quad + i\omega \rho c \lambda_g b_{og} dV_i dV_g dS_{og} + b_{ig} b_{og} \lambda_g dS_{ig} dV_g dS_{og} - \rho^2 c^2 u^2 b_{og} dV_i dV_o dS_{og}$$
$$\quad + b_{ig} b_{og} \lambda dS_{ig} dV_o dS_{og} + 2i\lambda b_{og} \omega \rho c dV_i dV_o dS_{og} + ib_{ig}\lambda_g \omega \rho c dV_g dV_o dS_{ig}$$
$$\quad + i\lambda \omega \rho_g c_g b_{og} dV_i dV_g dS_{og} - i\rho^2 c^2 u^2 \omega \rho_g c_g dV_i dV_o dV_g$$

$$a_1 = 2i\rho^2 c^2 u \omega^2 \rho_g c_g dV_i dV_o dV_g - ib_{ig} b_{og} \rho cu dV_o dS_{ig} dS_{og} + \rho cu \omega \rho_g c_g b_{og} dV_i dV_g dS_{og}$$
$$\quad + 2\rho^2 c^2 u \omega b_{ig} dV_i dV_o dS_{ig} + 2\rho^2 c^2 u \omega b_{og} dV_i dV_o dS_{og} + b_{ig}\omega \rho_g c_g \rho cu dV_o dV_g dS_{ig}$$
$$\quad - i\rho cu b_{ig} b_{og} dV_i dS_{ig} dS_{og}$$

$$a_0 = -i\omega^3 \rho^2 c^2 \rho_g c_g dV_i dV_o dV_g - \omega^2 \rho^2 c^2 b_{og} dV_i dV_o dS_{og} - \omega^2 \rho^2 c^2 b_{ig} dV_i dV_o dS_{ig}$$
$$\quad - \omega^2 \rho c \rho_g c_g b_{og} dV_i dV_g dS_{og} + i\omega \rho c b_{ig} b_{og} dV_i dS_{ig} dS_{og} + i\omega \rho c b_{ig} b_{og} dV_o dS_{ig} dS_{og}$$
$$\quad + ib_{ig}\omega \rho_g c_g b_{og} dV_g dS_{ig} dS_{og} - b_{ig}\omega^2 \rho_g c_g \rho c dV_o dV_g dS_{ig}$$

The convenient form of Eq. A9.1 was obtained using MAPLE software. Solution of Eq. A9.1 can be conducted using the IMSL mathematical library subroutine for finding the zeros of polynomials of complex coefficients, DZPOCC, (see IMSL).

CHAPTER 10

Spectral element model for borehole heat exchangers

In this chapter, we present a framework for deriving a semi-analytical model for transient heat transfer in a borehole heat exchanger using the spectral element method. The spectral element method is a powerful tool for solving transient problems in multi-layer and multi-component systems. It solves the governing field equations in a homogeneous medium representing a layer or a component analytically. Combining the system layers or components together is done using the matrix assembly technique of the finite element method. This chapter gives an introduction to the spectral element method and the basic procedure for formulating a spectral element. We formulate a two-node spectral element for general heat pipes, and a one-node spectral element for a single U-tube borehole heat exchanger. Latter, we verify the spectral element model against analytical and finite element solutions.

10.1 INTRODUCTION

The spectral element method (SEM) is a semi-numerical (semi-analytical) technique which combines the spectral analysis method (SAM), basically the discrete Fourier transform, with the finite element method. In the literature, the spectral element method corresponds to two different, and somewhat confusing, techniques. The first corresponds to the work introduced by Anthony Patera of MIT in 1984 (Patera 1984), and the second corresponds to the work introduced by James Doyle of Purdue University in 1988 (Doyle 1988a and 1997). Patera's spectral element method deals mainly with spectral formulations in the spatial domain. In this, the domain is discretized into a number of elements, and the field variable in each element is represented as a high-order Lagrangian interpolation through Chebyshev collocation points. It is thus a finite element method with high degree piecewise polynomial basis functions capable of producing high order accuracy.

Doyle's spectral element method, on the other hand, deals mainly with spectral formulations in the temporal domain. It is a combination of important features of the finite element method, the dynamic stiffness method and the spectral analysis method. Although Beskos (Narayanan and Beskos 1978) is the one who introduced the fundamental concept of the temporal domain SEM for the first time in 1978, the term "spectral element method" seemed to be coined first by Doyle (within his concept of SEM) in his 1992 work on wave propagation in layered systems (Rizzi and Doyle 1992). For more account of the historical and theoretical background of the spectral element method, the reader is referred to Lee (2009). In this book, we adopt the temporal SEM of Doyle, and any reference to the spectral element method is meant to be within this context.

The spectral element method is an elegant technique used mainly for solving wave propagation problems. One of the important features of this method is that its formulation leads to an algebraic set of equations, similar to that of the conventional finite element method. The fundamental difference, however, is that the spectral element stiffness matrix is exact and frequency dependent. Due to the exact description of the system, one element is sufficient to describe a whole homogenous domain. For a non-homogeneous medium consisting of several layers or members, the number of the spectral elements is equal to the number of the involved layers or members. This feature significantly reduces the size of the problem, and rendering this method computationally very efficient.

Similar to the spectral analysis, solving a space-time function using the spectral element method entails discretization of the field variable in the frequency domain and in the spatial domain. The discretization in the frequency domain is done using the fast Fourier transform (FFT) algorithm. The discretization in the spatial domain is done by eigenfunction expansion. By summing over

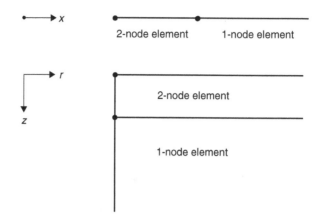

Figure 10.1   1D and 2D spectral element meshes.

all significant frequencies and eigenvalues, the general solution of the system can be obtained. Summing over the involved frequencies using FFT algorithm is, by definition, computationally very efficient. Summing over the eigenvalues, on the other hand, might be difficult and time consuming, since it requires solving for the roots of complex transcendental functions. This problem exuberates in infinite domains. However, it can be circumvented using the region-of-interest concept discussed in Chapter 9. In this, we postulate the presence of a fictitious homogeneous boundary, far away from the source, where it is known a priori that heat flux coming from the source vanishes. This allows a rather straight forward way for determining the eigenvalues. For finite domains, the eigenvalues can be obtained using standard algorithms for solving the roots of complex polynomial functions.

Similar to the finite element method, solving a non-homogeneous system consisting of several homogeneous layers or members requires a spectral element mesh. The difference, however, is that in the spectral element method each layer or member is described by one element. The assembly of the involved elements is conducted using any of the well-known finite element assembly techniques. Figure 10.1 shows typical one-dimensional and two-dimensional spectral element meshes describing a two-member domain, and a two-layer domain respectively. The mesh consists of two elements, one is two-node element, describing a homogeneous member or layer, and another is one-node element, describing a throw-off member or layer. Similar to the *boundary element method*, the spectral element method requires that only the boundary between different domains be partitioned into elements in order to determine the solution, whereas the finite element method, and also Patera's spectral element method, requires that the domain itself be discretized.

10.2   SPECTRAL ELEMENT FORMULATION

Formulation of a spectral element is to a great extent similar to the formulation of a conventional finite element. Techniques like the virtual method or the force-displacement method can be utilized for formulating a spectral element. The difference, however, is that in the spectral element method the governing equations and the boundary conditions are solved in the frequency domain. Another important difference between the two methods is that, in the spectral element method, the shape functions are exact.

In this section, we formulate a spectral element for transient heat conduction in a rod element subjected to heat fluxes at its ends. We adopt the force-displacement method in structural dynamics for formulating the element. The spectral element formulation procedure presented by Lee (2009) for wave propagation in solids will be closely followed. We develop a 1D two-node spectral element.

The partial differential equation governing heat conduction in a rod element subjected to heat fluxes at its ends can be described as

$$\rho c \frac{\partial T(x,t)}{\partial t} - \lambda \frac{\partial^2 T(x,t)}{\partial x^2} = q(x,t) \tag{10.1}$$

in which $q(x,t)$ is an external heat flux. Assume that the temperature and the heat flux can be expressed in terms of their Fourier transforms, as

$$q(x,t) = \frac{1}{N} \sum_{n=0}^{N-1} \hat{q}_n(x,\omega_n) e^{i\omega_n t} \tag{10.2}$$

$$T(x,t) = \frac{1}{N} \sum_{n=0}^{N-1} \hat{T}_n(x,\omega_n) e^{i\omega_n t} \tag{10.3}$$

in which $\hat{T}(x,\omega_n)$ is the spectral representation of the temperature, and $\hat{q}_n(x,\omega_n)$ is the spectral representation of the heat flux. Substituting Eqs. (10.2) and (10.3) into Eq. (10.1) yields

$$\rho c i\omega_n \hat{T}_n(x,\omega_n) - \lambda \frac{d^2 \hat{T}_n(x,\omega_n)}{dx^2} = \hat{q}_n(x,\omega_n) \tag{10.4}$$

Note that Eq. (10.4) is an ordinary differential equation, where the time derivative of the original partial differential equation, Eq. (10.1), is replaced by an algebraic term.

Having dealt with the time domain, we next need to deal with the spatial domain. In the finite element method, the spatial domain is described by a set of nodes constituting, together with the finite elements, the finite element mesh. The primary variables, termed degrees of freedom in the finite element method, are described at the element nodes. The eigenmodes (eigenvalues) of the system in this case are equal to the number of the degrees of freedom, and the roots of the system of equations representing the eigenmodes are set a priori at the mesh nodes. In the spectral element method, on the other hand, the nodes are only at the extreme boundaries of the element, no nodes exist in its interior, similar in that to the boundary element method. The eigenmodes of the system in this case are not known and need to be determined from the roots of the corresponding eigenfunctions. This can be achieved by solving for the system homogeneous equation analytically. The homogeneous equation of Eq. (10.4), ignoring the dependency between brackets for simplicity, is

$$\rho c i\omega_n \hat{T}_n - \lambda \frac{d^2 \hat{T}_n}{dx^2} = 0 \tag{10.5}$$

Solution of this ordinary differential equation might be written as

$$\hat{T} = A e^{-ikx} \tag{10.6}$$

in which A is an integration constant and k represents the eigenvalues of the system. In a vector form, Eq. (10.6) can be written as

$$\hat{\mathbf{T}} = \mathbf{A} e^{-ikx} \tag{10.7}$$

Substituting Eq. (10.7) into Eq. (10.5) yields an eigenvalue problem of the form

$$\mathbf{Q}_n(k;\omega_n)\mathbf{A} = \mathbf{0} \tag{10.8}$$

As it was explained in Chapter 6 and Chapter 9, the non-trivial solution of Eq. (10.8) can be achieved only if we let the determinant of matrix, $\mathbf{Q}_n(k;\omega_n)$, equal to zero, such that

$$\det \mathbf{Q}_n(k;\omega_n) = 0 \tag{10.9}$$

This equation is the characteristic equation of the system, which is a polynomial with a degree equal to the size of the system, expressed as

$$F(k) = a_m k^m + a_{m-1} k^{m-1} + \cdots + a_1 k + a_0 \tag{10.10}$$

where $a_j (j = 1, \ldots, m)$ are the polynomial coefficients and m is the degree of the polynomial. For a given frequency, ω_n, solving for the roots of Eq. (10.10) gives eigenvalues $k_j (j = 1, \ldots, m)$ of the system. The number of the eigenvalues depends on the number of degrees of freedom in the element. For a one-node element with two primary variables, such as $\hat{T}_i$ and $\hat{T}_g$, we obtain two pairs of complex conjugates. In comparison to the finite element method, the roots of Eq. (10.10) implicitly denote the positions of virtual internal nodes. The difference, however, is that in the spectral element method the positions of the virtual nodes are calculated exactly, giving optimal eigenmodes.

Once the eigenvalues are determined, the general solution of the spatial domain can be expressed by summing over all significant eigenmodes, as

$$\hat{T}(x) = \sum_{j=1}^{m} A_j e^{-ik_j x} \tag{10.11}$$

Having done that, the analytic shape functions, also known as the dynamic shape functions, can be formulated. This can be achieved by solving Eq. (10.11) at the element nodes. For a two-node element consisting of node 1 at $z = 0$ and node 2 at $z = h$, with h representing the element length, Eq. (10.11) reads

$$\hat{T}(x) = \sum_{j=1}^{m} A_j e^{-ik_j x} + \sum_{j=1}^{m} B_j e^{-ik_j (h-x)} \tag{10.12}$$

where in this case we have two heat fluxes or prescribed temperatures applied at the nodes. The first term on the left-hand side designates heat flux in the positive x direction, and the second term designates heat flux in the opposite direction, see Figure 10.1. In a matrix form, Eq. (10.12) can be described as

$$\hat{\mathbf{T}}(x) = \mathbf{E}(k_m, \omega_n) \mathbf{A} \tag{10.13}$$

where $\mathbf{E}(k_m, \omega_n)$ represents the exponential terms, and $\mathbf{A}$ represents the constant (coefficient) terms. Solving Eq. (10.12) at the nodes, yields

Node 1, $z = 0$

$$\hat{\mathbf{T}} = \hat{\mathbf{T}}_1 = \sum_{j=1}^{m} \mathbf{A}_j + \sum_{j=1}^{m} \mathbf{B}_j e^{-ik_j h} \tag{10.14}$$

Node 2, $z = h$

$$\hat{\mathbf{T}} = \hat{\mathbf{T}}_2 = \sum_{j=1}^{m} \mathbf{A}_j e^{-ik_j h} + \sum_{j=1}^{m} \mathbf{B}_j \tag{10.15}$$

where $\hat{\mathbf{T}}_1$ and $\hat{\mathbf{T}}_2$ are the nodal vectors which might involve more than one degree of freedom. Putting Eq. (10.14) and Eq. (10.15) together, leads symbolically to

$$\hat{\mathbf{T}}_{node}(k_m, \omega_n) = \mathbf{H}(k_m, \omega_n) \mathbf{A}(k_m, \omega_n) \tag{10.16}$$

Solving for $\mathbf{A}(k_m, \omega_n)$ and substituting into Eq. (10.13), we obtain the temperature distribution in terms of the nodal temperature, as

$$\hat{\mathbf{T}}(x) = \mathbf{N}(k_m, \omega_n) \hat{\mathbf{T}}_{node} \tag{10.17}$$

in which $\mathbf{N}(k_m, \omega_n) = \mathbf{E}(k_m, \omega_n) \mathbf{H}^{-1}(k_m, \omega_n)$ is the dynamic shape function of the spectral element.

To finalize the spectral element formulation we need to relate the heat flux to the temperature. The relationship between the heat flux and the temperature in a domain can be expressed as

$$\hat{\mathbf{q}}(x) = \mathbf{L}\hat{\mathbf{T}}(x) \tag{10.18}$$

where $\mathbf{L}$ is a linear differential operator for the natural boundary conditions. Substituting Eq. (10.17) into Eq. (10.18) and solving for the nodal boundary conditions, we obtain
Node 1, $z = 0$

$$\hat{\mathbf{q}} = \hat{\mathbf{q}}_1 = \mathbf{L}\,\mathbf{N}_1(k_m, \omega_n)\hat{\mathbf{T}}_1 \tag{10.19}$$

Node 2, $z = h$

$$\hat{\mathbf{q}} = \hat{\mathbf{q}}_2 = \mathbf{L}\,\mathbf{N}_2(k_m, \omega_n)\hat{\mathbf{T}}_2 \tag{10.20}$$

Putting Eqs. (10.19) and (10.20) in a matrix form gives

$$\mathbf{K}(k_m, \omega_n)\hat{\mathbf{T}}_{\text{node}} = \hat{\mathbf{q}}_{\text{node}} \tag{10.21}$$

Matrix $\mathbf{K}(k_m, \omega_n)$ is the spectral element stiffness matrix, in resemblance to that of the finite element method. However, the spectral element matrix is exact and frequency-dependent, usually termed the dynamic stiffness matrix.

The spectral element formulation procedure outlined in Eqs. (10.1)–(10.21) provides the means for developing spectral elements consisting of one or two nodes for any number of degrees of freedom. In an alternative way, we are able to formulate a spectral element without explicitly formulating the dynamic shape functions. Here we determine the constants vector $\mathbf{A}(k_m, \omega_n)$ directly from the natural boundary conditions. Continuing from Eq. (10.16), we apply a natural boundary condition of the form

$$\hat{\mathbf{q}}_{\text{node}} = \overline{h}(\hat{\mathbf{T}}_{\text{node}} - \hat{T}_\infty) \tag{10.22}$$

in which $\hat{\mathbf{q}}_{\text{node}}$ represents heat fluxes at the element boundaries, $\hat{T}_\infty$ is the temperature at the surrounding region, and $\overline{h}$ is a convection coefficient. Substituting Eqs. (10.14)–(10.15) into Eq. (10.22) we obtain, symbolically,

$$\hat{\mathbf{q}}_{\text{node}}(k_m, \omega_n) = \mathbf{M}(k_m, \omega_n)\mathbf{A}(k_m, \omega_n) \tag{10.23}$$

Solving Eq. (10.23) for $\mathbf{A}(k_m, \omega_n)$ and substituting into Eq. (10.16) yields

$$\hat{\mathbf{T}}_{\text{node}}(k_m, \omega_n) = \mathbf{G}(k_m, \omega_n)\,\hat{\mathbf{q}}_{\text{node}}(k_m, \omega_n) \tag{10.24}$$

in which

$$\mathbf{G}(k_m, \omega_n) = \mathbf{H}(k_m, \omega_n)\,\mathbf{M}^{-1}(k_m, \omega_n) \tag{10.25}$$

Matrix $\mathbf{G}(k_m, \omega_n)$ is the transfer function of the spectral element. This matrix is the reciprocal of the spectral element stiffness matrix, i.e. $\mathbf{G}(k_m, \omega_n) = \mathbf{k}^{-1}(k_m, \omega_n)$. Note that solving for the shape functions leads directly to the spectral element stiffness matrix, in resemblance to the finite element method. On the other hand, solving directly for the constants, we obtain the transfer function of the system. Nevertheless, both formulations give exactly the same results.

10.3   SPECTRAL ELEMENT FORMULATION FOR BOREHOLE HEAT EXCHANGERS

Here we employ the theory for formulating spectral elements, described in the previous section, to formulate a spectral element for heat flow in a borehole heat exchanger in contact with a soil mass.

Physically, heat flow in a typical borehole heat exchanger is dominated by a convective heat arises from a fluid coming from a heat pump via the inlet of pipe-in, and a conductive heat arises from the contact with the soil mass. From modeling point of view, one-node element is

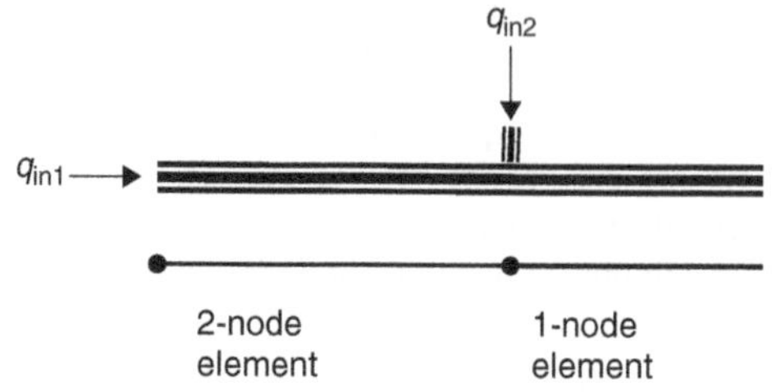

Figure 10.2   Spectral element mesh of a heat pipe subjected to two inlet fluxes.

sufficient to describe heat flow in the heat pipe. The pipe component temperatures are calculated at the element node. Then, temperatures at any point along the pipe can be calculated as a post processing. However, and in order to present a general spectral element formulation that can be useful to other applications, we introduce here, in addition to a one-node element, formulation of a two-node element. Such an element might be utilized for other types of heat exchangers which might be subjected to more than one inlet heat flux. Figure 10.2 shows a hypothetical example where the heat pipe is subjected to two inlet fluxes from two locations. This system can be modeled using two spectral elements: a two-node element, describing the two-inlets part, and a one-node element, describing the flow-off part.

Recalling the governing equations and the initial and boundary conditions of a single U-tube borehole heat exchanger, we have

$$\rho c \frac{\partial T_i}{\partial t} dV_i - \lambda \frac{\partial^2 T_i}{\partial z^2} dV_i + \rho c u \frac{\partial T_i}{\partial z} dV_i - b_{ig}(T_i - T_g) dS_{ig} = 0$$

$$\rho c \frac{\partial T_o}{\partial t} dV_o - \lambda \frac{\partial^2 T_o}{\partial z^2} dV_o - \rho c u \frac{\partial T_o}{\partial z} dV_o - b_{og}(T_o - T_g) dS_{og} = 0 \qquad (10.26)$$

$$\rho c_g \frac{\partial T_g}{\partial t} dV_g - \lambda_g \frac{\partial^2 T_g}{\partial z^2} dV_g - b_{ig}(T_g - T_i) dS_{ig} - b_{og}(T_g - T_o) dS_{og} = 0$$

with

$$T_i(z, 0) = T_o(z, 0) = T_g(z, 0) = T_s(z, 0)$$
$$T_i(0, t) = T_{in}(t)$$
$$-\lambda_g \frac{\partial T_g(z, t)}{\partial z} = b_{gs} [T_g(z, t) - T_s(z, t)] \qquad (10.27)$$
$$T_i(L, t) = T_o(L, t)$$

Alternatively, the initial and boundary conditions can be expressed as

$$T_i(z, 0) = T_o(z, 0) = T_g(z, 0) = T_s(z, 0)$$
$$q_i(0, t) = q_{in}(t)$$
$$-\lambda_g \frac{\partial T_g(z, t)}{\partial z} dS_{gs} - b_{ig} [T_g(z, t) - T_i(z, t)] dS_{ig} - b_{og} [T_g(z, t) - T_o(z, t)] dS_{og} \quad (10.28)$$
$$= b_{gs} [T_g(z, t) - T_s(z, t)] dS_{gs}$$
$$T_i(L, t) = T_o(L, t)$$

in which T_s is the soil temperature immediately around the pipe, and dS_{gs} is the surface area of the grout in contact with the soil. Other parameters are described in Chapter 9.

Applying the Fourier transform to Eq. (10.26) yields

$$-\lambda \frac{d^2 \hat{T}_i}{dz^2} dV_i + \rho cu \frac{d\hat{T}_i}{dz} dV_i + (i\omega\rho c dV_i - b_{ig}dS_{ig})\hat{T}_i = -b_{ig}\hat{T}_g dS_{ig}$$

$$-\lambda \frac{d^2 \hat{T}_o}{dz^2} dV_o - \rho cu \frac{d\hat{T}_o}{dz} dV_o + (i\omega\rho c dV_o - b_{og}dS_{og})\hat{T}_o = -b_{og}\hat{T}_g dS_{og} \qquad (10.29)$$

$$-\lambda_g \frac{d^2 \hat{T}_g}{dz^2} dV_g + (i\omega\rho c_g dV_g - b_{ig}dS_{ig} - b_{og}dS_{og})\hat{T}_g = -b_{ig}\hat{T}_i dS_{ig} - b_{og}\hat{T}_o dS_{og}$$

where the spectral representation of the time derivative is

$$\frac{\partial T}{\partial t} \Rightarrow i\omega\hat{T} \qquad (10.30)$$

and the spectral representation of the spatial derivatives is

$$\frac{\partial^m T}{\partial z^m} \Rightarrow \frac{\partial^m \hat{T}}{\partial z^m} \qquad (10.31)$$

As for typical spectral analysis, the resulting system of equations is frequency dependent and needs to be solved for every frequency, ω_n.

10.3.1 Two-node element

In Chapter 9, we have seen that the solution of Eq. (10.29) might be represented by two sub-systems, one representing pipe-in – grout, and other representing pipe-out – grout. The temperature at any point along the heat pipe is calculated by the superposition of incident fluxes from the two nodes. Following the spectral analysis procedure outlined in Chapter 9, the temperature in a spectral element spanning from $z = 0$ to $z = h$, with h representing its length, can be described as

Pipe-in – grout:

$$\hat{T}_i = A_1 e^{-ik_1 z} + A_2 e^{-ik_2 z} + A_3 e^{-ik_1(h-z)} + A_4 e^{-ik_2(h-z)}$$

$$\hat{T}_{gi} = \overline{Y}_1 A_1 e^{-ik_1 z} + \overline{Y}_2 A_2 e^{-ik_2 z} + \overline{Y}_3 A_3 e^{-ik_1(h-z)} + \overline{Y}_4 A_4 e^{-ik_2(h-z)} \qquad (10.32)$$

Pipe-out – grout:

$$\hat{T}_o = A_{o1} e^{ik_1 z} + A_{o2} e^{ik_2 z} + A_{o3} e^{ik_1(h-z)} + A_{o4} e^{ik_2(h-z)}$$

$$\hat{T}_{go} = \overline{\overline{Y}}_1 A_{o1} e^{-ik_1 z} + \overline{\overline{Y}}_2 A_{o2} e^{-ik_2 z} + \overline{\overline{Y}}_3 A_{o3} e^{-ik_1(h-z)} + \overline{\overline{Y}}_4 A_{o4} e^{-ik_2(h-z)} \qquad (10.33)$$

where k_1 and k_2 are the eigenvalues of the element.

In what follows, pipe-in – grout system of equations, Eq. (10.32), will be treated. The pipe-out – grout formulation is similar.

At node 1, $z = 0$, Eq. (10.32) becomes

$$\hat{T}_{i1} = A_1 + A_2 + A_3 e^{-ik_1 h} + A_4 e^{-ik_2 h}$$

$$\hat{T}_{gi1} = \overline{Y}_1 A_1 + \overline{Y}_2 A_2 + \overline{Y}_3 A_3 e^{-ik_1 h} + \overline{Y}_4 A_4 e^{-ik_2 h} \qquad (10.34)$$

At node 2, $z = h$, Eq. (10.32) becomes

$$\hat{T}_{i2} = A_1 e^{-ik_1 h} + A_2 e^{-ik_2 h} + A_3 + A_4$$
$$\hat{T}_{gi2} = \overline{Y}_1 A_1 e^{-ik_1 h} + \overline{Y}_2 A_2 e^{-ik_2 h} + \overline{Y}_3 A_3 + \overline{Y}_4 A_4 \tag{10.35}$$

Note that in Eq. (10.34), if there is no heat flux exerted at $z = h$, $A_3 = A_4 = 0$ and the nodal temperature reads $\hat{T}_{i1} = A_1 + A_2$, representing the amplitude of the temperatures of pipe-in and grout at node 1. Similarly, in Eq. (10.35), if there is no heat flux exerted at $z = 0$, $A_1 = A_2 = 0$ and the nodal temperature reads $\hat{T}_{i2} = A_3 + A_4$, representing the amplitude of the temperatures of pipe-in and grout at node 2. The same is valid for the grout temperature in both equations, Eqs. (10.34) and (10.35).

Writing Eqs. (10.34) and (10.35) in a matrix form gives

$$\begin{pmatrix} \hat{T}_{i1} \\ \hat{T}_{gi1} \\ \hat{T}_{i2} \\ \hat{T}_{gi2} \end{pmatrix} = \begin{pmatrix} 1 & 1 & e^{-ik_1 h} & e^{-ik_2 h} \\ \overline{Y}_1 & \overline{Y}_2 & \overline{Y}_3 e^{-ik_1 h} & \overline{Y}_4 e^{-ik_2 h} \\ e^{-ik_1 h} & e^{-ik_2 h} & 1 & 1 \\ \overline{Y}_1 e^{-ik_1 h} & \overline{Y}_2 e^{-ik_2 h} & \overline{Y}_3 & \overline{Y}_4 \end{pmatrix} \begin{pmatrix} A_1 \\ A_2 \\ A_3 \\ A_4 \end{pmatrix} \tag{10.36}$$

The constants in Eq. (10.36) can be expressed in terms of the nodal temperatures, in a compacted from, as

$$\mathbf{A} = \mathbf{Q}^{-1}\hat{\mathbf{T}} \tag{10.37}$$

The solution so far gave the temperatures in terms of the integration constants $\mathbf{A} = [A_1 \ A_2 \ A_3 \ A_4]^T$. These constants must be determined from the boundary conditions, Eq. (10.28). After solving for $\mathbf{A}$ in terms of nodal temperatures, the temperature in pipe-in at any arbitrary point in the pipe can then be obtained using Eq. (10.32), written symbolically as

$$\hat{T}_i(z, \omega) = \hat{N}_1(z, \omega)\hat{T}_{i1} + \hat{N}_2(z, \omega)\hat{T}_{i2} \tag{10.38}$$

in which $\hat{N}_1(z, \omega)$ and $\hat{N}_2(z, \omega)$ are the shape functions of the heat exchanger. The same is valid for the other pipe components.

The corresponding heat fluxes in pipe-in and grout are

$$q_i = -\lambda \frac{d\hat{T}_i}{dz} + \rho c u \hat{T}_i$$
$$q_{ig} = -\lambda_g \frac{dT_g(z,t)}{dz} dS_{gs} - b_{ig}(T_g - T_i) dS_{ig} \tag{10.39}$$

Letting $dS_{gs} = 2\pi r_{gs} z$ and $dS_{ig} = 2\pi r_{ig} z$, in which r_{gs} represents the radius of the grout in contact with the soil (the borehole radius) and r_{ig} represents the radius of the grout in contact with pipe-in, the boundary conditions can be described as

$$q_i(0, t) = q_{in1}(t)$$
$$q_i(h, t) = q_{in2}(t) \tag{10.40}$$
$$q_{ig} = -\lambda_g \frac{dT_g(z,t)}{dz} - b_{ig}(T_g - T_i)\frac{r_{ig}}{r_{gs}} = b_{gs}(T_g - T_s)$$

Substituting Eq. (10.32) and Eq. (10.40) into Eq. (10.39) yields

At $z = 0$:

$$\hat{q}_i = \hat{q}_{in1} = -\lambda\left(-ik_1 A_1 - ik_2 A_2 + ik_1 A_3 e^{-ik_1 h} + ik_2 A_4 e^{-ik_2 h}\right)$$

$$+\rho cu\left(A_1 + A_2 + A_3 e^{-ik_1 h} + A_4 e^{-ik_2 h}\right)$$

$$\hat{q}_{ig} = \hat{q}_{ig1} = -\lambda_g\left(-\overline{Y}_1 ik_1 A_1 - \overline{Y}_2 ik_2 A_2 + \overline{Y}_3 ik_1 A_3 e^{-ik_1 h} + \overline{Y}_4 ik_2 A_4 e^{-ik_2 h}\right) \quad (10.41)$$

$$-b_{ig}\left(\begin{array}{c}\overline{Y}_1 A_1 + \overline{Y}_2 A_2 + \overline{Y}_3 A_3 e^{-ik_1 h} + \overline{Y}_4 A_4 e^{-ik_2 h} \\ -\left(A_1 + A_2 + A_3 e^{-ik_1 h} + A_4 e^{-ik_2 h}\right)\end{array}\right)\frac{r_{ig}}{r_{gs}}$$

$$-b_{gs}\left(\overline{Y}_1 A_1 + \overline{Y}_2 A_2 + \overline{Y}_3 A_3 e^{-ik_1 h} + \overline{Y}_4 A_4 e^{-ik_2 h} - \hat{T}_s\right)$$

At $z = h$:

$$\hat{q}_i = \hat{q}_{in2} = -\lambda\left(-ik_1 A_1 e^{-ik_1 h} - ik_2 A_2 e^{-ik_2 h} + ik_1 A_3 + ik_2 A_4\right)$$

$$+\rho cu\left(A_1 e^{-ik_1 h} + A_2 e^{-ik_2 h} + A_3 + A_4\right)$$

$$\hat{q}_{ig} = \hat{q}_{ig2} = -\lambda_g\left(-\overline{Y}_1 ik_1 A_1 e^{-ik_1 h} - \overline{Y}_2 ik_2 A_2 e^{-ik_2 h} + \overline{Y}_3 ik_1 A_3 + \overline{Y}_4 ik_2 A_4\right) \quad (10.42)$$

$$-b_{ig}\left(\begin{array}{c}\overline{Y}_1 A_1 e^{-ik_1 h} + \overline{Y}_2 A_2 e^{-ik_2 h} + \overline{Y}_3 A_3 + \overline{Y}_4 A_4 \\ -\left(A_1 e^{-ik_1 h} + A_2 e^{-ik_2 h} + A_3 + A_4\right)\end{array}\right)\frac{r_{ig}}{r_{gs}}$$

$$-b_{gs}\left(\overline{Y}_1 A_1 e^{-ik_1 h} + \overline{Y}_2 A_2 e^{-ik_2 h} + \overline{Y}_3 A_3 + \overline{Y}_4 A_4 - \hat{T}_s\right)$$

Writing Eqs. (10.41) and (10.42) in a matrix form, gives

$$\hat{\mathbf{q}} = \hat{\mathbf{M}}\,\mathbf{A} = \begin{pmatrix} m_{11} & m_{12} & m_{13} & m_{14} \\ m_{21} & m_{22} & m_{23} & m_{24} \\ m_{31} & m_{32} & m_{33} & m_{34} \\ m_{41} & m_{42} & m_{43} & m_{44} \end{pmatrix}\mathbf{A} \quad (10.43)$$

in which

$$m_{11} = \lambda ik_1 + \rho cu$$

$$m_{12} = \lambda ik_2 + \rho cu$$

$$m_{13} = -\lambda ik_1 e^{-ik_1 h} + \rho cu e^{-ik_1 h}$$

$$m_{14} = -\lambda ik_2 e^{-ik_2 h} + \rho cu e^{-ik_2 h}$$

$$m_{21} = \lambda_g \overline{Y}_1 ik_1 - b_{ig}\left(\overline{Y}_1 - 1\right)\frac{r_{ig}}{r_{gs}} - b_{gs}\overline{Y}_1$$

$$m_{22} = \lambda_g \overline{Y}_2 ik_2 - b_{ig}\left(\overline{Y}_2 - 1\right)\frac{r_{ig}}{r_{gs}} - b_{gs}\overline{Y}$$

$$m_{23} = -\lambda_g \overline{Y}_3 ik_1 e^{-ik_1 h} - b_{ig}\left(\overline{Y}_3 e^{-ik_1 h} - e^{-ik_1 h}\right)\frac{r_{ig}}{r_{gs}} - b_{gs}\overline{Y}_3 e^{-ik_1 h}$$

$$m_{24} = -\lambda_g \overline{Y}_4 ik_2 e^{-ik_2 h} - b_{ig}\left(\overline{Y}_4 e^{-ik_2 h} - e^{-ik_2 h}\right)\frac{r_{ig}}{r_{gs}} - b_{gs}\overline{Y}_4 e^{-ik_2 h}$$

$$m_{31} = \lambda i k_1 e^{-ik_1 h} + \rho c u e^{-ik_1 h}$$

$$m_{32} = \lambda i k_2 e^{-ik_2 h} + \rho c u e^{-ik_2 h}$$

$$m_{33} = -\lambda i k_1 + \rho c u$$

$$m_{34} = -\lambda i k_2 + \rho c u$$

$$m_{41} = \lambda_g \overline{Y}_1 i k_1 e^{-ik_1 h} - b_{ig} \left(\overline{Y}_1 e^{-ik_1 h} - e^{-ik_1 h}\right) \frac{r_{ig}}{r_{gs}} - b_{gs} \overline{Y}_1 e^{-ik_1 h} \tag{10.44}$$

$$m_{42} = \lambda_g \overline{Y}_2 i k_2 e^{-ik_2 h} - b_{ig} \left(\overline{Y}_2 e^{-ik_2 h} - e^{-ik_2 h}\right) \frac{r_{ig}}{r_{gs}} - b_{gs} \overline{Y}_2 e^{-ik_2 h}$$

$$m_{43} = -\lambda_g \overline{Y}_3 i k_1 - b_{ig} \left(\overline{Y}_3 - 1\right) \frac{r_{ig}}{r_{gs}} - b_{gs} \overline{Y}_3$$

$$m_{44} = -\lambda_g \overline{Y}_4 i k_2 - b_{ig} \left(\overline{Y}_4 - 1\right) \frac{r_{ig}}{r_{gs}} - b_{gs} \overline{Y}_4$$

Substituting Eq. (10.37) into Eq. (10.43), yields

$$\hat{\mathbf{q}} = \hat{\mathbf{M}} \mathbf{Q}^{-1} \hat{\mathbf{T}} = \hat{\mathbf{K}} \hat{\mathbf{T}} \tag{10.45}$$

in which $\hat{\mathbf{K}}$ is the spectral element dynamic stiffness matrix.

Having formulated a spectral element, assembly of the elements constituting the heat pipe can be achieved in exactly the same way as for the finite element method. The resulting global stiffness matrix is complex, and hence requires a solver, which is capable of solving complex system of equations.

10.3.2   One-node element

In practice, the borehole heat exchanger has only one inlet located at the point where pipe-in is connected to the heat pump, usually placed inside the building. Moreover, unlike wave propagation problems, where reflection occurs at the boundary between two different media, in heat transfer problems, no reflection takes place. Hence, one-node spectral element is sufficient to describe heat flow in a borehole heat exchanger. In wave propagation problems, this kind of elements is referred to as *throw-off element*, usually used for modeling wave propagation in a semi-infinite medium. The system of equations is solved at the node. However, temperatures at any point within the element can be determined in the post processing.

In this case, the two sub-systems, pipe-in – grout and pipe-out – grout, are described as
Pipe-in – grout:

$$\hat{T}_i = A_i e^{-ik_1 \zeta} + B_i e^{-ik_2 \zeta}$$

$$\hat{T}_{gi} = \overline{Y}_1 A_i e^{-ik_1 \zeta} + \overline{Y}_2 B_i e^{-ik_2 \zeta} \tag{10.46}$$

Pipe-out – grout:

$$\hat{T}_o = A_o e^{ik_1 \zeta} + B_o e^{ik_2 \zeta}$$

$$\hat{T}_{go} = \overline{\overline{Y}}_1 A_o e^{-ik_1 \zeta} + \overline{\overline{Y}}_2 B_o e^{-ik_2 \zeta} \tag{10.47}$$

in which, k_1 and k_2 are the eigenvalues, and ζ represents the element local coordinate. This coordinate is useful if more than one element is used to describe the BHE. The spectral element is assumed to span from $\zeta = 0$ to $\zeta = h$, with h representing its length. In what follows, pipe-in – grout system, Eq. (10.46), will be treated. The pipe-out – grout formulation follows suit.

At the element node, $\zeta = 0$, Eq. (10.46) becomes

$$\begin{pmatrix} \hat{T}_i \\ \hat{T}_{gi} \end{pmatrix} = \begin{pmatrix} 1 & 1 \\ \overline{Y}_1 & \overline{Y}_2 \end{pmatrix} \begin{pmatrix} A_i \\ B_i \end{pmatrix}$$

(10.48)

In a compact form, Eq. (10.48) can be written as

$$\mathbf{A} = \mathbf{Q}^{-1} \hat{\mathbf{T}}$$

(10.49)

in which

$$\mathbf{Q}^{-1} = \frac{1}{\overline{Y}_2 - \overline{Y}_1} \begin{pmatrix} \overline{Y}_2 & -1 \\ -\overline{Y}_1 & 1 \end{pmatrix}$$

(10.50)

The corresponding heat fluxes in pipe-in and grout are

$$\hat{q}_i = -\lambda \frac{d\hat{T}_i}{dz} + \rho c u \hat{T}_i$$

(10.51)

$$\hat{q}_{gi} = -\lambda_g \frac{d\hat{T}_{gi}}{dz}$$

(10.52)

The boundary conditions, Eq. (10.27), can be expressed in the spectral domain as

$$\hat{q}_{gi} = -\lambda_g \frac{d\hat{T}_{gi}}{dz} = b_{gs}\left[\hat{T}_{gi} - \hat{T}_s(z, \omega)\right]$$

(10.53)

Note that we used the boundary condition, Eq. (10.27), instead of Eq. (10.28), just as an alternative. Substituting Eq. (10.46) into Eq. (10.51) and Eq. (10.53) results to

$$\hat{q}_i = \left(ik_1 \lambda e^{-ik_1\zeta} + \rho c u e^{-ik_1\zeta}\right) A_i + \left(ik_2 \lambda e^{-ik_2\zeta} + \rho c u e^{-ik_2\zeta}\right) B_i$$

(10.54)

$$\hat{q}_{gs} = \left(ik_1 \lambda_g \overline{Y}_1 e^{-ik_1\zeta} + \overline{Y}_1 e^{-ik_1\zeta} b_{gs}\right) A_i + \left(ik_2 \lambda_g \overline{Y}_2 e^{-ik_2\zeta} + \overline{Y}_2 e^{-ik_2\zeta} b_{gs}\right) B_i$$

(10.55)

in which $\hat{q}_{gs} = b_{gs}\hat{T}_s(z, \omega)$.

In the first element, where the element node is at $\zeta = 0$, $\hat{q}_i = \hat{q}_{in}$, the element heat flux equations, Eqs. (10.54) and (10.55), can be expressed as

$$\begin{pmatrix} \hat{q}_{in} \\ \hat{q}_{gs} \end{pmatrix} = \begin{pmatrix} ik_1 \lambda + \rho c u & ik_2 \lambda + \rho c u \\ ik_1 \lambda_g \overline{Y}_1 + \overline{Y}_1 b_{gs} & ik_2 \lambda_g \overline{Y}_2 + \overline{Y}_2 b_{gs} \end{pmatrix} \begin{pmatrix} A_i \\ B_i \end{pmatrix}$$

(10.56)

This equation might be written in a vector notation as

$$\hat{\mathbf{q}} = \hat{\mathbf{M}} \mathbf{A}$$

(10.57)

Substituting $\mathbf{A}$ from Eq. (10.49) into Eq. (10.57), leads to

$$\hat{\mathbf{q}} = \hat{\mathbf{M}} \mathbf{Q}^{-1} \hat{\mathbf{T}} = \hat{\mathbf{K}} \hat{\mathbf{T}}$$

(10.58)

in which $\hat{\mathbf{K}}$ is the spectral element stiffness matrix of a one-node element.

Solving Eq. (10.58), the nodal temperatures of pipe-in and grout can be determined. Once the temperatures at the node are obtained, the coefficients A_i and B_i can be determined from Eq. (10.49). Then the temperature at any point along the length of the element can be calculated using Eq. (10.46).

For a system consisting of more than one element, a sequential solution is required. This differs from the classical spectral element technique where element assembly, similar to that utilized in the finite element method, is necessary. Here, the continuity condition between two sequential elements, j and $j+1$ is imposed at $\zeta = h$, such that

$$\hat{q}_i^{j+1}(0,\omega) = \hat{q}_i^j(h,\omega) \tag{10.59}$$

The heat flux equations, Eq. (10.56), for element $j+1$ can then be described as

$$\begin{pmatrix} \hat{q}_i^j(h,\omega) \\ \hat{q}_{gs}^{j+1} \end{pmatrix} = \begin{pmatrix} ik_1^{j+1}\lambda + \rho c u & ik_2^{j+1}\lambda + \rho c u \\ ik_1^{j+1}\lambda_g \overline{Y}_1 + \overline{Y}_1 b_{gs} & ik_2^{j+1}\lambda_g \overline{Y}_2 + \overline{Y}_2 b_{gs} \end{pmatrix} \begin{pmatrix} A_i^{j+1} \\ B_i^{j+1} \end{pmatrix} \tag{10.60}$$

in which k_1^{j+1}, k_2^{j+1} are the eigenvalues of element $j+1$, need to be determined. The corresponding temperature distribution in pipe-in and grout is described as

$$\begin{pmatrix} \hat{T}_i^{j+1} \\ \hat{T}_g^{j+1} \end{pmatrix} = \begin{pmatrix} e^{-ik_1^{j+1}\zeta} & e^{-ik_2^{j+1}\zeta} \\ \overline{Y}_1 e^{-ik_1^{j+1}\zeta} & \overline{Y}_2 e^{-ik_2^{j+1}\zeta} \end{pmatrix} \begin{pmatrix} A_i^{j+1} \\ B_i^{j+1} \end{pmatrix} \tag{10.61}$$

By substituting the coefficients A_i^{j+1} and B_i^{j+1} from Eq. (10.60) into Eq. (10.61) the temperatures at any arbitrary point along element $j+1$ can be determined. The same procedure can be followed for all elements involved.

General solution of a system consisting of one or more spectral elements can be solved by imposing the boundary conditions. For instance, a prescribed temperature on a node or on an element can be described as

$$\hat{T}(\xi,\omega) = \hat{F}_m(\xi)\hat{F}_n(\omega) \tag{10.62}$$

where $\hat{F}_m(\xi)$ represents the spatial distribution of the force and $\hat{F}_n(\omega)$ represents its frequency spectrum. The general solution can then be obtained by summing over all significant spatial modes and frequencies, such that

$$T(z,t) = \sum_n \sum_m \hat{G}(k_m,\omega_n)\hat{F}_m\hat{F}_n e^{i\omega_n t} \tag{10.63}$$

in which $\hat{G}(k_m,\omega_n) = \hat{K}(k_m,\omega_n)^{-1}$ representing the transfer function of the system. In case of a prescribed temperature at the inlet of pipe-in, $\hat{T}_{in}$, $\hat{F}_m = 1$ and $\hat{F}_n$ is calculated by means of the forward FFT algorithm. For a time varying soil temperature and a constant distribution in space, the same is valid. However, for a varying soil temperature in space, summing over the spatial modes must be used for each element.

10.4   ELEMENT VERIFICATION

To verify the formulation and the computer implementation of the BHE spectral element, we use van Genuchten and Alves (1982) solution of heat flow in one-dimensional domain, described in Chapter 9. The BHE pipe is simulated using one spectral element. The involved material and geometrical properties are similar to those utilized in the verification example of Chapter 9. The calculation results obtained from the spectral element model together with the van Genuchten and Alves solution are shown in Figure 10.3. Apparently there is a quite good match between the two results.

The figure also shows two finite element calculation results obtained from the finite element package FEFLOW (Diersch *et al.* 2011a and 2011b). Part of the finite element mesh is shown in Figure 10.4, where a cut along the borehole heat exchanger is made. Two mesh coarseness along

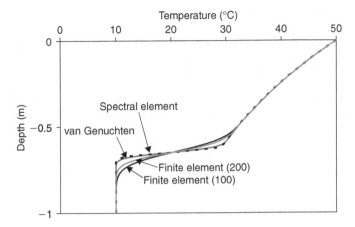

Figure 10.3   Spectral element model vs. van Genuchten model; and finite element model.

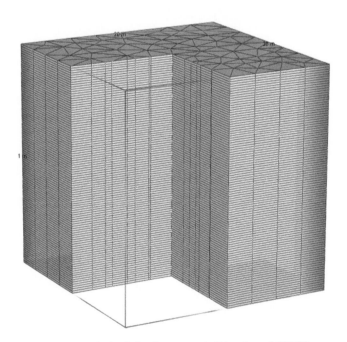

Figure 10.4   A cut along the BHE in the finite element mesh (Diersch *et al.* 2011b).

the vertical axis were utilized: one using 100 elements (slices), and another using 200 elements. The dimension of the mesh is 20 m × 20 m × 1 m. Thermal coefficients of pipe-in and grout are made equal, i.e. $b_{ig} = b_{gs}$, while the coefficient in pipe-out is set to zero, $b_{og} = 0$, to eliminate its thermal interaction with other pipe components. The temperature in the soil is kept constant.

The figure shows that even though the finite element results are in good agreement with the analytical solution, the spectral element result is more accurate. Adding to that, the spectral element model is remarkably efficient. The spectral element method required only one element to describe the whole pipe, while the finite element method needed 200 elements (in the vertical direction) to come close, but not too close, to the analytical solution.

10.5   CONCLUDING REMARKS

The spectral element model introduced in this chapter is generic and can be used as a framework for developing comprehensive models for shallow geothermal systems consisting of multiple borehole heat exchangers embedded in multilayer soil systems. The borehole heat exchanger can be modeled using one or more spectral elements. The soil mass can be modeled using the spectral element method or any of the available analytical, semi-analytical or numerical methods. The thermal interaction between the soil mass and the boreholes can be done using staggered (sequential) algorithm, wherein the system is divided into two sub-systems: one representing the borehole heat exchangers and the other representing the soil. The two systems are solved separately, but coupled by their external forces, i.e. heat fluxes at their boundary surfaces. The reader is encouraged to develop spectral elements for the soil mass and make the coupling to the borehole heat exchangers.

Part III

Numerical modeling

CHAPTER 11

Finite element methods for conduction-convection problems

In the last three chapters we introduced analytical and semi-analytical methods for solving initial and boundary value problems involving heat conduction-convection processes. We show that the utilization of such methods, in spite of their rigor, they are limited to some relatively simple geometry and initial and boundary conditions. This limitation can be circumvented by the use of numerical methods, among which the finite element method represents one of the most versatile technique in this category. In this chapter we introduce some basic finite element procedures suitable for solving steady-state and transient heat conduction-convection problems. Focus is placed on the discretization of the finite element in space and time.

11.1 INTRODUCTION

The finite element method (FEM) is a versatile numerical technique utilized for solving wide range of partial differential equations. It is particularly suitable for solving initial and boundary value problems occurring in engineering applications, constituting complicated geometry and initial and boundary conditions. The fundamental property of the finite element method is that it partitions a continuum of space and time into a set of discrete components, represented by elements and nodes. This property is shared by many other numerical techniques including the finite difference method.

During 1960's and early 1970's the finite element method was utilized to solve engineering problems related mainly to solid mechanics. The finite difference method was then the tool for solving heat and fluid flow problems. However, in mid-1970's, the finite element method was utilized for solving transient and steady-state problems dealing with advective-diffusive transport. Since then, it became an important tool for modeling wide range of thermo-hydro mechanical problems in engineering and geosciences.

This somewhat late utilization of the finite element method for solving flow problems was not merely due to preferences. Rather, it was a result of the inherited limitation of the standard finite element procedures to solve the type of differential equations dealing with convective-dominated processes. In retrospect, there are mainly three types of differential equations: *elliptic, parabolic* and *hyperbolic*. These equations cover wide range of engineering and physical applications. Mathematically, they arise from the quasilinear differential equation:

$$au_{xx} + 2bu_{xy} + cu_{yy} = F \tag{11.1}$$

in which $u_{xx} = \partial^2 u/\partial x^2$, etc. and a, b and c are some relevant coefficients. This equation is said to be

$$
\begin{array}{lll}
\text{elliptic} & \text{if} & ac - b^2 > 0 \\
\text{parabolic} & \text{if} & ac - b^2 = 0 \\
\text{hyperbolic} & \text{if} & ac - b^2 < 0
\end{array}
\tag{11.2}
$$

There are many important engineering and scientific applications of these equations. Applications involving elliptic equations are the well-known Laplace and Poisson problems. Applications involving parabolic equations are those dealing with species diffusion and heat conduction. Applications involving hyperbolic equations are wave propagation problems. Using the finite element method, these equations are replaced by a discrete set of algebraic equations. For the elliptic equation, the finite element discretization guarantees convergence. That is, the approximate solution converges to the exact solution as the element size tends to zero. This is not necessarily the case for the parabolic and the hyperbolic equations. For these two types, it is necessary to employ additional conditions to assure convergence and stability. Stability mean that small perturbation at any time remains small at latter times.

Diffusion-advection (or conduction-convection) applications are somewhat in the middle between parabolic and hyperbolic type equations. For a low convection case, the problem is close to being parabolic, but for a high convection case, the problem is close to being hyperbolic. We shall see later in this chapter that the convergence and stability of the first type can be improved by enhancing the discretization of the time domain. However, for the second type, we need to employ somewhat more complicated finite element procedure to enhance the spatial and the temporal domains.

In this chapter, focus will be placed on the discretization of heat flow problems in space and time. There are mainly two methods of discretization: *Rayleigh-Ritz variational* method, and *Galerkin method of weighted residuals*. The variational method solves problems by finding stationary values of some *functional* through an energy minimization procedure. Functional is an integral function which describes governing equations and boundary conditions of a physical system. The drawback of the variational method is that the functional is needed to be known a priori. For second order elasticity problems, the functional is well known, and therefore it is common to use the variational method to solve this kind of problems. However, for problems consisting of first order derivatives, such as convection problems, no functional is available, and thus the variational method is not directly applicable. (In some cases the variational method may be applied to pseudo-variational or quasi-variational functions.) For this kind of problems, the Galerkin weighted residual method is more suitable, since the governing equations can be directly discretized. Accordingly, we will utilize the weighted residual method to drive and solve transient and steady-state conductive-convective problems occurring in shallow geothermal systems. This will be covered in the next chapter. Here, we discuss the applicability of the Galerkin and the *upwind finite element* method for the discretization of such problems.

11.2   SPATIAL DISCRETIZATION

Appropriate spatial discretization of convective problems is important because there is a direct relationship between the element size and the fluid velocity. Inappropriate mesh size might lead to sever distortion of the finite element results. In this section we discuss this issue.

11.2.1   *Galrekin finite element method*

Galerkin finite element method is an approximation technique commonly utilized to solve complicated initial and boundary value problems. It is a piecewise application of the weighted residual method, where an exact field variable is approximated by a discrete number of degrees of freedom using some polynomial interpolation functions. As a result, the governing partial differential equations are replaced by a system of algebraic equations. However, and since the solution is approximate, the differential equations will not be satisfied, and there will be an inevitable error in the result. This error is known as the residual.

Consider a differential equation describing a one-dimensional steady-state heat conduction in a solid:

$$-\lambda \frac{d^2 T}{dx^2} + Q = 0, \quad \text{over } V \tag{11.3}$$

with a boundary condition of the form

$$-\lambda \frac{dT}{dn} + q_n = 0, \quad \text{over } S \tag{11.4}$$

where λ is the thermal conductivity, Q is a heat source, q_n is a heat flux normal to the boundary, n is the direction normal to Γ, and T is the primary variable, which is exact in its strong form, Eq. (11.3) and Eq. (11.4). A possible approximation to these equations can be obtained by expanding the unknown variable in terms of basis or trial functions as

$$\tilde{T} = \sum_{i=1}^{m} N_i T_i \tag{11.5}$$

in which $\tilde{T}$ is an approximation to the exact form T, T_i, $i = 1, \ldots, m$, represents the nodal values, with m, the number of the element nodes, and N_i, $i = 1, \ldots, m$ represents a set of m piecewise locally defined basis function, usually a complete set of mutually orthogonal normalized polynomials. In the context of the finite element method, the basis functions are referred to as the *shape functions*.

For the approximation to be convergent and sufficiently accurate, the *completeness condition* must be satisfied. This condition states that as the limit of $m \to \infty$, the exact solution must be retrieved, such that

$$T = \sum_{i=1}^{\infty} N_i T_i \tag{11.6}$$

Substituting Eq. (11.5) into Eq. (11.3) and Eq. (11.4), and since $\tilde{T}$ is an approximation to the exact form T, an error is introduced to the solution. This error is the residual, which is not zero, and defined as

$$R_V = -\lambda \frac{d^2\tilde{T}}{dx^2} + Q \neq 0$$

$$R_S = -\lambda \frac{d\tilde{T}}{dn} + q_n \neq 0 \tag{11.7}$$

The method of the weighted residual requires that the nodal values in Eq. (11.5) are computed by multiplying Eq. (11.7) by any arbitrary weighting function W, so that the integrals of the residual functions become zero over their intervals, such that

$$\int_V W_V R_V d\Omega + \int_S W_S R_S dS = 0 \tag{11.8}$$

For the integrals to be bounded, the *continuity condition* must be satisfied. This condition implies that the involved functions must be differentiable over some least possible orders. If nth-order derivatives exist in the original function, then its derivatives to the $(n-1)$th-order must be continuous. That is for a function with its first derivative (say $\partial T/\partial x$) continuous and its second order derivative unbounded, the function is said to be C_0. The associated finite element of this function is called the C_0-element. The C_0-element is the most popular type of finite elements, and is sufficient for heat flow problems. The continuity condition, together with the completeness condition, forms what is known as the *admissibility condition*.

The weighted residual equations, in their original forms, are expressed in what is known as the *strong form*. For instance, substituting Eq. (11.7) into Eq. (11.8), the strong form of the heat equation, Eq. (11.3), is obtained, in a vector form, as

$$\int_V \mathbf{W}_V \left(-\lambda \frac{d^2\tilde{T}}{dx^2} + Q \right) dV = 0 \tag{11.9}$$

As a consequence of the continuity condition, the order of the derivatives can be lowered using integration by parts. If, for example, a C_0-element is required for the above heat flow problem, then applying integration by parts to Eq. (11.9), using Eq. (11.5), we obtain

$$\int_V \left[\frac{d\mathbf{W}_V}{dx} \lambda \frac{d\mathbf{N}}{dx} \tilde{\mathbf{T}} + \mathbf{W}_V Q \right] dV = 0 \tag{11.10}$$

This form is called the *weak form*. Not that, in this form, the order of the derivative of the approximation function is the same as for the derivative of the weighting function. This means that the function is *self-adjoint*, and the resulting weak form has the best arrangement of the admissibility requirement of both the approximation function and the weighting function. In many other cases, successive integrations by parts are needed. The latter case is usually needed in boundary methods of approximation.

The choice of the weighting function is very important. Basically, any weighting function can be used. However, different weighting functions produce totally different methods of approximation. In the Galerkin method, the weighting function is chosen to be equal to the basis function, i.e. $\mathbf{W} = \mathbf{N}$. In this form, the Galerkin method is known as the Bubnov-Galerkin method. There are of course many other weighting functions. Commonly used functions are:

1. Point collocation: The weighting function of this method is $\mathbf{W} = \delta$, with δ is the Dirac delta function of the form

$$\delta (x - x_i) = \begin{cases} \infty & x = x_i \\ 0 & \text{else} \end{cases}$$

$$\int_V \mathbf{W} dV = \mathbf{I}$$

(11.11)

in which $\mathbf{I}$ is a unit matrix. The finite difference method is a particular case of this approximation technique.
2. Subdomain collocation: The weighting function of this method is $\mathbf{W} = \mathbf{I}$ in the specified domain and zero elsewhere. The finite volume method is a particular case of this technique.
3. Petrov-Galerkin: The weighting function of this method is any arbitrary function that is not equal to the shape function, i.e. $\mathbf{W} \neq \mathbf{N}$. In the next section we discuss this approximation technique in more details.

In many engineering applications, the choice $\mathbf{W} = \mathbf{N}$ produces matrices which are symmetric. This makes the Galerkin method one of the most commonly utilized finite element procedure. However, the symmetry is by no means valid for all problems. Finite element matrices of heat conduction-convection problems, which we will be dealing with, are non-symmetric.

11.2.2 *Upwind finite element method*

The Galerkin finite element method is a powerful tool for solving parabolic problems, such as diffusive/conductive fluid/heat flow problems. However, for hyperbolic-type problems, such as advective/convective problems, the Galerkin method ceases, in many situations, to produce sufficiently accurate and stable results. This problem is not unexpected since the Galerkin formulation gives rise to a central-difference type approximation. In such a case, the results are under diffused and might be corrupted with spurious oscillation, the degree of which depends on the amount of convection. In fluid mechanics, the amount of convection is characterized by the *Peclet number*, defined as

$$\text{Pe} = \frac{uh}{D}$$

(11.12)

where D is a general representation of diffusion, u is the fluid velocity and h is a characteristic length, equal to the element length in one-dimensional problems. For Pe ≈ 0, the flow is dominated by diffusion, but for Pe $>> 1$, the flow is dominated by convection. Figure 11.1 shows a qualitative computational result of temperature distribution in an object subjected to convective heat flow, with a relatively high Peclet number. The conventional Galerkin finite element method was utilized for the discretization of the problem. Obviously, the result exhibits spurious oscillations.

One possibility to eliminate spurious oscillations is to utilize excessively fine meshes. Evidently, fine meshes are affordable for small geometries. However, for large geometries, such

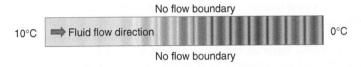

Figure 11.1   Galerkin finite element method for a highly convective problem.

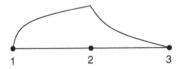

Figure 11.2   Petrov-Galerkin weighting function of a quadratic element.

as those representing geothermal systems, fine meshes are difficult to afford. This problem and other convection related problems have motivated researchers to develop alternative formulations, which are oscillation-free regardless of the mesh size. Researchers in the field of finite difference method found that the spurious oscillations can be eliminated by the use of *upwind differencing* in the convection term. The upwind convection term can be constructed by adding an artificial diffusion to the central difference formulation. This simple modification was sufficient to obtain a spurious-free solution. However, the solution is, for many applications, overly diffused, in contrast to the Galerkin method, which is under diffused. The reason for this is that the upwind difference is only first-order accurate. Over diffusion and inaccuracy of the upwind formulation brought many criticisms to this method. However, it was found that putting the upwind difference and the central difference together, a balanced diffusion can be obtained. This led to the development of the *optimal upwind* method, which incorporates an adaptive diffusion term in the central difference equations.

In the finite element method, there are three basic techniques capable of achieving the upwind effect (Brooks and Hughes 1982):

1. Artificial diffusion: This technique, also known as the *balancing diffusion* method, is made to balance the negative diffusion of the Galerkin formulation. It has been proved that by adding an artificial diffusion term given by

$$\tilde{k} = \frac{uh}{2}\tilde{\xi}$$

$$\tilde{\xi} = \left(\coth \alpha - \frac{1}{\alpha} \right), \quad \alpha = \frac{uh}{2\lambda} \qquad (11.13)$$

to the conventional Galerkin finite element discretization, an exact solution at the nodes can be obtained, particularly for 1D elements.
2. Quadrature: This technique is proposed by Hughes (1978). It is constructed by modifying the numerical quadrature rule of the convection term to achieve the upwind effect. In a one-dimensional case, for a linear element, a single quadrature point, $\tilde{\xi}$, is positioned within the element according to Eq. (11.13).
3. Petrov-Galerkin: This technique is constructed by modifying the weighting function for the upwind node of an element more heavily than the downwind node. An example of such a weighting function is shown in Figure 11.2.

All of the above methods when applied to a steady-state one-dimensional case with a prescribed flow direction, give similar computational results. However, they produce different results for multidimensional and transient cases. In multidimensional problems, the solutions exhibit excessive diffusion in the direction perpendicular to the flow. Overly diffused results also appeared in many transient problems. Consequently, this shortcoming has been a subject for arguments between different groups questioning the reliability of the upwind finite element method. The topic of the upwind finite element method has been covered in many papers and text books, see for example Yu and Heinrich (1986), Heinrich and Pepper (1999), and Lewis *et al.* (2004), among others. Here, we follow the outlines introduced by Heinrich and Pepper (1999).

11.2.2.1   *Upwind formulation*
As mentioned earlier, the upwind method is developed to eliminate spurious oscillations that normally occur when we utilize the conventional Galerkin finite element method to model highly convective heat and fluid flow. The basic principle of the upwind method is to weight the upstream nodes with a higher value than that for the downstream.

Consider a one-dimensional bar of length, L, containing a fluid moving with a constant velocity, u, in the x-direction. The fluid is subjected to prescribed temperatures at both ends. The balance equation describing steady-state heat transfer in the object, in the absence of a source or a sink, is

$$-\frac{d}{dx}\left(\kappa\frac{dT}{dx}\right) + u\frac{dT}{dx} = 0; \qquad 0 \le x \le L \tag{11.14}$$

where κ is the thermal diffusivity. For $u = 0$, the problem is purely conductive, and the equation is parabolic. For $u >> \kappa$ the problem is convective and behaves like a hyperbolic equation, which requires some sort of damping to keep it under control. However, the amount of damping must be chosen in such a way that it does not alter the physics of the problem.

Using the weighted residual method to discretize Eq. (11.14) for a 1D element, and upon ignoring the boundary conditions, gives

$$\int_0^L \kappa\frac{d\mathbf{W}}{dx}\frac{d\mathbf{N}}{dx}dx\overline{\mathbf{T}} + \int_0^L u\mathbf{W}\frac{d\mathbf{N}}{dx}dx\overline{\mathbf{T}} = 0 \tag{11.15}$$

in which $\mathbf{W}$ denotes the weighting function, and $\mathbf{N}$ the shape function. We apply the Galerkin finite element method, i.e. $\mathbf{W} = \mathbf{N}$. If we utilize a two-node linear element with shape functions of the form

$$N_1 = 1 - \frac{x}{h}; \qquad N_2 = \frac{x}{h} \tag{11.16}$$

where h is the length of the element, the element stiffness matrix (also known as the *conductance* matrix) can readily be obtained as

$$\mathbf{k} = \begin{pmatrix} \dfrac{\kappa}{h} - \dfrac{u}{2} & -\dfrac{\kappa}{h} + \dfrac{u}{2} \\[2mm] -\dfrac{\kappa}{h} - \dfrac{u}{2} & \dfrac{\kappa}{h} + \dfrac{u}{2} \end{pmatrix} \tag{11.17}$$

Letting $u = 0$, the solution for a purely conductive problem is obtained, as

$$\mathbf{k} = \frac{\kappa}{h}\begin{pmatrix} 1 & -1 \\ -1 & 1 \end{pmatrix} \tag{11.18}$$

The difference between the two matrices in Eq. (11.17) and Eq. (11.18) is that the former is un-symmetric, and the diagonal terms are not necessarily dominant, especially for $u >> \kappa$. These two features constitute the main source of numerical nuisance in highly convective problems.

Assume that the geometry is described by a mesh of two elements of equal size, as shown in Figure 11.3.

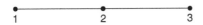

Figure 11.3   Two line-elements.

Assembling the global stiffness matrix gives:

$$\begin{pmatrix} \dfrac{\kappa}{h} - \dfrac{u}{2} & -\dfrac{\kappa}{h} + \dfrac{u}{2} & 0 \\[2mm] -\dfrac{\kappa}{h} - \dfrac{u}{2} & 2\dfrac{\kappa}{h} & -\dfrac{\kappa}{h} + \dfrac{u}{2} \\[2mm] 0 & -\dfrac{\kappa}{h} - \dfrac{u}{2} & \dfrac{\kappa}{h} + \dfrac{u}{2} \end{pmatrix} \begin{Bmatrix} T_1 \\ T_2 \\ T_3 \end{Bmatrix} = 0 \tag{11.19}$$

For the interior node (node 2), the difference equation can be written as

$$\left(-1 - \frac{Pe}{2}\right) T_1 + 2T_2 + \left(-1 + \frac{Pe}{2}\right) T_3 = 0 \tag{11.20}$$

in which $Pe = uh/\kappa$ is the element Peclet number. Eq. (11.20) is identical to the central difference equation of the finite difference method, which is normally written as

$$\left(-1 - \frac{Pe}{2}\right) T_{i-1} + 2T_i + \left(-1 + \frac{Pe}{2}\right) T_{i+1} = 0 \tag{11.21}$$

Solution of this difference equation can be obtained by assuming a solution of the form $T_i = Cr^i$ where C is some arbitrary constant (Heinrich and Pepper 1999). Substitute T_i into Eq. (11.21), a characteristic equation is obtained of the form

$$ar^2 + br + c = 0 \tag{11.22}$$

in which a is the coefficient of T_{i+1}, b is the coefficient of T_i and c is the coefficient of T_{i-1}. Applying the quadratic rule to Eq. (11.22) leads to

$$T = Cr = C\left(\frac{2 + Pe}{2 - Pe}\right) \tag{11.23}$$

Accordingly, for a general difference equation, the solution can be written as

$$T_i = A + B\left(\frac{2 + Pe}{2 - Pe}\right)^i \tag{11.24}$$

This equation reveals that for $Pe > 2$ the dominator becomes negative and the numerical solution will be oscillatory. This problem has led to the development of the upwind finite difference method. Applying the backward derivative approximation to Eq. (11.14) leads to the difference equation (see Sub-section 11.3.1.2 for details)

$$-\frac{\kappa}{h^2} (T_{i+1} - 2T_i + T_{i-1}) + \frac{u}{h} (T_i - T_{i-1}) = 0 \tag{11.25}$$

which can be expressed as

$$T_{i+1} - (2 + Pe) T_i + (1 + Pe) T_{i-1} = 0 \tag{11.26}$$

As for Eqs. (11.21)–(11.24), solution of Eq. (11.26) yields

$$T_i = A + B(1 + \text{Pe})^i \tag{11.27}$$

which shows, in contrast to Eq. (11.24), it is stable for all values of Pe.

Heinrich and Pepper (1999) have shown that the truncation error analysis of Eq. (11.25) gives

$$\underbrace{\left(-\kappa \frac{d^2 T}{dx^2} + u \frac{dT}{dx}\right)_{x=x_i}}_{\text{exact derivative}} - \underbrace{\left\{-\frac{\kappa}{h^2}(T_{i+1} - 2T_i + T_{i-1}) + \frac{u}{h}(T_i - T_{i-1}) = 0\right\}}_{\text{approximated derivative}} \tag{11.28}$$

$$= \frac{uh}{2} \frac{d^2 T}{dx^2} + \text{HOT}$$

where HOT stands for high order terms. So, in order to reduce the truncation error, the leading term on the right-hand side must be moved to the left-hand side (Roache 1976). By ignoring the high order terms, the conductive-convective heat transfer equation may now be written as

$$-\left(\kappa + \frac{uh}{2}\right) \frac{d^2 T}{dx^2} + u \frac{dT}{dx} = 0 \tag{11.29}$$

The term $uh/2$ is called the *artificial numerical diffusion*, which is in some aspects similar to the numerical damping used in structural dynamics. As we shall see later, the upwind differencing method results to an unconditionally stable algorithm, simply by adding an artificial numerical diffusion term to the system. However, this approach produces solutions which are over diffused and, in certain physical and boundary conditions, might significantly reduce the accuracy of the problem. Several research works have shown that this shortcoming can be eliminated or its effect reduced if the upwind differencing approximation is constructed through a proper Petrov-Galerkin weighted residual formulation.

11.2.2.2   *Petrov-Galerkin formulation*

Petrov-Galerkin method is an important generalization of the Galerkin method and has been utilized successfully for modeling highly convective transport problems. The main objective of this method is to eliminate spurious oscillations and numerical instability in the system by introducing artificially diffused weighting functions consisting of standard basis functions and polynomials of higher order. Several Petrov-Galerkin algorithms have been introduced. Among others, are the works given by Westernik and Sheta (1989), Idelsohn *et al.* (1996a, 1996b), and Heinrich and Pepper (1999). Each algorithm has distinct features which make it accurate in some cases and less in others. Here again we follow the outlines presented by Heinrich and Pepper (1999).

In order to control the amount of numerical diffusion in the system, the modified convective-diffusion equation, Eq. (11.29), is supplemented by a relaxation parameter, α, such that

$$-\left(\kappa + \frac{\alpha uh}{2}\right) \frac{d^2 T}{dx^2} + u \frac{dT}{dx} = 0, \qquad 0 \le \alpha \le 1 \tag{11.30}$$

where the term $\alpha uh/2$ is now referred to as the *balancing diffusion*. It can be noticed that if we set $\alpha = 0$ we obtain the original formulation, which is under diffused, and if we set $\alpha = 1$, we obtain the upwind formulation, which is over diffused. So, it is reasonable to assume that at some value between 0 and 1, we expect to obtain the right solution.

As for Eq. (11.14), the differencing equation of Eq. (11.30) can be derived to yield

$$\left(1 - \frac{\text{Pe}}{2}(\alpha + 1)\right) T_{i-1} + 2\left(1 + \frac{\alpha \text{Pe}}{2}\right) T_i + \left(1 + \frac{\text{Pe}}{2}(\alpha + 1)\right) T_{i+1} = 0 \tag{11.31}$$

The exact difference solution of this differencing equation is:

$$T_i = A + B \left[\frac{2 + Pe(\alpha + 1)}{2 + Pe(\alpha - 1)} \right]^i \tag{11.32}$$

from which it can readily be found that for $\alpha = 0$, the Galerkin solution is obtained, indicating that the scheme exhibits oscillation for all $Pe > 2$. Hence, in order to avoid oscillation in the system, α must be greater than some critical value, α_{cr}, which can be determined by letting the denominator equal to zero and solving for α to give

$$\alpha \geq \alpha_{cr} \equiv 1 - \frac{2}{Pe} \tag{11.33}$$

It is important to find an optimal value, α_o such that the diffusion is neither over stated nor under stated. To quantify α_o Heinrich and Pepper (1999) conducted a truncation error analysis of the heat equation, Eq. (11.14), and compared it with the discretized form. They obtained:

$$TE \equiv \underbrace{\left(-\kappa \frac{d^2 T}{dx^2} + u \frac{dT}{dx} \right)_{x=x_i}}_{\text{exact derivative}}$$

$$\underbrace{- \frac{1}{h^2} \left\{ -\left(1 + \frac{Pe}{2}(\alpha_o + 1) \right) T_{i-1} + 2 \left(1 + \frac{\alpha_o Pe}{2} \right) T_i - \left(1 + \frac{Pe}{2}(\alpha_o - 1) \right) T_{i+1} \right\}}_{\text{approximated derivative}}$$

$$\tag{11.34}$$

Applying Taylor series expansion to T_{i+1}, yields

$$T_{i+1} = T_i(x_i + h)$$
$$= T_i(x_i) + hT_i'(x_i) + \frac{h^2}{2!}T_i''(x_i) + \frac{h^3}{3!}T_i'''(x_i) + \frac{h^4}{4!}T_i''''(x_i) + \cdots \tag{11.35}$$

Taylor series expansion of T_{i-1} yields

$$T_{i-1} = T_i(x_i - h)$$
$$= T_i(x_i) - hT_i'(x_i) + \frac{h^2}{2!}T_i''(x_i) - \frac{h^3}{3!}T_i'''(x_i) + \frac{h^4}{4!}T_i''''(x_i) - \cdots \tag{11.36}$$

in which $()''$ indicates the second derivative of the term, etc. Substituting Eqs. (11.35)–(11.36) into Eq. (11.34) leads to

$$TE \equiv \frac{\alpha_o Pe}{2} T_i''(x_i) + \frac{2}{h^2} \left(1 + \frac{\alpha_o Pe}{2} \right) \left[\frac{h^4}{4!} T_i''''(x_i) + \frac{h^6}{6!} T_i''''''(x_i) + \cdots \right]$$

$$- \frac{Pe}{h^2} \left[\frac{h^3}{3!} T_i'''(x_i) + \frac{h^5}{5!} T_i'''''(x_i) + \cdots \right] \tag{11.37}$$

Returning to the heat equation, Eq. (11.14), we can readily write the first order derivative in terms of the second order derivative, as

$$T_i' = \left(\frac{Pe}{h} \right)^{-1} T_i'' \tag{11.38}$$

This can be used to express the higher order derivatives of T_i by means of a recursive relationship of the form

$$T_i^{(n)} = \left(\frac{\text{Pe}}{h}\right)^{n-2} T_i^{(2)}$$ (11.39)

Substituting this relationship into Eq. (11.37), Heinrich and Pepper have shown that the truncation error in Eq. (11.37) can be expressed as

$$TE \equiv \left[\frac{1}{\text{Pe}^2}\left\{2\left(1+\frac{\alpha_o\text{Pe}}{2}\right)\tanh\frac{\text{Pe}}{2} - \text{Pe}\right\}\sinh\text{Pe}\right]T_i''$$ (11.40)

Now letting $TE = 0$ and solving for α_o, we obtain

$$\alpha_o = \coth\frac{\text{Pe}}{2} - \frac{2}{\text{Pe}}$$ (11.41)

α_o provides the exact amount of balancing diffusion which counteracts the numerical diffusion resulting from the discretization of the heat equation. It is usually referred to as the optimal diffusion, indicated by the subscript o. For one-dimensional problems, this value produces the exact solution if the fluid velocity is constant and the mesh is uniform. For simplicity of notation, in what follow we use α instead of α_o.

Having derived the optimal amount of diffusion, we are now able to derive an optimal weighting function. There are several ways to derive the Petrov-Galerkin weighting function, but here we utilize the one introduced by Heinrich and Pepper (1999). The Galerkin weak form of Eq. (11.30) can be expressed as

$$\int_0^L \left\{\kappa\frac{d\mathbf{N}}{dx}\frac{d\mathbf{T}}{dx} + u\left(\mathbf{N} + \frac{\alpha h}{2}\frac{d\mathbf{N}}{dx}\right)\frac{d\mathbf{T}}{dx}\right\}dx = 0$$ (11.42)

We choose the term containing the optimal diffusion as the optimal weighting function, i.e.

$$\mathbf{W} = \mathbf{N} + \frac{\alpha h}{2}\frac{d\mathbf{N}}{dx}$$ (11.43)

In a more general form, the Petrov-Galerkin weighting function can be written, for a one-dimensional element, as

$$\mathbf{W}(x) = \mathbf{N}(x) + \alpha F(x)$$ (11.44)

in which $F(x)$ is some linear function of space. Note that this form of Petrov-Galerkin weighting function is very much similar to the one utilized recently using the Extended finite element method (X-FEM) and the partition of unity method (PUM). Apparently, the work on the upwind Petrov-Galerkin in the late seventies of the last century has preceded the X-FEM and PUM for more than a decade.

Using Eq. (11.43), the Petrov-Galerkin approximation of Eq. (11.42) might then be written as

$$\int_0^L \left\{\kappa\frac{d\mathbf{W}}{dx}\frac{d\mathbf{T}}{dx} + u\mathbf{W}\frac{d\mathbf{T}}{dx}\right\}dx = 0$$ (11.45)

So far the Petrov-Galerkin weighting function was derived for a linear element, where, as we found, only one dissipative parameter, α, was necessary. For a quadratic element, two dissipative parameters are necessary. Heinrich and Zienkiewicz (reported in Heinrich 1980) suggested weighting functions for a one-dimensional three-nodded element of the form

$$\mathbf{W}_i(x,\alpha) = \mathbf{N}_i(x) - \alpha F(x)$$ (11.46)

for i corresponds to a corner node, and

$$\mathbf{W}_i(x, \beta) = \mathbf{N}_i(x) + \beta\, F(x) \tag{11.47}$$

for i corresponds to a middle node. For an element spanning from 0 to h, $F(x)$ is chosen cubic as

$$F(x) = \frac{5}{2}\frac{x}{h}\left[2\left(\frac{x}{h}\right)^2 - 3\frac{x}{h} + 1\right]; \qquad 0 \le x \le h \tag{11.48}$$

shown in Figure 11.2. Similar to linear elements, values of α and β can be determined from the truncation error analysis. The exact difference solution of the corresponding difference equation yields

$$T_i = A + B\left[\frac{1 + \dfrac{\text{Pe}}{4}(\beta + 1)}{1 + \dfrac{\text{Pe}}{4}(\beta - 1)}\right]^i \tag{11.49}$$

This equation shows that for $\beta = 0$, the Galerkin solution is obtained, which shows that the scheme exhibits oscillation for Pe > 4. Hence, in order to avoid oscillations, β must fulfill the following condition:

$$\beta \ge \beta_{\text{cr}} \equiv 1 - \frac{4}{\text{Pe}} \tag{11.50}$$

That is β must be greater than or equal to some critical value, depending on the Peclet number of the element. Christie *et al.* (1976 and 1978) have shown that the optimal value of β is given by

$$\beta_0 = \coth\frac{\text{Pe}}{4} - \frac{4}{\text{Pe}} \tag{11.51}$$

The critical value of α must fulfill this condition:

$$\alpha \ge \alpha_{\text{cr}} \equiv 1 - \frac{2}{\text{Pe}} \tag{11.52}$$

and the optimal value of α is

$$\alpha_0 = 2\tanh\frac{\text{Pe}}{2}\left(1 + 3\frac{\beta_0}{\text{Pe}} + \frac{12}{\text{Pe}^2}\right) - \frac{12}{\text{Pe}} - \beta_0 \tag{11.53}$$

For $\alpha = \alpha_0$ and $\beta = \beta_0$ with constant fluid velocity and a uniform 1D mesh, exact solution can be obtained.

For transient problems, the Petrov-Galerkin scheme is constructed using enhanced weighting functions, commonly obtained by perturbing the shape functions by their derivatives. This results to an implicit *space-time finite element* algorithm. Similar to classical bilinear space-time shape functions, the Petrov-Galerkin linear weighting functions produce solutions, which are overly diffused. To reduce the diffusion, a weighting function that is parabolic in time is introduced. The time function is given by

$$M(t) = \frac{4t}{\Delta t}\left(1 - \frac{t}{\Delta t}\right) \tag{11.54}$$

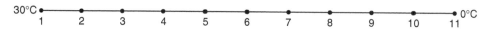

Figure 11.4   Finite element mesh for 1D bar.

in which Δt is a time step, representing the length of the element in the time domain. The time function is multiplied by the standard shape functions in space. For a linear element, $0 \le x \le h$, this results to

$$N_1(x,t) = 4\left(1 - \frac{x}{h}\right)\frac{t}{\Delta t}\left(1 - \frac{t}{\Delta t}\right)$$

$$N_2(x,t) = 4\frac{x}{h}\frac{t}{\Delta t}\left(1 - \frac{t}{\Delta t}\right)$$

(11.55)

for node 1 and 2, respectively. The Petrov-Galerkin weighting function is proposed to be of the form (Heinrich and Pepper 1999)

$$\mathbf{W} = \mathbf{N}(x,t) + \frac{\alpha h}{2}\frac{\partial \mathbf{N}(x,t)}{\partial x} + \frac{\beta h \Delta t}{4}\frac{\partial^2 \mathbf{N}(x,t)}{\partial x \partial t}$$

(11.56)

As we have seen for the steady-state, the diffusive parameters, α and β, can be determined from the truncation error analysis, by first writing the difference equation for a middle node, compare it with the exact form of the heat equation and then expand it using Taylor series. Applying this procedure, it was found that in order to eliminate the diffusion error, α and β must have the following forms

$$\alpha = \coth\frac{Pe}{2} - \frac{2}{Pe}$$

$$\beta = \frac{c}{3} - \frac{2\alpha}{cPe}$$

(11.57)

in which c is the *Courant number* defined as

$$c = \frac{u \Delta t}{h}$$

(11.58)

This type of weighting functions gives rise to an algorithm which is third-order accurate in space and second-order in time. If $\beta = 0$ it becomes second-order accurate in space and time, and is unconditionally stable for all α. If $\beta \neq 0$ the algorithm is conditionally stable, except for $c \le 1$.

11.2.3   Numerical example

In this example we examine the numerical performance of the three approximation procedures: Galerkin method, upwind differencing, and Petrov-Galerkin method. We utilize them to simulate a steady-state conductive-convective problem in an insulated one-dimensional bar element. The bar is uniform, 10 m in length and subjected to 30°C at the left end, and 0°C at the other end, as shown in Figure 11.4.

For this, the boundary value problem to be solved is:

$$-\lambda\frac{d^2 T}{dx^2} + \rho c u\frac{dT}{dx} = 0, \quad 0 \le x \le L$$

$$T(x = 0) = 30; \quad T(x = L) = 0$$

(11.59)

Ten linear finite elements with 1 m in length are used to simulate the bar. Using the weighted residual method, the stiffness matrix obtained from the Galerkin and the upwind differencing, using Eqs. (11.16) and (11.30), can be expressed as

$$
\mathbf{k} = \begin{pmatrix} \dfrac{\tilde{D}}{h} - \dfrac{u}{2} & -\dfrac{\tilde{D}}{h} + \dfrac{u}{2} \\[2mm] -\dfrac{\tilde{D}}{h} - \dfrac{u}{2} & \dfrac{\tilde{D}}{h} + \dfrac{u}{2} \end{pmatrix}, \qquad \tilde{D} = \dfrac{\lambda}{\rho c} + \dfrac{\alpha u h}{2} \qquad (11.60)
$$

where with $\alpha = 0$, the Galerkin matrix is obtained, and with $\alpha = 1$, the upwind matrix is obtained. The stiffness matrix obtained from the Petrov-Galerkin formulation, using Eqs. (11.16), (11.43), and (11.45), can be expressed as

$$
\mathbf{k} = \begin{pmatrix} \dfrac{D}{h} + (\alpha - 1)\dfrac{u}{2} & -\dfrac{D}{h} - (\alpha - 1)\dfrac{u}{2} \\[2mm] -\dfrac{D}{h} - (\alpha + 1)\dfrac{u}{2} & \dfrac{D}{h} + (\alpha + 1)\dfrac{u}{2} \end{pmatrix}, \qquad D = \dfrac{\lambda}{\rho c}; \quad \alpha = \coth\dfrac{\mathrm{Pe}}{2} - \dfrac{2}{\mathrm{Pe}} \qquad (11.61)
$$

For simplicity, we assume that the material parameters are of unit magnitude, i.e. $\lambda = \rho = c = 1$. The exact solution of this problem can readily be obtained as

$$
T = T_1 + (T_2 - T_1)\dfrac{1}{e^{\mathrm{Pe}}-1}\left[e^{(\mathrm{Pe}\cdot x/L)-1}\right] \qquad (11.62)
$$

in which T_1 is the temperature at the left-hand side of the bar, and T_2 is on the right-hand side. Computational results obtained from the three numerical algorithms, together with the analytical solution, for different fluid velocity, u, are presented in Figure 11.5. In Figure 11.5(a) the fluid velocity is zero, resembling a pure conduction case. Apparently, the three algorithms produced the exact solution. In Figure 11.5(b) the fluid velocity is 2 m/s. This figure shows the followings:

- The Petrov-Galerkin solution matches perfectly the exact solution.
- The upwind differencing solution is over diffused.
- The Galerkin algorithm is under diffused, but not to the degree of oscillation. This velocity leads to an element Peclet number equal 2, just at the critical point, see Eq. (11.24).

Figure 11.5(c) and Figure 11.5(d) show the computational results for fluid velocity 4 m/s and 10 m/s respectively. As for the previous case, the Petrov-Galerkin solution matches the exact solution, the upwind solution is over diffused, but the results are to some extent reliable. However, the Galerkin solution is spurious and the computational results are not reliable, especially for higher fluid velocity. Note that using the Petrov-Galerkin algorithm, the utilization of the optimal diffusion, α_o, in a uniform mesh with constant fluid velocity has resulted to the exact solution. However, if the problem has been simulated using a non-uniform mesh, the exact solution can no longer be obtained, but an accurate solution might be obtained, if α_o is calculated for each element.

11.3 TIME DISCRETIZATION

In steady-state heat transfer, the governing equation is described in space, and the solution is governed by prescribed boundary conditions at the extreme boundaries of the domain. This kind of engineering analysis is important in describing heat transfer for long terms. However, most engineering applications, including heat flow in shallow geothermal systems, are transient, and it is necessary to study heat flow in time. In such cases, the solution is constrained by the initial state of the system and the boundary conditions. Hence, the transient problem is often referred to as an initial and boundary value problem. The finite element discretization of such a problem is done in space and time. Typically, the finite element discretization of the transient heat conduction problem

$$
\rho c \dfrac{\partial T}{\partial t} - \nabla \cdot (\lambda \nabla T) = Q \qquad (11.63)
$$

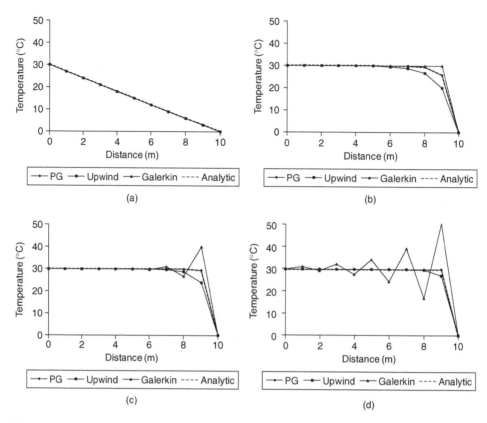

Figure 11.5   Finite element results for (a) $u = 0$, (b) $u = 2$, (c) $u = 4$ and (d) $u = 10$.

can be written as

$$\mathbf{M\dot{T} + KT = F} \tag{11.64}$$

where $\mathbf{M}$ is the *capacitance* (mass) matrix, $\mathbf{K}$ is the conductance (stiffness) matrix, and $\mathbf{F}$ is the flux (force) vector; the dot represents derivative with time. In finite element analysis, Eq. (11.64) is a first order differential equation of time, termed as the *semi-discrete* equation.

There are basically two methods to solve semi-discrete equations: the finite difference method, and the finite element method. In this section we briefly present most commonly used time integration schemes for solving first order differential equations. For more details, the reader is refereed to Lewis *et al.* (2004), Hughes (2000), Zienkiewicz *et al.* (2005), Wood (1990), among others.

11.3.1   *Finite difference time integration schemes*

Finite difference time integration schemes are most widely used. The finite difference method approximates the solution of a differential equation without the need for calculus, simply by replacing derivative expressions by approximately equivalent difference equations. Given a discrete function of uniformly spaced axis, h, it is convenient to express a function in terms of three differencing schemes: *forward difference*, *backward difference* and *central difference*.

Based on the finite difference method, we present here the most widely used time marching schemes for solving the first order semi-discrete equation, Eq. (11.64).

11.3.1.1   *Forward difference method (Forward Euler, Explicit)*
Mathematically, the forward difference is expressed as

$$\Delta f(x) = f(x + h) - f(x) \tag{11.65}$$

where Δ refers to the forward differencing, and h is the spacing between two grid points. For uniformly spaced grid, Δ_h is also used. This approximation is made forward, and hence it is convenient to evaluate the independent variable at x_n, x_{n+1}, etc.

In this time scheme, the heat equation is solved at the current time step, t_n, such that

$$\mathbf{M\dot{T}}_n + \mathbf{KT}_n = \mathbf{F}_n \tag{11.66}$$

and the approximation of the time derivative is made forward, i.e. from t_n towards t_{n+1}. To find a value of the unknown variable at a future time step, t_{n+1}, from a known value at a current time step, t_n, we employ Taylor series expansion. Taylor expansion of the temperature at time t_{n+1} can be expressed as

$$T(t_n + \Delta t) = T(t_n) + \Delta t\, \dot{T}(t_n) + \frac{\Delta t^2}{2!} \ddot{T}(t_n) + \cdots \tag{11.67}$$

Omitting the higher order terms, Eq. (11.67) can be written as

$$T_{n+1} = T_n + \Delta t \dot{T}_n \tag{11.68}$$

where we used the simplified notation T_{n+1} instead of $T(t_n + \Delta t)$, etc. Solving for $\dot{T}_n$, Eq. (11.68) can be expressed as

$$\dot{T}_n = \frac{T_{n+1} - T_n}{\Delta t} \tag{11.69}$$

Substituting Eq. (11.69) into Eq. (11.66) yields

$$\mathbf{MT}_{n+1} = (\mathbf{M} - \Delta t \mathbf{K})\mathbf{T}_n + \Delta t \mathbf{F}_n \tag{11.70}$$

This equation is algebraic, and hence suitable for numerical calculations. However, and as for discretization in space, error is inevitable. The amount of error can be estimated by the truncation error analysis of Taylor expansion. The truncation error can be determined by first writing Eq. (11.67) in terms of $\dot{T}_n$ such that

$$\dot{T}(t_n) = \frac{T(t_n + \Delta t) - T(t_n)}{\Delta t} - \frac{\Delta t}{2!} \ddot{T}(t_n) + \cdots \tag{11.71}$$

Then, comparing between the exact form of the time derivative, $\partial T(t_n)/\partial t$ and the expanded form, Eq. (11.71), gives

$$\underbrace{\frac{\partial T(t_n)}{\partial t}}_{\text{exact form}} - \underbrace{\frac{T(t_n + \Delta t) - T(t_n)}{\Delta t}}_{\text{expanded form}} = \underbrace{\frac{\Delta t}{2} \ddot{T}(t_n) + \cdots}_{\text{truncation error}} \tag{11.72}$$

This equation reveals that the error involved in the approximation of the time derivative is of the order of $O(\Delta t)$. Therefore, the forward difference scheme is a first-order time integration algorithm.

An important feature of the forward difference scheme, is that Eq. (11.70) can be solved directly, without inversion of the conductive matrix, $\mathbf{K}$; provided that the mass matrix is lumped. This scheme is therefore *explicit*. This property is proved to be efficient for many computational procedures, especially those dealing with small time steps.

11.3.1.2   *Backward difference method (Backward Euler, Implicit)*

Mathematically, the backward difference is expressed as

$$\nabla f(x) = f(x) - f(x - h) \tag{11.73}$$

where ∇ indicates the backward differencing (not to be confused with the spatial differential operator, grad f). The approximation is made backward, i.e. it uses the function value at x_n and x_{n-1}, etc.

In this time scheme, the heat equation is conveniently expressed at t_{n+1} time step such that

$$\mathbf{M\dot{T}}_{n+1} + \mathbf{KT}_{n+1} = \mathbf{F}_{n+1} \tag{11.74}$$

and the time derivative is made backward, i.e. from t_{n+1} towards t_n. Using Taylor series to expand the independent variable at time $t_n = t_{n+1} - \Delta t$, in terms of t_{n+1} leads to

$$T_n = T(t_{n+1} - \Delta t) = T(t_{n+1}) - \Delta t \dot{T}(t_{n+1}) + \frac{\Delta t^2}{2!} \ddot{T}(t_{n+1}) + \cdots \tag{11.75}$$

Omitting the higher order terms, Eq. (11.75) can be written as

$$T_n = T_{n+1} - \Delta t \dot{T}_{n+1} \tag{11.76}$$

Solving for $\dot{T}_{n+1}$, gives

$$\dot{T}_{n+1} = \frac{T_{n+1} - T_n}{\Delta t} \tag{11.77}$$

Substituting Eq. (11.77) into Eq. (11.74) gives

$$(\mathbf{M} + \Delta t \mathbf{K})\mathbf{T}_{n+1} = \mathbf{MT}_n + \Delta t \mathbf{F}_{n+1} \tag{11.78}$$

Similar to the forward difference scheme, the truncation error analysis yields

$$\underbrace{\frac{\partial T (t_{n+1})}{\partial t}}_{\text{exact form}} - \underbrace{\frac{T (t_{n+1}) - T (t_{n+1} - \Delta t)}{\Delta t}}_{\text{expanded form}} = \underbrace{\frac{\Delta t}{2} \ddot{T}(t_{n+1}) + \cdots}_{\text{truncation error}} \tag{11.79}$$

This reveals that the error involved in the approximation of the time derivative is of the order of $O(\Delta t)$. Therefore, the backward difference scheme is also a first-order algorithm.

Unlike the forward differencing, solution of Eq. (11.78) is not direct and requires inversion of the coefficient matrix, $\mathbf{M} + \Delta t \, \mathbf{K}$, in order to calculate $\mathbf{T}_{n+1}$. The scheme is therefore *implicit*.

11.3.1.3   *Central difference method (Crank-Nicolson, Trapezoidal rule)*

Mathematically, the central difference is given by

$$\delta f(x) = f\left(x + \frac{1}{2}h\right) - f\left(x - \frac{1}{2}h\right) \tag{11.80}$$

where δ indicates the central differencing. The approximation is made backward at x_n, x_{n-1}, etc., and forward at x_n, x_{n+1}, etc. In this time scheme, the heat equation can be expressed at t_n time step such that

$$\mathbf{M\dot{T}}_n + \mathbf{KT}_n = \mathbf{F}_n \tag{11.81}$$

and the time derivative is made to span one-step backward, towards t_{n-1}; and one step forward, towards t_{n+1}. Using Taylor series to expand the independent variable at time $t_{n+1} = t_n + \Delta t$, and $t_{n-1} = t_n - \Delta t$ in terms of t_n leads to

$$T_{n+1} = T(t_n + \Delta t) = T(t_n) + \Delta t \dot{T}(t_n) + \frac{\Delta t^2}{2!}\ddot{T}(t_n) + \frac{\Delta t^3}{3!}\dddot{T}(t_n) + \cdots \tag{11.82}$$

$$T_{n-1} = T(t_n - \Delta t) = T(t_n) - \Delta t \dot{T}(t_n) + \frac{\Delta t^2}{2!}\ddot{T}(t_n) - \frac{\Delta t^3}{3!}\dddot{T}(t_n) + \cdots \tag{11.83}$$

Adding the above two equations and omitting the higher order terms, yields

$$\dot{T}_n = \frac{T_{n+1} - T_{n-1}}{2\Delta t} \tag{11.84}$$

Substituting this equation into Eq. (11.81) gives

$$\mathbf{M}\mathbf{T}_{n+1} = \mathbf{M}\mathbf{T}_{n-1} - 2\Delta t\mathbf{K}\mathbf{T}_n + 2\Delta t\mathbf{F}_n \tag{11.85}$$

The truncation error analysis yields

$$\underbrace{\frac{\partial T(t_n)}{\partial t}}_{\text{exact form}} - \underbrace{\frac{T(t_n + \Delta t) - T(t_n - \Delta t)}{2\Delta t}}_{\text{expanded form}} = \underbrace{\frac{\Delta t^2}{3!}\dddot{T}(t_n) + \cdots}_{\text{truncation error}} \tag{11.86}$$

which reveals that the error involved in the approximation of the time derivative is of the order of $O(\Delta t^2)$. Therefore, the central difference scheme is a second-order algorithm. As for the forward scheme, the central difference scheme, in the form presented in Eq. (11.85), is explicit, provided **M** is lumped.

11.3.1.4 *The theta-method*
This method is one of the most commonly used time integration schemes since it represents different time schemes put together in an elegant way. In this scheme, the time derivative is approximated using the backward differencing, as

$$\dot{T} = \frac{T_{n+1} - T_n}{\Delta t} \tag{11.87}$$

The independent variable is defined as

$$T = \theta T_{n+1} + (1 - \theta)T_n \tag{11.88}$$

in which $0 \le \theta \le 1$, denoting a relaxation parameter. Substituting Eq. (11.88) and Eq. (11.87) into the finite element equation:

$$\mathbf{M}\dot{T} + \mathbf{K}T = \mathbf{F} \tag{11.89}$$

results to

$$(\mathbf{M} + \theta\Delta t\mathbf{K})\mathbf{T}_{n+1} = (\mathbf{M} - (1-\theta)\Delta t\mathbf{K})\mathbf{T}_n + \Delta t(\theta\mathbf{F}_{n+1} + (1-\theta)\mathbf{F}_n) \tag{11.90}$$

where the force vector, **F**, is defined as that for **T**, see Eq. (11.88). The essence of this time integration scheme is that by varying θ, it is possible to conduct any of the above-mentioned time integration schemes, such that

- Forward difference scheme: $\theta = 0$
- Backward difference scheme: $\theta = 1$
- Central difference scheme: $\theta = 1/2$

11.3.1.5 *Convergence*

One of the important aspects of any time integration scheme is its capability to converge to the right solution. Convergence implies that the numerical solution of the differencing procedure tends to the exact solution of the original partial differential equation as the grid spacing tends to zero. There are two important conditions which need to be satisfied in order to determine the convergence of a particular finite difference scheme, namely *consistency* and *stability*.

Consistency entails that the differencing scheme is capable of approximating the solution of the partial differential equation such that the truncation error must tend to zero as the numerical mesh size tends to zero. The stability is a measure of the capability of the scheme for solving boundary value problems, which might inheritably contain errors due to discretization. A time scheme is said to be stable if the error introduced at any stage of the computation either decreases or remains constant during subsequent steps. Unstable schemes produce unbounded results, and the error increases indefinitely. More details concerning consistency and stability of time integration schemes can be found in Zienkiewicz *et al.* (2005) and Hughes (2000).

There are several mathematical methods which can be utilized to evaluate the stability of a finite difference time integration scheme. Here we utilize the eigenvalue analysis for examining the stability of the theta-method. Consider the finite element equation of a transient heat conduction problem, Eq. (11.90). For a homogeneous case, i.e. $\mathbf{F} = 0$, this equation can be written in a compact form as

$$\mathbf{T}_{n+1} = \mathbf{B}\mathbf{T}_n \tag{11.91}$$

in which $\mathbf{B}$ is known as the *amplification matrix*, described in this case as

$$\mathbf{B} = (\mathbf{M} + \theta \Delta t \mathbf{K})^{-1}(\mathbf{M} - (1 - \theta)\Delta t \mathbf{K}) \tag{11.92}$$

In general, a recursive relationship between temperatures at different time steps can be expressed. For instance, the relationship between time step n and time step $n+1$ can be described as

$$\mathbf{T}_{n+1} = \mu \mathbf{T}_n \tag{11.93}$$

Substituting Eq. (11.93) into Eq. (11.91) leads to

$$(\mathbf{B} - \mu \mathbf{I})\mathbf{T}_n = 0 \tag{11.94}$$

in which $\mathbf{I}$ is a unit matrix. Apparently, Eq. (11.94) is the eigenfunction of the system, and μ is the corresponding eigenvalue. The *spectral radius* of matrix $\mathbf{A} = (\mathbf{B} - \mu \mathbf{I})$ is defined as

$$\rho(\mathbf{A}) = \max_i (|\mu_i|) \tag{11.95}$$

in which $|\mu_i| = (\mu_i \overline{\mu_i})^{1/2}$, with $\overline{\mu_i}$ is the complex conjugate of μ_i. Spectral radius of a matrix is a contour of the extremes of absolute values of the matrix elements. The basic properties of the spectral radius (also known as *spectral radii*) of a matrix are:

$$\begin{array}{llll} \rho(\mathbf{A}) > 1, & |\mu_i| > 1, & \mathbf{A} \text{ is not bounded; unstable} \\ \rho(\mathbf{A}) < 1, & |\mu_i| < 1, & \mathbf{A} \text{ is bounded; stable} \end{array} \tag{11.96}$$

Therefore, evaluating the stability of any time integration scheme requires the determination of the system eigenvalues, i.e. solving Eq. (11.94). For a large system of equations, which is typical for most finite element analyses, determination of the system eigenvalues is difficult. However, Irons and Treharne (reported in Zienkiewicz *et al.* 2005) have shown that the system eigenvalues are bounded by the eigenvalues of individual elements. The stability analysis can thus be conducted for a single element using a scalar equation. For a single degree of freedom, the amplification matrix **B** becomes the amplification factor, and by means of Eq. (11.92), it can be expressed as

$$B = \frac{m - (1 - \theta)\Delta t\, k}{m + \theta \Delta tk} \tag{11.97}$$

The eigenvalues can then be obtained from Eq. (11.94), and the non-trivial solution is given by

$$\mu = B = \frac{1 - (1 - \theta)\omega\Delta t}{1 + \theta\omega\Delta t} \tag{11.98}$$

where $\omega = k/m$; the eigenfrequency of the element. Following the stability condition, Eq. (11.96), we observe that in order to insure stability of an algorithm, the amplification factor must satisfy the following condition

$$-1 < \frac{1 - (1 - \theta)\omega\Delta t}{1 + \theta\omega\Delta t} < 1 \tag{11.99}$$

The right-hand side inequality is satisfied for all allowable values. For the left-hand side, the inequality yields for

- $\theta \geq 1/2$, $|\mu_i| \leq 1$, the algorithm is *unconditionally stable*.
- $\theta < 1/2$, $|\mu_i| < 1$, the algorithm is *conditionally stable*. This condition restricts the choice of the time step to the element properties, such that

$$\Delta t = \Delta t_{cr} \leq \frac{2}{(1 - 2\theta)\omega} \tag{11.100}$$

with Δt_{cr} indicates a critical time step beyond which the scheme becomes unstable. Note that the critical time step depends on ω, which is the maximum eigenfrequency of the system.

To give an impression on the value of the critical time step for a single degree of freedom heat conduction problem, we use an example given by Zienkiewicz *et al.* (2005). For a linear element with length h, the finite element coefficients are

$$m = \int_V N\rho c N dV$$
$$k = \int_V \frac{dN}{dx} \lambda \frac{dN}{dx} dV \tag{11.101}$$

with a single degree of freedom spatial shape function

$$N = \frac{h - x}{h} \tag{11.102}$$

Integrating Eq. (11.101) yields

$$m = \int_0^h \rho c N^2 dx = \frac{1}{3}\rho ch$$
$$k = \int_0^h \lambda \left(\frac{dN}{dx}\right)^2 dx = \frac{\lambda}{h} \tag{11.103}$$

Therefore the corresponding eigenfrequency of this system is

$$\omega = \frac{3\lambda}{\rho ch^2} \tag{11.104}$$

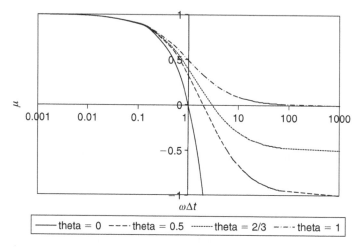

Figure 11.6   Stability of some time schemes.

This gives

$$\Delta t_{cr} \le \frac{2}{(1-2\theta)} \frac{\rho c h^2}{3\lambda} \tag{11.105}$$

This relationship shows that the critical time step is proportional to h^2, and thus decreases rapidly with decreasing element size. In many practical cases, this property exerts computational difficulties. Using explicit time integration schemes for an unstructured finite element mesh, the smallest element size must be considered for the calculation of the critical time step, making such schemes computationally inefficient.

Figure 11.6 shows a plot of the amplification factor, μ versus $\omega\Delta t$ for different time integration schemes. As can be inferred from Eq. (11.100), the forward difference scheme ($\theta=0$) becomes unstable for $\omega\Delta t \ge 2$. For the backward difference scheme ($\theta=1$), the amplification factor is always positive, resulting to stable solutions for all values of $\omega\Delta t$. Time schemes with $1/2 \le \theta < 1$, the amplification factor becomes negative at some stages, leading to oscillatory solutions, though stable. This property of the differencing schemes is controversial and confusing. The time scheme that is said to be unconditionally stable, does not necessarily produce accurate results for all conditions. Example of such a time scheme is the central difference method or the crank-Nicolson method. Assume

$$T_n = \mu T_{n-1} \tag{11.106}$$

This gives

$$T_{n+1} = \mu^2 T_{n-1} \tag{11.107}$$

Applying Eqs. (11.106) and (11.107) to the central difference equation, Eq. (11.85), yields a quadratic equation of the form

$$\mu^2 + (2\Delta t\omega)\mu - 1 = 0 \tag{11.108}$$

The roots of this equation are

$$-\Delta t\,\omega + \sqrt{1+(\Delta t\,\omega)^2}, \quad -\Delta t\,\omega - \sqrt{1+(\Delta t\,\omega)^2} \tag{11.109}$$

which are real with modulus (absolute value) of the second root gives always $|\mu| > 1$. This indicates that the central difference method is generally not stable (Lewis *et al.* 2004).

11.3.2 *Finite element time integration schemes*

Another important class of time integration schemes is the one which is based on the finite element method. In this scheme, the time is discretized in the same way as for the primary variables in space. For the heat equation, for example, the time dependent temperature is discretized as

$$T(t) = \sum_i N_i(t)T_i \tag{11.110}$$

in which $N_i(t)$ is the temporal shape function. For a first order derivative of time, such as that of the heat equation, a linear shape function is sufficient.

Recall the semi-discrete finite element formulation of the heat equation:

$$\mathbf{M}\dot{\mathbf{T}} + \mathbf{K}\mathbf{T} = \mathbf{F} \tag{11.111}$$

Assume that the time dependent temperature is discretized as in Eq. (11.110). Using a two-node linear element in time $N(t) \in (t_n, t_{n+1}]$, the temperature can be approximated as

$$T(t) = N_n \mathbf{T}_n + N_{n+1}\mathbf{T}_{n+1} \tag{11.112}$$

where

$$N_n = 1 - \frac{\tau}{\Delta t}, \quad N_{n+1} = \frac{\tau}{\Delta t} \tag{11.113}$$

with $\tau = t - t_n$ and $\Delta t = t_{n+1} - t_n$. Differentiating Eq. (11.112) with time yields

$$\dot{\mathbf{T}}(t) = -\frac{1}{\Delta t}\mathbf{T}_n + \frac{1}{\Delta t}\mathbf{T}_{n+1} \tag{11.114}$$

Using the weighted residual method, Eq. (11.111) can be discretized in the time domain as

$$\int_t \mathbf{W}(\mathbf{M}\dot{\mathbf{T}} + \mathbf{K}\mathbf{T} - \mathbf{F})dt = 0 \tag{11.115}$$

with $\mathbf{W}$ is any arbitrary weighting function in time. Substituting Eq. (11.112) and Eq. (11.114) into Eq. (11.115), leads to

$$\int_{t_n}^{t_{n+1}} \mathbf{W}\left(\mathbf{M}\left(-\frac{1}{\Delta t}\mathbf{T}_n + \frac{1}{\Delta t}\mathbf{T}_{n+1}\right) + \mathbf{K}(N_n\mathbf{T}_n + N_{n+1}\mathbf{T}_{n+1}) - \mathbf{F}\right)dt = 0 \tag{11.116}$$

The forcing vector, $\mathbf{F}$, can be discretized in the same way as the temperature, that is

$$\mathbf{F}(t) = N_n \mathbf{F}_n + N_{n+1}\mathbf{F}_{n+1} \tag{11.117}$$

Working out Eq. (11.116), yields

$$\begin{aligned}
\frac{\mathbf{M}}{\Delta t}(-\mathbf{T}_n + \mathbf{T}_{n+1})\int_{t_n}^{t_{n+1}} \mathbf{W}dt &+ \mathbf{K}\mathbf{T}_n\int_{t_n}^{t_{n+1}} \mathbf{W}dt - \frac{\mathbf{K}}{\Delta t}\mathbf{T}_n\int_{t_n}^{t_{n+1}} \mathbf{W}\tau dt \\
&+ \frac{\mathbf{K}}{\Delta t}\mathbf{T}_{n+1}\int_{t_n}^{t_{n+1}} \mathbf{W}\tau dt - \mathbf{F}_n\int_{t_n}^{t_{n+1}} \mathbf{W}dt + \frac{\mathbf{F}_n}{\Delta t}\int_{t_n}^{t_{n+1}} \mathbf{W}\tau dt \\
&- \frac{\mathbf{F}_{n+1}}{\Delta t}\int_{t_n}^{t_{n+1}} \mathbf{W}\tau dt = 0
\end{aligned} \tag{11.118}$$

Dividing Eq. (11.118) by $\Delta t \int_{t_n}^{t_{n+1}} \mathbf{W} dt$ and rearranging, gives

$$\left(\frac{\mathbf{M}}{\Delta t} + \theta \mathbf{K}\right) \mathbf{T}_{n+1} = \left(\frac{\mathbf{M}}{\Delta t} - (1 - \theta)\mathbf{K}\right) \mathbf{T}_n + (1 - \theta)\mathbf{F}_n + \theta \mathbf{F}_{n+1} \qquad (11.119)$$

in which

$$\theta = \frac{1}{\Delta t} \frac{\int_{t_n}^{t_{n+1}} \mathbf{W}\tau dt}{\int_{t_n}^{t_{n+1}} \mathbf{W} dt} \qquad (11.120)$$

which is referred to as the *algorithmic parameter* of the time integration scheme. This scheme is also known as SS11; one of the time integration scheme family of SSpj, which stands for single step with approximation of degree p for equations of order j, (Zienkiewicz *et al.* 2005).

It is evident that the weighted residual time integration scheme gives similar formulation as that for the θ-method. For $\theta = 1$ we obtain the fully implicit backward difference scheme; $\theta = 1/2$ we obtain the central difference scheme; $\theta = 0$ we obtain the explicit forward difference scheme. The main advantage, however, is that the weighted residual finite element time integration scheme produces an analytical algorithmic parameter, which depends on the weighting function and the time step. Hence, by selecting a proper weighting function, we are able to produce a time integration scheme which is stable and its numerical dissipation can be effectively controlled.

The stability analysis of Eq. (11.119) is similar to that of the θ-method, which has been shown that in order to have an unconditionally stable algorithm, θ must be $\geq 1/2$. It is therefore important to choose a weighting function that leads to $\theta \geq 1/2$ in Eq. (11.120). For $\mathbf{W} = w = N_n$, Eq. (11.113), upon solving the integration in Eq. (11.120), gives $\theta = 1/3$, which is unstable. However, for $\mathbf{W} = w = N_{n+1}$, the integration gives $\theta = 2/3$, which is unconditionally stable. The amplification factor for this weighting function is shown in Figure 11.6, which shows that it is more stable than schemes with $\theta \leq 1/2$, and less stable than that for $\theta = 1$.

In choosing the time integration scheme, it is important to satisfy certain aspects, including:

- Accuracy, preferably, second order accurate.
- Stability, in most cases it is preferable to employ unconditionally stable algorithms.
- Self-starting, i.e. a single step algorithm.
- Controllable damping in the high frequency range.
- Low period error.
- Low damping error.
- Computational efficiency.

These aspects have been discussed intensively in literature, see for example Wood (1990) and Zienkiewicz *et al.* (2005).

11.3.3   *Numerical example*

In this example we employ the finite difference and the finite element time integration schemes to solve heat conduction and conduction-convection problems in a one-dimensional fluid bar. The bar is assumed 10 m in length, subjected to a prescribed temperature of 30°C at one end and 0°C at the other. The fluid is assumed to flow with the direction of the heat flow, i.e. from high to low temperature. All thermal parameters, λ, ρ, c, are assumed equal to 1.

Discretization in space is done using 10 linear elements with $h = 1$ m in length, see Figure 11.4. Discretization in time is done using finite difference and finite element time integration schemes. The finite difference scheme is represented by the theta-method, given by the forward difference

($\theta = 0$), the central difference ($\theta = 1/2$), and the backward difference ($\theta = 1$) algorithms. The finite element time integration scheme is represented by the Galerkin finite element algorithm ($\theta = 2/3$), and the space-time Petrov-Galerkin algorithm, defined by Eqs. (11.55)–(11.56).

Various time step sizes and fluid velocities are utilized. The forward difference algorithm is conditionally stable and, hence, the time step size must be calculated according to Eq. (11.100); to recall:

$$\Delta t_{cr} \leq \frac{2}{(1 - 2\theta)\omega}$$

(11.121)

where ω is the largest eigenfrequency of the eigenvalue problem:

$$(\mathbf{K} - \omega\mathbf{M})\mathbf{T} = 0$$

(11.122)

Using the weighted residual method, the element stiffness matrix is

$$\mathbf{k} = \begin{pmatrix} \dfrac{\lambda}{h} - \dfrac{1}{2}v\rho c & -\dfrac{\lambda}{h} + \dfrac{1}{2}v\rho c \\[2mm] -\dfrac{\lambda}{h} - \dfrac{1}{2}v\rho c & \dfrac{\lambda}{h} + \dfrac{1}{2}v\rho c \end{pmatrix}$$

(11.123)

The element mass matrix is

$$\mathbf{m} = \begin{pmatrix} \dfrac{1}{3}h\rho c & \dfrac{1}{6}h\rho c \\[2mm] \dfrac{1}{6}h\rho c & \dfrac{1}{3}h\rho c \end{pmatrix}$$

(11.124)

This mass matrix is known as the *consistent mass* matrix. However, for an explicit time integration scheme, the mass must be lumped. This can be done by summing rows of the consistent matrix and lump them on the associated diagonal term, as

$$\overline{M}_{ij} = \begin{cases} \displaystyle\sum_k M_{ik} & \text{if } i = j \\ 0 & \text{if } i \neq j \end{cases}$$

(11.125)

For a linear element, the *lumped mass* is:

$$\overline{\mathbf{m}} = \begin{pmatrix} \dfrac{1}{2}h\rho c & 0 \\[2mm] 0 & \dfrac{1}{2}h\rho c \end{pmatrix}$$

(11.126)

Using the lumped mass, the system of equations of the explicit algorithm can be solved by simple algebraic inversion of the mass matrix, as

$$\overline{\mathbf{M}}_{ij}^{-1} = \frac{1}{\overline{\mathbf{M}}_{ii}}$$

(11.127)

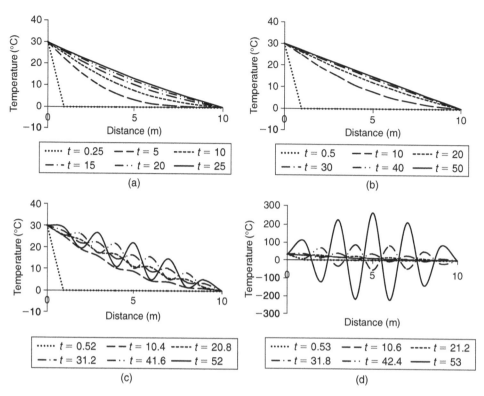

Figure 11.7   Theta = 0, conduction, (a) $\Delta t = 0.25$, (b) $\Delta t = 0.5$, (c) $\Delta t = 0.52$ and (d) $\Delta t = 0.53$.

Substituting Eqs. (11.123) and (11.126) into Eq. (11.122), the eigenvalues, for the lumped mass system, yields

$$0, \quad \frac{4}{h^2}\frac{\lambda}{\rho c} \tag{11.128}$$

For the consistent mass system, the eigenvalues are

$$0, \quad \frac{12}{h^2}\frac{\lambda}{\rho c} \tag{11.129}$$

Accordingly, the critical time step of the lumped mass system is 0.5 s, and of the consistent mass system is 0.166 s. That is, using a lumped mass matrix, the time step can be three times larger than that required by a consistent mass matrix. Lumped mass matrices are simple to form, require less memory and less computational efforts. However, care must be taken when choosing between the two matrices, because it cannot be stated that either lumped or consistent matrices are suitable for all problems. For more details on this issue, the reader is referred to Cook *et al.* (1989).

In Figure 11.7 to Figure 11.16 we show the performance of different time integration schemes for simulating heat conduction and heat conduction-convection in the medium. Figure 11.7 shows the temperature distribution as a result of heat conduction along the fluid bar at several times using time step sizes: $\Delta t = 0.25, 0.5, 0.52, 0.53$ s. The forward time integration scheme is utilized. The figures show clearly that for time steps less than the critical, the results are accurate. However, once the critical time step has been violated, spurious oscillation occurs. With larger time steps, the results become unbounded.

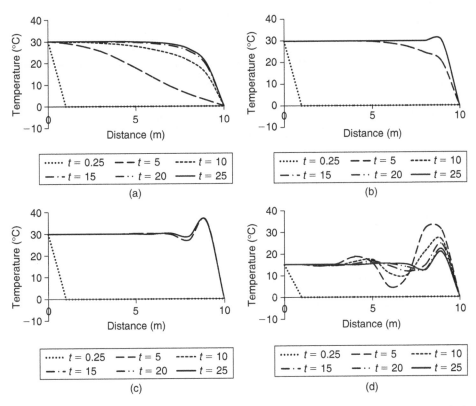

Figure 11.8   Theta $= 0$, conduction-convection, $\Delta t = 0.25$, (a) $v = 1$, (b) $v = 2$, (c) $v = 3$ and (d) $v = 4$ m/s.

Figure 11.8 shows the temperature distribution as a result of heat conduction and convection, using the forward time integration scheme. The time step is fixed at $\Delta t = 0.25$ s, but varying fluid velocity, $v = 1, 2, 3, 4$ m/s. Studying these figures, two issues can be observed:

1. Physically: the fluid flow has a significant effect on the heat flow.
2. Numerically: in spite of using a time step size which is less than the critical, the computational results for the high velocity cases exhibit spurious oscillations. This occurs despite the fact that the fluid velocity does not appear in the eigenvalues of the system, Eqs. (11.128) and (11.129). Hence, in principle, the fluid velocity should not affect the choice of the time step size. However, it can be noticed that when the fluid crosses more than half the length of the element in one time step, oscillation occurs. This is apparent in Figure 11.8(c), and Figure 11.8(d) where the fluid crossed $3/4\,h$ and h respectively.

Figure 11.9 to Figure 11.14 show temperature distributions due heat conduction and conduction-convection using the central difference, the Galerkin finite element and the backward difference algorithms. They are unconditionally stable, and, in principle, capable of producing results using any time step size. However, the figures show some oscillations, especially at the beginning; at time $t = 0.1$–0.25 s. They also exhibit oscillation for high fluid velocity, but still less than that for the forward scheme.

Figure 11.15 and Figure 11.16 show the performance of the space-time Petrov-Galerkin method for solving heat conduction and conduction-convection problems. Apparently, this method produces results which are accurate and smooth for all time step sizes and fluid velocities. In fact, this method produces exact results, but only for 1D problems. For 2D and 3D, the method accuracy declines.

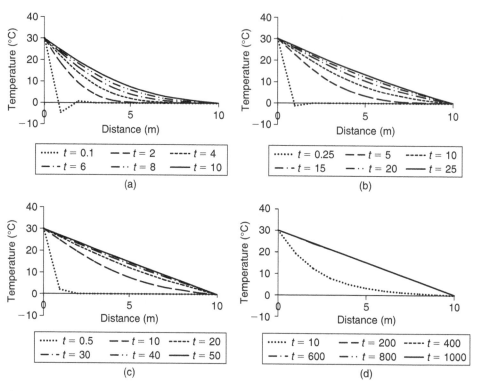

Figure 11.9   Theta $= 1/2$, conduction, (a) $\Delta t = 0.1$, (b) $\Delta t = 0.25$, (c) $\Delta t = 0.5$ and (d) $\Delta t = 10$.

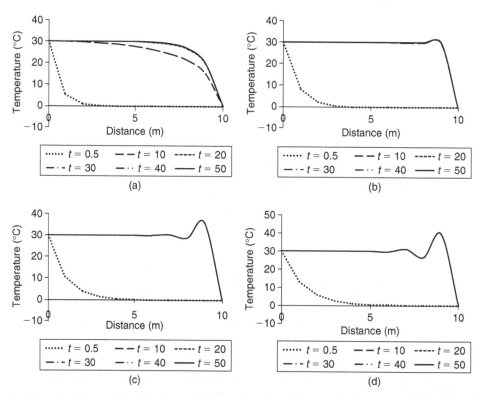

Figure 11.10   Theta $= 1/2$, conduction-convection, $\Delta t = 0.5$, (a) $v = 1$, (b) $v = 2$, (c) $v = 3$ and (d) $v = 4$ m/s.

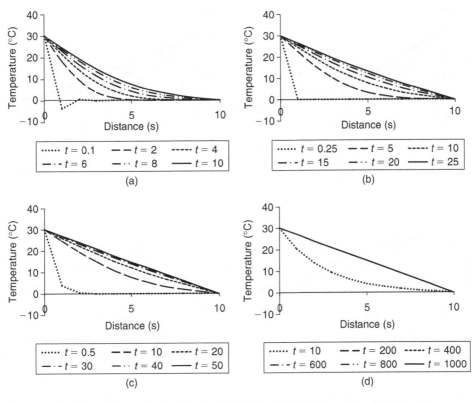

Figure 11.11   Theta $= 2/3$, conduction, (a) $\Delta t = 0.1$, (b) $\Delta t = 0.25$, (c) $\Delta t = 0.5$ and (d) $\Delta t = 10$.

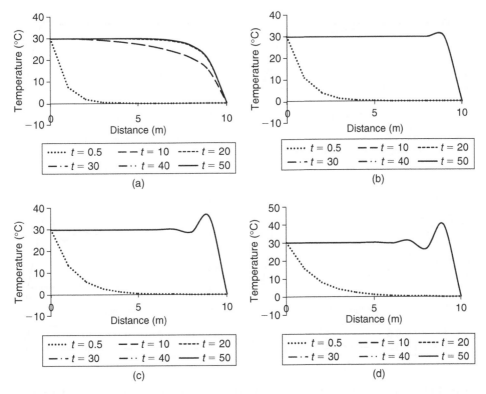

Figure 11.12   Theta $= 2/3$, conduction-convection, $\Delta t = 0.5$, a) $v = 1$, b) $v = 2$, c) $v = 3$ and d) $v = 4$ m/s.

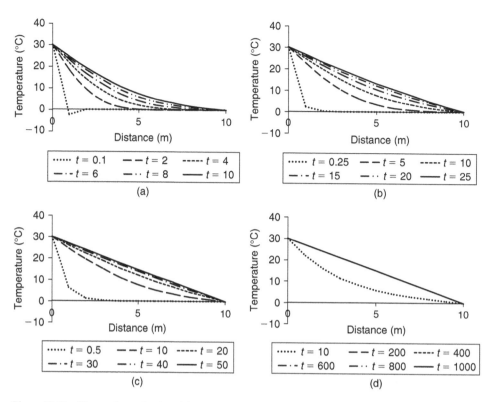

Figure 11.13 Theta = 1, conduction, (a) $\Delta t = 0.1$, (b) $\Delta t = 0.25$, (c) $\Delta t = 0.5$ and (d) $\Delta t = 10$.

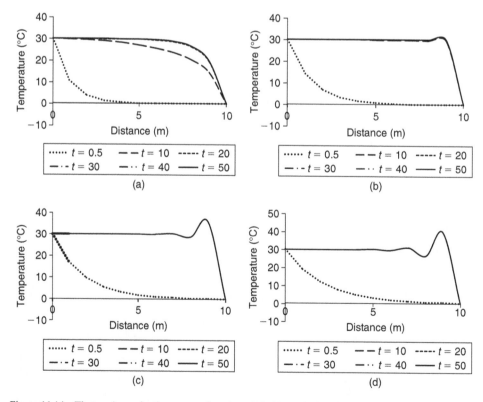

Figure 11.14 Theta = 1, conduction-convection, $\Delta t = 0.5$, (a) $v = 1$, (b) $v = 2$, (c) $v = 3$ and (d) $v = 4$ m/s.

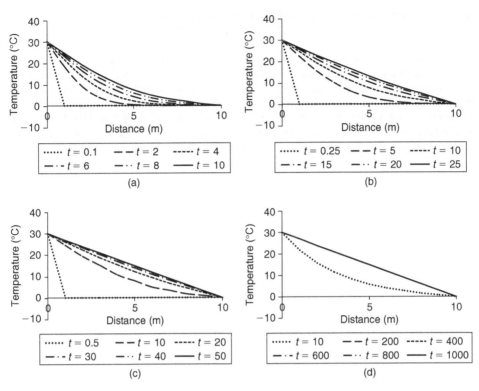

Figure 11.15 Petrov-Galerkin, conduction, (a) $\Delta t = 0.1$, (b) $\Delta t = 0.25$, (c) $\Delta t = 0.5$ and (d) $\Delta t = 10$.

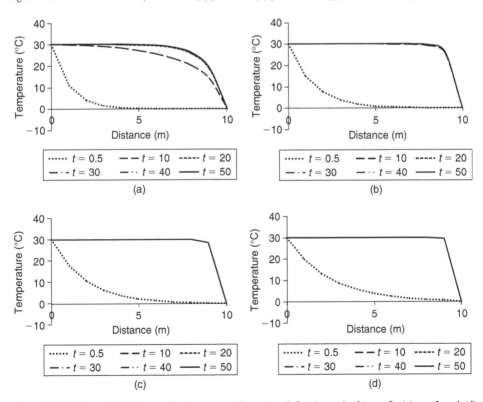

Figure 11.16 Petrov-Galerkin, conduction-convection, $\Delta t = 0.5$, (a) $v = 1$, (b) $v = 2$, (c) $v = 3$ and (d) $v = 4$ m/s.

An important point that needs to be mentioned in this context is that these schemes have apparent stability because of their inherited numerical diffusion, which eliminates some of the high frequencies from the system. With increasing θ, the numerical diffusion increases. This property is somewhat controversial since in one hand, it provides smooth and stable results, but on the other hand, it eliminates some of the physical behavior of the system. The choice of the time integration scheme depends on the problem, and the user must be aware of their advantages and disadvantages before applying them for solving engineering problems.

CHAPTER 12

Finite element modeling of shallow geothermal systems

Based on the finite element techniques introduced in Chapter 11, in this chapter, we formulate finite elements for conductive-convective heat flow in a soil mass and borehole heat exchangers. Galerkin, upwind and Petrov-Galerkin methods will be utilized for this purpose. Focus is placed on developing an efficient computational model for the simulation of heat flow in a borehole heat exchanger and its thermal interaction with the surrounding soil mass. As it will be shown in this chapter, this model is capable of simulating complicated BHE configurations and geometry using coarse finite element meshes. Furthermore, we introduce two numerical procedures for solving the resulting finite element equations: sequential and static condensation.

12.1 INTRODUCTION

In Part II of this book we presented several commonly utilized analytical and semi-analytical solution techniques for the simulation of heat flow in shallow geothermal systems, particularly ground source heat pumps. We show that in order to obtain a feasible solution to the governing equations, many simplifications are necessary to model the involved geometry and initial and boundary conditions. Only limited level of complexities is possible. These limitations make the analytical and the semi-analytical models fall short of the versatility of the numerical models. The numerical models are quite effective in describing complicated geometry and initial and boundary conditions.

The finite difference and the finite element (FEM) methods have been utilized to model shallow geothermal systems. The finite difference method seems to dominate this research field, as it is traditionally utilized to solve heat and fluid flow problems. Some of the known finite difference and its associated finite volume models are, among others, Eskilson and Claesson (1988), Clauser (2003), Lee and Lam (2008), Sliwa and Gonet (2005), and Yavuzturk (1999). The finite element method is utilized to a lesser extent, but recently this method is gaining momentum. One of the leading works in this field is that of Muraya (1994). Later, several finite element models have been introduced including, Al-Khoury *et al.* (2005, 2006 and 2010), van der Meer *et al.* (2009), and Diersch *et al.* (2011a and 2011b). Here, we focus on the use of the finite element method to model shallow geothermal systems, particularly the ground source heat pump.

In the field of shallow geothermal systems, the finite element method is yet considered of limited practical applications. The primary hurdle in this context is the computational inefficiency of FEM to model the borehole heat exchanger. The slenderness of the BHE exerts enormous challenges to the finite element developers. Mesh sizes of the order of millions of elements are not unusual to simulate three-dimensional multiple borehole heat exchangers embedded in multilayer systems. Even with current advent in computational power, this problem is pretty demanding. Consequently, models based on the finite element method cannot be directly incorporated into building energy simulation software, and thus have limited applicability in engineering practice.

As a result of the difficulty to model the borehole heat exchanger, the geothermal system is usually simulated in a two-dimensional space. Most of the existing models describe the geothermal system from the top view. That is, heat flow describing the thermal interaction between the BHE and the surrounding soil mass is calculated in the radial direction. Other finite element models simulate heat flow along the axis of the BHE in an axially symmetric system. In both cases, the standard Galerkin finite element method is utilized, and hence fine meshes are necessary to model the involved heat convection processes.

In standard finite element formulations, heat flow in each individual BHE component is described separately, as

$$\mathbf{k}_i \mathbf{T}_i = \mathbf{F}_i, \quad \text{Pipe-in}$$
$$\mathbf{k}_o \mathbf{T}_o = \mathbf{F}_o, \quad \text{Pipe-out} \tag{12.1}$$
$$\mathbf{k}_g \mathbf{T}_g = \mathbf{F}_g, \quad \text{Grout}$$

in which $\mathbf{k}_i$ is the element stiffness matrix of pipe-in, $\mathbf{T}_i$ is the nodal temperature vector of pipe-in, and $\mathbf{F}_i$ is the nodal heat flux vector of pipe-in, etc. Coupling these equations in space is done via standard finite element assembly technique. As a consequence, and due to the slenderness of the BHE, together with the presence of convective processes, this way of modeling requires excessively fine meshes, and thus demands an enormous computational power for conducting calculations, especially for those of engineering application scales.

One possible remedy to this problem is to couple the BHE components in a single finite element, circumventing thus the discretization in space. This can be done by coupling the BHE governing partial differential equations using thermal interaction terms between components in direct contact. As such, the discretization of the individual components in space is no longer necessary, and a simple line element will suffice. This idea has been introduced by Al-Khoury *et al.* (2005 and 2006), who developed a pseudo three-dimensional finite element model for heat flow in borehole heat exchangers using a line element. In this chapter, we focus on this model, and develop a finite element modeling procedure for heat flow in different types of shallow geothermal systems.

12.2 SOIL FINITE ELEMENT

Heat flow in a shallow geothermal system arises from the thermal interaction between the borehole heat exchangers and the soil mass. In a one-phase soil mass, constituting a solid skeleton, heat is merely conductive. In the presence of groundwater flow, the soil mass constitutes a multiphase porous medium and the heat flow in this case is conductive-convective. In typical shallow geothermal systems, it is realistic to assume that the groundwater flow is steady-state and occurs in confined fully saturated soil layers. Also, groundwater flow is independent of temperature. However, the temperature of the soil is highly dependent on the groundwater flow.

12.2.1 *Basic heat equation*

In Chapter 2 we derived a conductive-convective heat equation of a three-dimensional domain based on Taylor series expansion and Fourier's law. We show that, considering a rectangular parallelepiped element with volume $dx\,dy\,dz$ shown in Figure 12.1, the net flow of heat conduction into the element along the x-axis can be expressed as

$$q_{x1} - q_{x2} = \frac{\partial}{\partial x}\left(\lambda_x \frac{\partial T}{\partial x}\right) dx dy dz \tag{12.2}$$

The same is valid for the other axes. For a material with density ρ and a specific heat c, the rate of gain of heat is given by

$$\rho c \frac{\partial T}{\partial t} dx dy dz \tag{12.3}$$

Collecting terms and including a heat source, we saw that the heat equation can be written in a compact form as

$$\rho c \frac{\partial T}{\partial t} = \nabla \cdot (\lambda \nabla T) + Q(x, y, z) \tag{12.4}$$

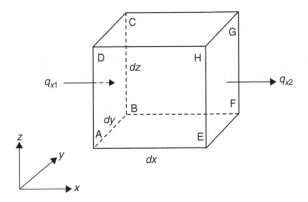

Figure 12.1　A parallelepiped element with volume $dx\,dy\,dz$.

in which λ is the heat conductivity tensor. In practice, for three dimensional problems, the conductivity is described as anisotropic in the principal axes, that is

$$\lambda = \mathrm{diag}(\lambda_x, \lambda_y, \lambda_z) \tag{12.5}$$

For an isotropic material $\lambda_x = \lambda_y = \lambda_z = \lambda$.

If heat convection due to groundwater flow also exists, the net heat flow in x-direction then becomes

$$q_{x1} - q_{x2} = \frac{\partial}{\partial x}\left(\lambda_x \frac{\partial T}{\partial x}\right) dxdydz + \frac{\partial}{\partial x}(\rho c v_x T)dxdydz \tag{12.6}$$

Collecting terms and including a heat source, we saw that the heat conduction-convection equation can be written as

$$\rho c \frac{\partial T}{\partial t} = \nabla \cdot (\lambda \nabla T) + \nabla(\rho c v T) + Q(x, y, z) \tag{12.7}$$

in which $\mathbf{v}$ is the Darcy velocity tensor.

12.2.2　*Governing equations of heat flow in a fully saturated porous medium*

Heat flow in a shallow geothermal system arises due to heat conduction and convection occurring in the borehole heat exchangers and the surrounding soil mass. Heat conduction in a soil mass occurs as a result of temperature gradients between the bottom of the earth and the surface of soil in contact with the air. Heat convection arises as a result of diffusion and advection of heat due to the presence of groundwater flow. The coupling between heat flow and groundwater flow is manifested in the fluid velocity tensor $\mathbf{v}$ in Eq. (12.7). This tensor is determined using Darcy's law which can be expressed in terms of pressure gradient as

$$\mathbf{v} = -\frac{\overline{\mathbf{k}}}{\mu} \cdot (\mathbf{grad}\,P - \rho_w \mathbf{g}) \tag{12.8}$$

where $\overline{\mathbf{k}}$ (m^2) is the *intrinsic permeability* tensor, μ (kg/ms) is the dynamic viscosity, and $\mathbf{g}$ (m/s^2) is the gravity vector.

In a wet soil layer, pore pressure generates due to hydrostatic and/or hydraulic conditions. The hydrostatic pore pressure occurs as a result of the weight of water above the point of interest, described as

$$P = \rho_w g h = \gamma_w h \tag{12.9}$$

in which γ_w is the unit weight of water and h is the depth of the point where the pressure is measured. In this condition, as the term indicates, the water resides in the pores and the pressure is static.

The hydraulic pore pressure, on the other hand, arises due to pressure gradients, and as the term indicates, the hydraulic condition involves movement of the fluid. In natural geothermal aquifers and soil layers the pressure gradient is relatively low. Therefore it is reasonable to assume that the fluid flow is laminar and obeys Darcy's law, Eq. (12.8). Another assumption, which is also specific for such a system, is that the soil, in the presence of groundwater, can be considered as fully saturated two phase material, constituting a solid matrix and water. Furthermore, the temperature and the temperature gradient in shallow geothermal systems are relatively low, in the order of −5°C to 30°C. It is therefore realistic to assume that all involved material parameters are not function of pressure or temperature.

The two phases of the soil, solid and liquid, are assumed locally in a state of thermodynamic equilibrium, i.e. $T_s = T_w = T$, where the subscripts s and w refer to the solid and liquid (water) phases, respectively. It is also assumed that there is no net heat transfer from one phase to another. For this, and following the derivation in Chapter 3, the macroscopic energy balance equation for the solid phase can be described as

$$(1-n)\rho c_s \frac{\partial T_s}{\partial t} - (1-n)\nabla \cdot (\lambda_s \nabla T_s) = 0 \tag{12.10}$$

and for the fluid phase,

$$n\rho c_w \frac{\partial T_w}{\partial t} + \rho c_w \mathbf{v} \cdot \nabla T_w - n\nabla \cdot (\lambda_w \nabla T_w) = 0 \tag{12.11}$$

Setting $T_s = T_w = T$ and adding Eq. (12.10) to Eq. (12.11), the energy balance equations for a fully saturated two-phase medium can then be described as

$$\rho c \frac{\partial T}{\partial t} + \rho c_w \mathbf{v} \cdot \nabla T - \nabla \cdot (\lambda \nabla T) = 0 \tag{12.12}$$

in which

$$\begin{aligned} \lambda &= (1-n)\lambda_s + n\lambda_w \\ \rho c &= (1-n)\rho c_s + n\rho c_w \end{aligned} \tag{12.13}$$

Eq. (12.12) contains two unknowns, the temperature and the fluid velocity. For this to be complete, we need to introduce the momentum conservation equation. The derivation of this equation follows exactly that for the heat equation derived in the previous section. The continuity condition enforces that the total flow into a differential element is equal to the total flow leaving the element. In x-direction, this can be expressed as

$$J_{(\text{net})} = v_x dy - \left(v_x + \frac{\partial v_x}{\partial x}\right) dy \tag{12.14}$$

The constitutive relationship that governs the fluid flow is Darcy's law, Eq. (12.8). In practice, Eq. (12.8) is written in terms of the total head

$$\varphi = \frac{P}{\rho_w g} + z \tag{12.15}$$

where φ is the total head, $P/\rho_w g$ is the pressure head and z is the elevation head. Hence, Darcy's law can be expressed as

$$\mathbf{v} = -\mathbf{k}\nabla\varphi \tag{12.16}$$

in which $\mathbf{k} = \bar{\mathbf{k}}/\mu \rho_w g$ is the *hydraulic conductivity* tensor, commonly termed *permeability* (m/s). For three dimensional problems, the permeability is described as anisotropic in the principal axes, as

$$\mathbf{k} = \text{diag}(k_x, k_y, k_z) \tag{12.17}$$

For an isotropic material $k_x = k_y = k_z = k$. Introducing Eq. (12.16) to Eq. (12.14), and equating the net flow to a storage term and a source, the fluid flow balance equation can be expressed as

$$S\frac{\partial \varphi}{\partial t} = -\nabla \cdot (\mathbf{k}\varphi) + Q_p(x, y, z) \tag{12.18}$$

in which S is the *specific storage coefficient*, which represents the amount of water released per unit area per head gradient.

12.2.3   *Initial and boundary conditions*

To formulate an initial and boundary value problem of a shallow geothermal system, relevant initial and boundary conditions must be specified for both: the fluid flow and the heat flow. For the fluid flow, the initial condition in the soil mass, at time $t = 0$, can be described in terms of the hydrostatic head

$$\varphi(x, y, z, 0) = \varphi_0(x, y, z) \tag{12.19}$$

The soil boundary conditions for groundwater flow is most probably associated with a head difference between an upper stream and a lower stream occurring in a confined aquifer that might be described, for example, as

$$
\begin{aligned}
\varphi(x = 0, y, z) &= \varphi_1, & &\text{on } x = 0 \text{ surface} \\
\varphi(x = L, y, z) &= \varphi_2, & &\text{on } x = L \text{ surface} \\
\mathbf{k}\nabla\varphi \cdot \mathbf{n} &= J, & &\text{on any of the boundary surfaces}
\end{aligned}
\tag{12.20}
$$

in which L is the length (along the x-axis) between two known hydraulic heads, and J is a fluid flux.

For the heat flow, the initial condition of the soil mass, at time $t = 0$, can be described as a steady-state condition:

$$T(x, y, z, 0) = F(x, y, z) \tag{12.21}$$

The soil boundary conditions associated with a shallow geothermal field might be:

$$
\begin{aligned}
T(P', t) &= F(P', t), & &\text{on a point or a surface } P' \\
\lambda \nabla T \cdot \mathbf{n} + b_{as}(T_s - T_a) &= 0, & &\text{on the surface in contact with the air} \\
\lambda_z \frac{\partial T}{\partial n} + b_{gs}(T_s - T_g) &= 0, & &\text{on the surface in contact with a BHE}
\end{aligned}
\tag{12.22}
$$

in which T_a is the air temperature, b_{as} is the convective heat transfer coefficient at the surface in contact with the air, T_g is the pipe (grout) temperature and b_{gs} is the reciprocal of the thermal resistance between the soil and the grout (borehole).

12.2.4   *Finite element discretization*

Recall the balance equations governing fluid and heat flow in a fully saturated porous medium:

$$S\frac{\partial\varphi}{\partial t} - \nabla\cdot(\mathbf{k}\varphi) + Q_p = 0 \tag{12.23}$$

$$\mathbf{v} = -\mathbf{k}\nabla\varphi \tag{12.24}$$

$$\rho c\frac{\partial T}{\partial t} + \rho c_w\mathbf{v}\cdot\nabla T - \nabla\cdot(\lambda\nabla T) + Q_T = 0 \tag{12.25}$$

As mentioned earlier, in shallow geothermal problems it is reasonable to assume that the fluid velocity influences the temperature distribution, but the temperature does not influence the fluid velocity. Hence, the solution of this problem can be achieved in two steps. First, the fluid velocity, **v**, is computed using Eqs. (12.23) and (12.24). Then it is utilized as an input to the heat equation, Eq. (12.25). That is no coupling is necessary.

We apply the weighted residual method to discretize the governing equations. The residual of the two flow fields, fluid and heat, can be described as

$$\begin{aligned} R_p &= S\frac{\partial\varphi}{\partial t} - \nabla\cdot(\mathbf{k}\varphi) + Q_p \neq 0 \\ R_T &= \rho c\frac{\partial T}{\partial t} + \rho c_w\mathbf{v}\cdot\nabla T - \nabla\cdot(\lambda\nabla T) + Q_T \neq 0 \end{aligned} \tag{12.26}$$

Multiplying the residual by a weighting function, **W**, integrating the product over the domain, and setting the product equal to zero, we obtain

Fluid flow:

$$\int_\Omega \mathbf{W}^T\left[S\frac{\partial\varphi}{\partial t} - \frac{\partial}{\partial x}\left(k_x\frac{\partial\varphi}{\partial x}\right) + \frac{\partial}{\partial y}\left(k_y\frac{\partial\varphi}{\partial y}\right) + \frac{\partial}{\partial z}\left(k_z\frac{\partial\varphi}{\partial x}\right) + Q_p\right]d\Omega = 0 \tag{12.27}$$

Heat flow:

$$\int_\Omega \mathbf{W}^T\left[\begin{array}{l} \rho c\frac{\partial T}{\partial t} - \frac{\partial}{\partial x}\left(\lambda_x\frac{\partial T}{\partial x}\right) + \frac{\partial}{\partial y}\left(\lambda_y\frac{\partial T}{\partial y}\right) + \frac{\partial}{\partial z}\left(\lambda_z\frac{\partial T}{\partial x}\right) + \\ \rho c_w\left(v_x\frac{\partial T}{\partial x} + v_y\frac{\partial T}{\partial y} + v_z\frac{\partial T}{\partial z}\right) + Q_T(x,y,z) \end{array}\right]d\Omega = 0 \tag{12.28}$$

These two equations are in their strong form since they contain high order derivatives. To convert them to a weak form, the second order derivative in Eqs. (12.27) and (12.28) must be replaced by a first order derivative using integration by parts. For example

$$\int_\Omega \mathbf{W}^T\lambda_x\frac{\partial^2 T}{\partial x^2}d\Omega = \int_\Omega\left(\lambda_x\frac{\partial}{\partial x}\left(\mathbf{W}^T\frac{\partial T}{\partial x}\right) - \lambda_x\frac{\partial\mathbf{W}^T}{\partial x}\frac{\partial T}{\partial x}\right)d\Omega \tag{12.29}$$

The first term on the right-hand side can be replaced by an integration over the surface using the Green's theorem, such that

$$\int_\Omega \lambda_x\frac{\partial}{\partial x}\left(\mathbf{W}^T\frac{\partial T}{\partial x}\right)d\Omega = \int_\Gamma \lambda_x\mathbf{W}^T\frac{\partial T}{\partial x}\cdot n_x d\Gamma \tag{12.30}$$

in which n_x is direction cosine of the normal to the surface in the x-direction. Applying this procedure to Eqs. (12.27) and (12.28), using vector notation, leads to

Fluid flow:

$$
\int_{\Omega} \left[\mathbf{W}^T S \frac{\partial \varphi}{\partial t} - \left(k_x \frac{\partial \mathbf{W}^T}{\partial x} \frac{\partial \varphi}{\partial x} + k_y \frac{\partial \mathbf{W}^T}{\partial y} \frac{\partial \varphi}{\partial y} + k_z \frac{\partial \mathbf{W}^T}{\partial z} \frac{\partial \varphi}{\partial z} \right) \right] d\Omega
$$
$$
- \int_{\Gamma} \mathbf{W}^T \left(\frac{\partial \varphi}{\partial x} n_x + \frac{\partial \varphi}{\partial y} n_y + \frac{\partial \varphi}{\partial z} n_z \right) d\Gamma - \int_{\Omega} \mathbf{W}^T Q_p d\Omega = 0
$$

(12.31)

Heat flow:

$$
\int_{\Omega} \left[\mathbf{W}^T \rho c \frac{\partial \mathbf{T}}{\partial t} - \left(\begin{array}{l} \lambda_x \dfrac{\partial \mathbf{W}^T}{\partial x} \dfrac{\partial \mathbf{T}}{\partial x} + \lambda_y \dfrac{\partial \mathbf{W}^T}{\partial y} \dfrac{\partial \mathbf{T}}{\partial y} + \lambda_z \dfrac{\partial \mathbf{W}^T}{\partial z} \dfrac{\partial \mathbf{T}}{\partial z} \\[6pt] - c_w \rho_w \mathbf{W}^T \left(v_x \dfrac{\partial \mathbf{T}}{\partial x} + v_y \dfrac{\partial \mathbf{T}}{\partial y} + v_z \dfrac{\partial \mathbf{T}}{\partial z} \right) \end{array} \right) \right] d\Omega
$$
$$
- \int_{\Gamma} \mathbf{W}^T \left(\frac{\partial \mathbf{T}}{\partial x} n_x + \frac{\partial \mathbf{T}}{\partial y} n_y + \frac{\partial \mathbf{T}}{\partial z} n_z \right) d\Gamma + \int_{\Omega} \mathbf{W}^T Q_T d\Omega = 0
$$

(12.32)

These formulations represent the weak form of the flow field finite element equations. Using Galerkin's finite element method, the unknown variables φ and **T** can be approximated by

$$
\varphi = \mathbf{N}_p \overline{\varphi}
$$
$$
\mathbf{T} = \mathbf{N}_T \overline{\mathbf{T}}
$$

(12.33)

where $\overline{\varphi}$ and $\overline{\mathbf{T}}$ are the total head and temperature nodal vectors, respectively, and $\mathbf{N}_p$ and $\mathbf{N}_T$ are the corresponding finite element shape functions. In practice the two shape function are made equal, i.e. $\mathbf{N} = \mathbf{N}_p = \mathbf{N}_T$. Substituting these approximations into Eqs. (12.31) and (12.32), and using the boundary conditions, Eqs. (12.20) and (12.22), yields

Fluid flow:

$$
\int_{\Omega} \mathbf{W}^T S \mathbf{N} \frac{\partial \overline{\varphi}}{\partial t} d\Omega - \int_{\Omega} \left(k_x \frac{\partial \mathbf{W}^T}{\partial x} \frac{\partial \mathbf{N}}{\partial x} + k_y \frac{\partial \mathbf{W}^T}{\partial y} \frac{\partial \mathbf{N}}{\partial y} + k_z \frac{\partial \mathbf{W}^T}{\partial z} \frac{\partial \mathbf{N}}{\partial z} \right) \overline{\varphi} \, d\Omega
$$
$$
- \int_{\Gamma} \mathbf{W}^T q_p d\Gamma - \int_{\Omega} \mathbf{W}^T Q_p d\Omega = 0
$$

(12.34)

Heat flow:

$$
\int_{\Omega} \mathbf{W}^T \rho c \mathbf{N} \frac{\partial \overline{\mathbf{T}}}{\partial t} d\Omega - \int_{\Omega} \left(\begin{array}{l} \lambda_x \dfrac{\partial \mathbf{W}^T}{\partial x} \dfrac{\partial \mathbf{N}}{\partial x} + \lambda_y \dfrac{\partial \mathbf{W}^T}{\partial y} \dfrac{\partial \mathbf{N}}{\partial y} + \\[6pt] \lambda_z \dfrac{\partial \mathbf{W}^T}{\partial z} \dfrac{\partial \mathbf{N}}{\partial z} + \\[6pt] \rho_w c_w \mathbf{W}^T \left(v_x \dfrac{\partial \mathbf{N}}{\partial x} + v_y \dfrac{\partial \mathbf{N}}{\partial y} + v_z \dfrac{\partial \mathbf{N}}{\partial z} \right) \end{array} \right) \overline{\mathbf{T}} d\Omega
$$
$$
- \int_{\Gamma} \mathbf{W}^T q_T d\Gamma + \int_{\Omega} \mathbf{W}^T Q_T d\Omega = 0
$$

(12.35)

As indicated in Eq. (12.22), heat flux in a geothermal system arises from thermal interactions between the soil surface and the air, and from the contact with the borehole heat exchanger. The finite element discretization of these boundary conditions is

Soil-air contact:

$$
\int_{\Gamma} \mathbf{W}^T b_{as} (\mathbf{T} - T_a) d\Gamma = \int_{\Gamma} \mathbf{W}^T b_{as} \mathbf{N} d\Gamma \overline{\mathbf{T}} - \int_{\Gamma} \mathbf{W}^T b_{as} T_a d\Gamma
$$

(12.36)

Soil-BHE contact:

$$\int_{\Gamma} \mathbf{W}^T b_{gs} (\mathbf{T} - \mathbf{T}_g) d\Gamma = \int_{\Gamma} \mathbf{W}^T b_{gs} \mathbf{N} d\Gamma \overline{\mathbf{T}} - \int_{\Gamma} \mathbf{W}^T b_{gs} \mathbf{T}_g d\Gamma \qquad (12.37)$$

For a typically slow groundwater flow, the standard Galerkin method is practically adequate for formulating the convective-conductive heat flow in a porous medium, i.e. $\mathbf{W} = \mathbf{N}$. The final finite element relationships can then be expressed as

Fluid flow:

$$\mathbf{M}_p \frac{\partial \overline{\varphi}}{\partial t} + \mathbf{K}_p \overline{\varphi} = \mathbf{F}_p \qquad (12.38)$$

where

$$\mathbf{M}_p = \int_{\Omega} \mathbf{N}^T S \mathbf{N} d\Omega$$

$$\mathbf{K}_p = \int_{\Omega} \mathbf{B}^T \mathbf{k} \mathbf{B} d\Omega \qquad (12.39)$$

$$\mathbf{F}_p = \int_{\Gamma} \mathbf{W}^T J d\Gamma - \int_{\Omega} \mathbf{W}^T Q_p d\Omega$$

with $\mathbf{B} = d\mathbf{N}/dz$.

Heat flow:

$$\mathbf{M}_T \frac{\partial \overline{\mathbf{T}}}{\partial t} + \mathbf{K}_T \overline{\mathbf{T}} = \mathbf{F}_T \qquad (12.40)$$

where

$$\mathbf{M}_T = \int_{\Omega} \mathbf{N}^T \rho c \mathbf{N} d\Omega$$

$$\mathbf{K}_T = \int_{\Omega} (\mathbf{B}^T \boldsymbol{\lambda} \mathbf{B} + \rho_w c_w \mathbf{N}^T \mathbf{v} \mathbf{B}) d\Omega + \int_{\Gamma} \mathbf{N}^T b_{as} \mathbf{N} d\Gamma + \int_{\Gamma} \mathbf{N}^T b_{sg} \mathbf{N} d\Gamma \qquad (12.41)$$

$$\mathbf{F}_T = \int_{\Omega} Q_t \mathbf{N}^T d\Omega + \int_{\Gamma} \mathbf{N}^T b_{gs} \overline{\mathbf{T}}_g d\Gamma + \int_{\Gamma} \mathbf{N}^T b_{as} T_a d\Gamma$$

The above set of equations, Eqs. (12.38)–(12.41), represents the finite element equation of heat flow in a saturated soil mass. It can be utilized to formulate simple, complex or multiplex finite elements using linear, quadratic, cubic or higher-order interpolation polynomials. To be compatible with the finite element equations of the borehole heat exchanger that will be derived in the next section, the nodal temperature of the soil will be denoted as $\overline{\mathbf{T}}_s$ and Eq. (12.41) will be expressed as

$$\mathbf{M}_T \frac{\partial \overline{\mathbf{T}}_s}{\partial t} + \mathbf{K}_T \overline{\mathbf{T}}_s = \mathbf{F}_T \qquad (12.42)$$

It is worth mentioning that this element is suitable for slow changes of boundary air or soil surface temperatures, and for this a relatively large element might be utilized. However, if a fast change in the boundary temperature occurs, large elements fail to capture the temperature profile in the soil, leading to numerical oscillations. To tackle this problem, van der Meer *et al.* (2009), developed a soil element, tailored specifically for highly transient geothermal heat flow problems. The basic idea is to incorporate time-dependent shape functions with their time dependency synchronized with the time variation of the problem. The quality of the shape functions is enhanced by incorporating basis functions of which it is known a priori to approximately match the solution. They introduced two time-dependent shape functions: 1) Iterative shape function, in which the time dependent variable is optimized iteratively, depending on the changes of the temperature gradients with time. 2) Analytically-based shape functions, in which the approximation functions are derived from an analytic solution of the basic problem. The proposed element is proved to be very efficient in capturing the high gradient temperatures even when coarse meshes are used.

12.3   BOREHOLE HEAT EXCHANGER FINITE ELEMENT

In Chapter 4 we derived the governing equations of heat flow in different types of borehole heat exchangers. Here we give a detailed formulation of the finite element equations of typical borehole heat exchangers. We solve the governing equations for both steady-state and transient conditions, using different finite element methods: the Galerkin method, the upwind method and the Petrov-Galerkin method.

12.3.1   *Governing equations of heat flow in a borehole heat exchanger*

Consider a single U-tube borehole heat exchanger. Heat flow in such a domain can be described using a control volume of length dz, consisting of three pipe components: pipe-in, denoted as i; pipe-out, denoted as o; and grout, denoted as g, shown in Figure 12.2. Equating the rate of energy entering the control volume to the rate of energy leaving it (see Chapter 4), the net heat flow into each of the pipe components can be expressed, for a transient condition, as

Pipes-in:

$$\rho c_f \frac{\partial T_i}{\partial t} - \lambda_f \frac{\partial^2 T_i}{\partial z^2} + \rho c_f u \frac{\partial T_i}{\partial z} = b_{ig}(T_i - T_g) \tag{12.43}$$

Pipes-out:

$$\rho c_f \frac{\partial T_o}{\partial t} - \lambda_f \frac{\partial^2 T_o}{\partial z^2} - \rho c_f u \frac{\partial T_o}{\partial z} = b_{og}(T_o - T_g) \tag{12.44}$$

Grout:

$$\rho c_g \frac{\partial T_g}{\partial t} - \lambda_g \frac{\partial^2 T_g}{\partial z^2} = b_{ig}(T_g - T_i) + b_{og}(T_g - T_o) \tag{12.45}$$

in which the subscript f represents the working fluid, u denotes the fluid velocity. Other parameters are defined earlier. For simplicity of notation the subscript f will not be used unless necessary.

This formulation emphasizes that, as manifested physically, coupling between the BHE components occurs via the grout, which works as an intermediate medium that transfers heat from

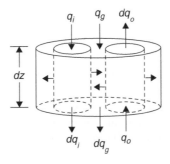

Figure 12.2   Control volume of a single U-Tube.

one pipe to another and vice versa. Other models, such as that of Eskilson and Claesson (1988), assume that there is a direct interaction between pipe-in and pipe-out, controlled by some thermal resistance.

12.3.2   *Initial and boundary conditions*

Initially, at $t = 0$, the temperature in the borehole heat exchanger is equal to the steady-state temperature in the soil before the heating/cooling operation starts, i.e.

$$T_i(z,0) = T_o(z,0) = T_g(z,0) = T_s(z,0) \tag{12.46}$$

in which T_s is the soil temperature immediately around the borehole.

Boundary conditions typically involved in an operating BHE are of two types: essential (Dirichlet) and natural (Neumann). In the first condition, at the inlet of pipe-in, $z = 0$, the temperature is equal to the fluid temperature at the moment it enters into pipe-in, that is

$$T_i(0,t) = T_{in}(t) \tag{12.47}$$

In the second condition, along the borehole, there is heat flow between the borehole and the neighboring soil mass, described as

$$-\lambda_g \frac{\partial T_g}{\partial n} = b_{gs}(T_g - T_s) \tag{12.48}$$

in which b_{gs} is the reciprocal of the thermal resistance between the soil and the grout. In the literature, T_g is usually denoted as T_b, denoting borehole temperature, and T_s as T_g, denoting ground temperature.

12.3.3   *Steady-state formulation*

The governing equations for the steady-state condition are

$$-\lambda \frac{d^2 T_i}{dz^2} + \rho c u \frac{dT_i}{dz} = b_{ig}(T_i - T_g) \tag{12.49}$$

$$-\lambda \frac{d^2 T_o}{dz^2} - \rho c u \frac{dT_o}{dz} = b_{og}(T_o - T_g) \tag{12.50}$$

$$-\lambda_g \frac{d^2 T_g}{dz^2} = b_{ig}(T_g - T_i) + b_{og}(T_g - T_o) \tag{12.51}$$

The boundary conditions are as stated in Eqs. (12.47) and (12.48).

12.3.3.1   *Discretization using the Galerkin method*
Here we utilize the Galerkin weighted residual method to discretize the steady-state differential equations, Eqs. (12.49), (12.50) and (12.51). We solve for

$$\int_\Omega \mathbf{N}^T \mathbf{R} \, d\Omega = 0 \tag{12.52}$$

where **N** is the Galerkin weighting function, equal to the conventional finite element shape function of an element, and **R** is the residual function, described, in this case, as

$$R_i = -\lambda \frac{d^2 T_i}{dz^2} + \rho c u \frac{dT_i}{dz} - b_{ig}(T_i - T_g)$$

$$R_o = -\lambda \frac{d^2 T_o}{dz^2} - \rho c u \frac{dT_o}{dz} - b_{og}(T_o - T_g) \tag{12.53}$$

$$R_g = -\lambda_g \frac{d^2 T_g}{dz^2} - b_{ig}(T_g - T_i) - b_{og}(T_g - T_o)$$

Including Eq. (12.53) into Eq. (12.52) yields

$$\int_V \left[\mathbf{N}^T \left(-\lambda \frac{d^2 T_i}{dz^2} \right) + \mathbf{N}^T \left(\rho c u \frac{dT_i}{dz} \right) \right] dV - \int_S \mathbf{N}^T b_{ig}(T_i - T_g) dS = 0$$

$$\int_V \left[\mathbf{N}^T \left(-\lambda \frac{d^2 T_o}{dz^2} \right) - \mathbf{N}^T \left(\rho c u \frac{dT_o}{dz} \right) \right] dV - \int_S \mathbf{N}^T b_{og}(T_o - T_g) dS = 0 \tag{12.54}$$

$$\int_V \mathbf{N}^T \left(-\lambda_g \frac{d^2 T_g}{dz^2} \right) dV - \int_S [\mathbf{N}^T b_{ig}(T_g - T_i) - \mathbf{N}^T b_{og}(T_g - T_o)] dS = 0$$

where heat conduction and convection terms are evaluated over the element volume, and heat flux between the BHE components are evaluated over the contact surface areas. Using integration by parts and the Green's theorem, as described in Eqs. (12.29)–(12.30), the second order derivatives in Eq. (12.54) can be replaced, for example, by

$$\int_V \mathbf{N}^T \lambda \frac{d^2 T}{dz^2} dV = -\int_V \frac{d\mathbf{N}^T}{dz} \lambda \frac{dT}{dz} dV + \int_S \mathbf{N}^T \lambda \frac{dT}{dz} \cdot \mathbf{n} dS \tag{12.55}$$

Substituting Eq. (12.55) into the corresponding term in Eq. (12.54), we obtain

$$\int_V \left[\frac{d\mathbf{N}^T}{dz} \lambda \frac{dT_i}{dz} dV + \mathbf{N}^T \left(\rho c u \frac{dT_i}{dz} \right) \right] dV - \int_S \mathbf{N}^T b_{ig}(T_i - T_g) dS = 0$$

$$\int_V \left[\frac{d\mathbf{N}^T}{dz} \lambda \frac{dT_o}{dz} dV - \mathbf{N}^T \left(\rho c u \frac{dT_o}{dz} \right) \right] dV - \int_S \mathbf{N}^T b_{og}(T_o - T_g) dS = 0 \tag{12.56}$$

$$\int_V \frac{d\mathbf{N}^T}{dz} \lambda_g \frac{dT_g}{dz} dV - \int_S \mathbf{N}^T \lambda \frac{dT_g}{dz} \cdot \mathbf{n} dS$$

$$- \int_S [\mathbf{N}^T b_{ig}(T_g - T_i) - \mathbf{N}^T b_{og}(T_g - T_o)] dS = 0$$

Note that, for pipe-in and pipe-out, the second term on the right-hand side of Eq. (12.55) does not appear explicitly in the first two equations of Eq. (12.56) because it is already included in the second integrals of these two equations. For the grout, this term is represented by the boundary condition, Eq. (12.48), such that

$$\int_S \mathbf{N}^T \lambda_g \frac{dT_g}{dz} \cdot \mathbf{n} dS = \int_S \mathbf{N}^T b_{gs}(T_g - T_s) dS \tag{12.57}$$

Using the conventional Galerkin method entails:

$$\begin{aligned} T_i &= \mathbf{N}_i \overline{\mathbf{T}}_i \\ T_o &= \mathbf{N}_o \overline{\mathbf{T}}_o \\ T_g &= \mathbf{N}_g \overline{\mathbf{T}}_g \end{aligned} \tag{12.58}$$

in which $\mathbf{N}_i$, $\mathbf{N}_o$, $\mathbf{N}_g$ are the interpolation functions (shape functions) for pipe-in, pipe-out and grout respectively. For simplicity, the interpolation functions are assumed equal, i.e $\mathbf{N}_i = \mathbf{N}_o = \mathbf{N}_g = \mathbf{N}$. Substituting Eq. (12.58) into Eq. (12.56) together with Eq. (12.57) gives
Pipe-in:

$$\int_V \left(\frac{d\mathbf{N}^T}{dz} \lambda \frac{d\mathbf{N}}{dz} dV + \mathbf{N}^T \rho c u \frac{d\mathbf{N}}{dz} \right) dV \overline{\mathbf{T}}_i - \int_S \mathbf{N}^T b_{ig} \mathbf{N} dS \overline{\mathbf{T}}_i$$
$$+ \int_S \mathbf{N}^T b_{ig} \mathbf{N} dS \overline{\mathbf{T}}_g = 0 \tag{12.59}$$

Pipe-out:

$$\int_V \left(\frac{d\mathbf{N}^T}{dz} \lambda \frac{d\mathbf{N}}{dz} dV - \mathbf{N}^T \rho c u \frac{d\mathbf{N}}{dz} \right) dV \overline{\mathbf{T}}_o - \int_S \mathbf{N}^T b_{og} \mathbf{N} dS \overline{\mathbf{T}}_o$$
$$+ \int_S \mathbf{N}^T b_{og} \mathbf{N} dS \overline{\mathbf{T}}_g = 0 \tag{12.60}$$

Grout:

$$\int_V \frac{d\mathbf{N}^T}{dz} \lambda_g \frac{d\mathbf{N}}{dz} dV \overline{\mathbf{T}}_g - \int_S \mathbf{N}^T b_{ig} \mathbf{N} dS \overline{\mathbf{T}}_g + \int_S \mathbf{N}^T b_{ig} \mathbf{N} dS \overline{\mathbf{T}}_i$$
$$- \int_S \mathbf{N}^T b_{og} \mathbf{N} dS \overline{\mathbf{T}}_g + \int_S \mathbf{N}^T b_{og} \mathbf{N} dS \overline{\mathbf{T}}_o \tag{12.61}$$
$$- \int_S \mathbf{N}^T b_{gs} \mathbf{N} dS \overline{\mathbf{T}}_g = - \int_S \mathbf{N}^T b_{gs} dS \overline{\mathbf{T}}_s$$

Putting these equations in a matrix form, leads to the standard finite element equation

$$\begin{pmatrix} \mathbf{K}_{ii} & \mathbf{K}_{ig} & 0 \\ \mathbf{K}_{gi} & \mathbf{K}_{gg} & \mathbf{K}_{go} \\ 0 & \mathbf{K}_{og} & \mathbf{K}_{oo} \end{pmatrix} \begin{pmatrix} \overline{\mathbf{T}}_i \\ \overline{\mathbf{T}}_g \\ \overline{\mathbf{T}}_o \end{pmatrix} = \begin{pmatrix} q_{in} \\ \mathbf{F}_{gs} \\ 0 \end{pmatrix} \tag{12.62}$$

where

$$\mathbf{K}_{ii} = \int_V \left(\mathbf{B}^T \lambda \mathbf{B} dV + \mathbf{N}^T \rho c u \mathbf{B} \right) dV - \int_S \mathbf{N}^T b_{ig} \mathbf{N} dS$$

$$\mathbf{K}_{ig} = \int_S \mathbf{N}^T b_{ig} \mathbf{N} dS$$

$$\mathbf{K}_{oo} = \int_V \left(\mathbf{B}^T \lambda \mathbf{B} dV - \mathbf{N}^T \rho c u \mathbf{B} \right) dV - \int_S \mathbf{N}^T b_{og} \mathbf{N} dS$$

$$\mathbf{K}_{og} = \int_S \mathbf{N}^T b_{og} \mathbf{N} dS$$

$$\mathbf{K}_{gg} = \int_V \mathbf{B}^T \lambda_g \mathbf{B} dV - \int_S \mathbf{N}^T b_{ig} \mathbf{N} dS - \int_S \mathbf{N}^T b_{og} \mathbf{N} dS - \int_S \mathbf{N}^T b_{gs} \mathbf{N} dS$$

$$\mathbf{K}_{gi} = \int_S \mathbf{N}^T b_{ig} \mathbf{N} dS$$

$$\mathbf{K}_{go} = \int_S \mathbf{N}^T b_{og} \mathbf{N} dS$$

$$\mathbf{F}_{gs} = -\int_S \mathbf{N}^T b_{gs} dS \overline{\mathbf{T}}_s \tag{12.63}$$

with $\mathbf{B} = d\mathbf{N}/dz$ and q_{in} is the inlet heat flux.

12.3.3.2 *Discretization using the balancing diffusion upwind method*

In Chapter 11, we have seen that by using the upwind method, we will be able to obtain a non-oscillating algorithm for problems with a large value of Peclet number. For a 1D element with length h, the energy equation can be modified by including a balancing diffusion term, such that

$$- \left(\lambda + \rho c \frac{\alpha u h}{2} \right) \frac{d^2 T}{dz^2} + \rho c u \frac{dT}{dz} = 0 \tag{12.64}$$

with

$$\alpha = \coth \frac{Pe}{2} - \frac{2}{Pe}, \quad Pe = \frac{uh}{\lambda} \tag{12.65}$$

If we let $\alpha = 0$, we obtain the Galerkin formulation, and if we let $\alpha = 1$, we get the upwind formulation. In the latter, $uh/2$ is known as the artificial numerical diffusion. Introducing the balancing diffusion term into Eq. (12.49) and Eq. (12.50) leads to

$$- \left(\lambda + \rho c \frac{\alpha u h}{2} \right) \frac{d^2 T_i}{dz^2} + \rho c u \frac{dT_i}{dz} - b_{ig}(T_i - T_g) = 0$$

$$- \left(\lambda - \rho c \frac{\alpha u h}{2} \right) \frac{d^2 T_o}{dz^2} - \rho c u \frac{dT_o}{dz} - b_{og}(T_o - T_g) = 0 \tag{12.66}$$

Applying the Galerkin finite element procedure to Eq. (12.66), yields

$$\int_V \left(\frac{d\mathbf{N}^T}{dz}\lambda\frac{d\mathbf{N}}{dz} + \frac{d\mathbf{N}^T}{dz}\rho c\frac{\alpha u h}{2}\frac{d\mathbf{N}}{dz} + \mathbf{N}^T\rho c u\frac{d\mathbf{N}}{dz} \right) dV \overline{\mathbf{T}}_i$$

$$- \int_S \mathbf{N}^T b_{ig}\mathbf{N}dS\overline{\mathbf{T}}_i + \int_S \mathbf{N}^T b_{ig}\mathbf{N}dS\overline{\mathbf{T}}_g = 0$$

$$\int_V \left(\frac{d\mathbf{N}^T}{dz}\lambda\frac{d\mathbf{N}}{dz} - \frac{d\mathbf{N}^T}{dz}\rho c\frac{\alpha u h}{2}\frac{d\mathbf{N}}{dz} - \mathbf{N}^T\rho c u\frac{d\mathbf{N}}{dz} \right) dV \overline{\mathbf{T}}_o$$

$$- \int_S \mathbf{N}^T b_{og}\mathbf{N}dS\overline{\mathbf{T}}_o + \int_S \mathbf{N}^T b_{og}\mathbf{N}dS\overline{\mathbf{T}}_g = 0$$

(12.67)

The discretization of the grout heat equation, Eq. (12.51), is the same as that in Eq. (12.61). The finite element equation of such a system can then be written as

$$\begin{pmatrix} \mathbf{K}_{ii} & \mathbf{K}_{ig} & 0 \\ \mathbf{K}_{gi} & \mathbf{K}_{gg} & \mathbf{K}_{go} \\ 0 & \mathbf{K}_{og} & \mathbf{K}_{oo} \end{pmatrix} \begin{pmatrix} \overline{\mathbf{T}}_i \\ \overline{\mathbf{T}}_g \\ \overline{\mathbf{T}}_o \end{pmatrix} = \begin{pmatrix} q_{in} \\ \mathbf{F}_{gs} \\ 0 \end{pmatrix}$$

(12.68)

where

$$\mathbf{K}_{ii} = \int_V \left(\mathbf{B}^T\lambda\mathbf{B} + \mathbf{B}^T\rho c\frac{\alpha u h}{2}\mathbf{B} + \mathbf{N}^T\rho c u\mathbf{B} \right) dV - \int_S \mathbf{N}^T b_{ig}\mathbf{N}dS$$

$$\mathbf{K}_{ig} = \int_S \mathbf{N}^T b_{ig}\mathbf{N}dS$$

$$\mathbf{K}_{oo} = \int_V \left(\mathbf{B}^T\lambda\mathbf{B} - \mathbf{B}^T\rho c\frac{\alpha u h}{2}\mathbf{B} - \mathbf{N}^T\rho c u\mathbf{B} \right) dV - \int_S \mathbf{N}^T b_{og}\mathbf{N}dS$$

$$\mathbf{K}_{og} = \int_S \mathbf{N}^T b_{og}\mathbf{N}dS$$

$$\mathbf{K}_{gg} = \int_V \mathbf{B}^T\lambda_g\mathbf{B}dV - \int_S \mathbf{N}^T b_{ig}\mathbf{N}dS - \int_S \mathbf{N}^T b_{og}\mathbf{N}dS - \int_S \mathbf{N}^T b_{gs}\mathbf{N}dS$$

(12.69)

$$\mathbf{K}_{gi} = \int_S \mathbf{N}^T b_{ig}\mathbf{N}dS$$

$$\mathbf{K}_{go} = \int_S \mathbf{N}^T b_{og}\mathbf{N}dS$$

$$\mathbf{F}_{gs} = -\int_S \mathbf{N}^T b_{gs}dS\overline{\mathbf{T}}_s$$

12.3.3.3 *Discretization using the Petrov-Galerkin method*

Formally, the weighted residual method states that

$$\int_\Omega \mathbf{W}\mathbf{R}d\Omega = 0$$

(12.70)

where $\mathbf{W}$ is any arbitrary weighting function and $\mathbf{R}$ is the residual. In Petrov-Galerkin method, the weighting function can be determined from the modified heat equations, Eq. (12.66), which include the balancing diffusion term. Eq. (12.67) can be written as

$$\int_V \frac{d\mathbf{N}_i^T}{dz} \lambda \frac{d\mathbf{T}_i}{dz} dV + \int_V \rho c u \left(\frac{\alpha h}{2} \frac{d\mathbf{N}_i^T}{dz} + \mathbf{N}_i^T \right) \frac{d\mathbf{T}_i}{dz} dV$$

$$- \int_S \mathbf{N}_i^T b_{ig} \mathbf{T}_i dS + \int_S \mathbf{N}_i^T b_{ig} \mathbf{T}_g dS = 0$$

$$\int_V \frac{d\mathbf{N}_o^T}{dz} \lambda \frac{d\mathbf{T}_o}{dz} dV + \int_V \rho c u \left(-\frac{\alpha h}{2} \frac{d\mathbf{N}_o^T}{dz} + \mathbf{N}_o^T \right) \frac{d\mathbf{T}_o}{dz} dV$$

$$- \int_S \mathbf{N}_o^T b_{og} \mathbf{T}_o dS + \int_S \mathbf{N}_o^T b_{og} \mathbf{T}_g dS = 0$$

(12.71)

Following Heinrich and Pepper (1999), the Petrov-Galerkin weighting function can be chosen to be the terms between brackets in Eq. (12.71), such that
Pipe-in:

$$\mathbf{W}_i = \mathbf{N}_i + \frac{\alpha h}{2} \frac{d\mathbf{N}_i}{dz}$$

(12.72)

Pipe-out:

$$\mathbf{W}_o = \mathbf{N}_o - \frac{\alpha h}{2} \frac{d\mathbf{N}_o}{dz}$$

(12.73)

Grout:

$$\mathbf{W}_g = \mathbf{N}_g$$

(12.74)

Accordingly, the discretization of the temperature in pipe-in, pipe-out and grout can be expressed as
Pipe-in:

$$\int_V \left(\frac{d\mathbf{W}_i^T}{dz} \lambda \frac{d\mathbf{N}}{dz} + \mathbf{W}_i^T \rho c u \frac{d\mathbf{N}}{dz} \right) dV \overline{\mathbf{T}}_i - \int_S \mathbf{W}_i^T b_{ig} \mathbf{N} dS \overline{\mathbf{T}}_i$$

$$+ \int_S \mathbf{W}_i^T b_{ig} \mathbf{N} dS \overline{\mathbf{T}}_g = 0$$

(12.75)

Pipe-out:

$$\int_V \left(\frac{d\mathbf{W}_o^T}{dz} \lambda \frac{d\mathbf{N}}{dz} - \mathbf{W}_o^T \rho c u \frac{d\mathbf{N}}{dz} \right) dV \overline{\mathbf{T}}_o - \int_S \mathbf{W}_o^T b_{og} \mathbf{N} dS \overline{\mathbf{T}}_o$$

$$+ \int_S \mathbf{W}_o^T b_{og} \mathbf{N} dS \overline{\mathbf{T}}_g = 0$$

(12.76)

Grout:

$$\int_V \frac{d\mathbf{N}^T}{dz} \lambda_g \frac{d\mathbf{N}}{dz} dV \overline{\mathbf{T}}_g - \int_S \mathbf{N}^T b_{ig} \mathbf{N} dS \overline{\mathbf{T}}_g + \int_S \mathbf{N}^T b_{ig} \mathbf{N} dS \overline{\mathbf{T}}_i$$

$$- \int_S \mathbf{N}^T b_{og} \mathbf{N} dS \overline{\mathbf{T}}_g + \int_S \mathbf{N}^T b_{og} \mathbf{N} dS \overline{\mathbf{T}}_o$$

(12.77)

$$- \int_S \mathbf{N}^T b_{gs} \mathbf{N} dS \overline{\mathbf{T}}_g = - \int_S \mathbf{N}^T b_{gs} dS \overline{\mathbf{T}}_s$$

The finite element system of equations of the Petrov-Galerkin formulation can then be expressed as

$$\begin{pmatrix} \mathbf{K}_{ii} & \mathbf{K}_{ig} & 0 \\ \mathbf{K}_{gi} & \mathbf{K}_{gg} & \mathbf{K}_{go} \\ 0 & \mathbf{K}_{og} & \mathbf{K}_{oo} \end{pmatrix} \begin{pmatrix} \overline{\mathbf{T}}_i \\ \overline{\mathbf{T}}_g \\ \overline{\mathbf{T}}_o \end{pmatrix} = \begin{pmatrix} q_{in} \\ \mathbf{F}_{gs} \\ 0 \end{pmatrix} \tag{12.78}$$

where

$$\mathbf{K}_{ii} = \int_V \left(\mathbf{P}_i^T \lambda \mathbf{B} + \mathbf{W}_i^T \rho c u \mathbf{B} \right) dV - \int_S \mathbf{W}_i^T b_{ig} \mathbf{N} dS$$

$$\mathbf{K}_{ig} = \int_S \mathbf{W}_i^T b_{ig} \mathbf{N} dS$$

$$\mathbf{K}_{oo} = \int_V \left(\mathbf{P}_o^T \lambda \mathbf{B} - \mathbf{W}_o^T \rho c u \mathbf{B} \right) dV - \int_S \mathbf{W}_o^T b_{og} \mathbf{N} dS$$

$$\mathbf{K}_{og} = \int_S \mathbf{W}_o^T b_{og} \mathbf{N} dS$$

$$\mathbf{K}_{gg} = \int_V \mathbf{B}^T \lambda_g \mathbf{B} dV - \int_S \mathbf{N}^T b_{ig} \mathbf{N} dS - \int_S \mathbf{N}^T b_{og} \mathbf{N} dS - \int_S \mathbf{N}^T b_{gs} \mathbf{N} dS$$

$$\mathbf{K}_{gi} = \int_S \mathbf{N}^T b_{ig} \mathbf{N} dS$$

$$\mathbf{K}_{go} = \int_S \mathbf{N}^T b_{og} \mathbf{N} dS$$

$$\mathbf{F}_{gs} = - \int_S \mathbf{N}^T b_{gs} dS \overline{\mathbf{T}}_s \tag{12.79}$$

with $\mathbf{P}_i = d\mathbf{W}_i/dz$, etc.

12.3.3.4 *1D steady-state line element*

A borehole heat exchanger is typically a uniform cylinder constituting a borehole embedded in which a U-tube plastic pipe, and filled with grout. Other pipe configurations, such as the coaxial BHE, share similar property of uniform cylindrical shape. Finite element modeling of such a shape can be made elegantly using a simple 1D line element. The theoretical model for heat flow in the borehole heat exchanger, Eqs. (12.49)–(12.51), describes full coupling between the BHE components. This would circumvent the need to spatially discretize the individual pipe components, and thus, with the use of the line element, saves enormous computational demands.

In the previous sections we described the finite element equations of a single U-tube BHE using the Galerkin method, the upwind method and the Petrov-Galerkin method. Here we derive the finite element formulation for a 1D line element on the basis of the Petrov-Galerkin method. Either linear or quadratic elements can be employed for this purpose. Here we use a three-node quadratic element, which is compatible with many 3D finite elements including the 15-node prism elements and the 20-node brick elements.

BHE element of a single U-tube is designated by three degrees of freedom at each node, representing the temperatures in pipe-in, pipe-out and grout. The geometry of the element is described by a pipe-in with cross sectional area A_i and surface area P_i; a pipe-out with cross sectional area A_o and surface area P_o; and a grout with cross sectional area A_g and surface area

P_g. For a three-node quadratic element with $h = 2L$, the shape functions are

$$N_1 = \frac{1}{2L^2}(z^2 - zL)$$

$$N_2 = \frac{1}{L^2}(L^2 - z^2) \qquad (12.80)$$

$$N_3 = \frac{1}{2L^2}(z^2 + zL)$$

The derivatives of the shape functions are

$$\frac{dN_1}{dz} = \frac{1}{2L^2}(2z - L)$$

$$\frac{dN_2}{dz} = -\frac{2}{L^2}z \qquad (12.81)$$

$$\frac{dN_3}{dz} = \frac{1}{2L^2}(2z + L)$$

Petrov-Galerkin weighting functions for pipe-in, using Eq. (12.72), can be expressed as

$$W_1 = \frac{1}{2L^2}(z^2 - zL) + \frac{\alpha}{2L}(2z - L)$$

$$W_2 = \frac{1}{L^2}(L^2 - z^2) - \frac{2\alpha}{L}z \qquad (12.82)$$

$$W_3 = \frac{1}{2L^2}(z^2 + zL) + \frac{\alpha}{2L}(2z + L)$$

Their derivatives are

$$\frac{dW_1}{dz} = \frac{1}{2L^2}(2z - L) + \frac{\alpha}{L}$$

$$\frac{dW_2}{dz} = -\frac{2}{L^2}z - \frac{2\alpha}{L} \qquad (12.83)$$

$$\frac{dW_3}{dz} = \frac{1}{2L^2}(2z + L) + \frac{\alpha}{L}$$

The same can be done for pipe-out. Applying these equations into Eq. (12.79) and integrating analytically over the corresponding volume and surface areas of the pipe components, we obtain

$$\int_V \mathbf{P}_i^T \mathbf{B} dV = \frac{A_i}{3L}\begin{bmatrix} \frac{7}{2} - 3\alpha & -4 & \frac{1}{2} + 3\alpha \\ -4 + 6\alpha & 8 & -4 - 6\alpha \\ \frac{1}{2} - 3\alpha & -4 & \frac{7}{2} + 3\alpha \end{bmatrix} \qquad (12.84)$$

$$\int_V \mathbf{P}_o^T \mathbf{B} dV = \frac{A_o}{3L}\begin{bmatrix} \frac{7}{2} + 3\alpha & -4 & \frac{1}{2} - 3\alpha \\ -4 - 6\alpha & 8 & -4 + 6\alpha \\ \frac{1}{2} + 3\alpha & -4 & \frac{7}{2} - 3\alpha \end{bmatrix} \qquad (12.85)$$

$$\int_V \mathbf{W}_i^T \mathbf{B} dV = A_i \begin{bmatrix} -\dfrac{1}{2}+\dfrac{7}{6}\alpha & \dfrac{2}{3}-\dfrac{4}{3}\alpha & -\dfrac{1}{6}+\dfrac{1}{6}\alpha \\ -\dfrac{2}{3}-\dfrac{4}{3}\alpha & \dfrac{8}{3}\alpha & \dfrac{2}{3}-\dfrac{4}{3}\alpha \\ \dfrac{1}{6}+\dfrac{1}{6}\alpha & -\dfrac{2}{3}-\dfrac{4}{3}\alpha & \dfrac{1}{2}+\dfrac{7}{6}\alpha \end{bmatrix} \tag{12.86}$$

$$\int_V \mathbf{W}_o^T \mathbf{B} dV = A_o \begin{bmatrix} -\dfrac{1}{2}-\dfrac{7}{6}\alpha & \dfrac{2}{3}+\dfrac{4}{3}\alpha & -\dfrac{1}{6}-\dfrac{1}{6}\alpha \\ -\dfrac{2}{3}+\dfrac{4}{3}\alpha & -\dfrac{8}{3}\alpha & \dfrac{2}{3}+\dfrac{4}{3}\alpha \\ \dfrac{1}{6}-\dfrac{1}{6}\alpha & -\dfrac{2}{3}+\dfrac{4}{3}\alpha & \dfrac{1}{2}-\dfrac{7}{6}\alpha \end{bmatrix} \tag{12.87}$$

$$\int_S \mathbf{W}_i^T \mathbf{N} dS = P_i L \begin{bmatrix} \dfrac{4}{15}-\dfrac{1}{2}\alpha & \dfrac{2}{15}-\dfrac{2}{3}\alpha & -\dfrac{1}{15}+\dfrac{1}{6}\alpha \\ \dfrac{2}{15}+\dfrac{2}{3}\alpha & \dfrac{16}{15} & \dfrac{2}{15}-\dfrac{2}{3}\alpha \\ -\dfrac{1}{15}-\dfrac{1}{6}\alpha & \dfrac{2}{15}+\dfrac{2}{3}\alpha & \dfrac{4}{15}+\dfrac{1}{2}\alpha \end{bmatrix} \tag{12.88}$$

$$\int_S \mathbf{W}_o^T \mathbf{N} dS = P_o L \begin{bmatrix} \dfrac{4}{15}+\dfrac{1}{2}\alpha & \dfrac{2}{15}+\dfrac{2}{3}\alpha & -\dfrac{1}{15}-\dfrac{1}{6}\alpha \\ \dfrac{2}{15}-\dfrac{2}{3}\alpha & \dfrac{16}{15} & \dfrac{2}{15}+\dfrac{2}{3}\alpha \\ -\dfrac{1}{15}+\dfrac{1}{6}\alpha & \dfrac{2}{15}-\dfrac{2}{3}\alpha & \dfrac{4}{15}-\dfrac{1}{2}\alpha \end{bmatrix} \tag{12.89}$$

$$\int_S \mathbf{N}^T \mathbf{N} dS = PL \begin{bmatrix} \dfrac{4}{15} & \dfrac{2}{15} & -\dfrac{1}{15} \\ & \dfrac{16}{15} & \dfrac{2}{15} \\ \text{sym} & & \dfrac{4}{15} \end{bmatrix} \tag{12.90}$$

$$\int_V \mathbf{B}^T \mathbf{B} dV = \dfrac{A_g}{3L} \begin{bmatrix} \dfrac{7}{2} & -4 & \dfrac{1}{2} \\ & 8 & -4 \\ \text{sym} & & \dfrac{7}{2} \end{bmatrix} \tag{12.91}$$

12.3.4 *Transient formulation*

In this section we present the finite element formulation of single and double U-tube borehole heat exchangers. Recall the transient heat equations of a single U-tube BHE

$$\rho c \frac{\partial T_i}{\partial t} - \lambda \frac{\partial^2 T_i}{\partial z^2} + \rho c u \frac{\partial T_i}{\partial z} - b_{ig}(T_i - T_g) = 0 \tag{12.92}$$

$$\rho c \frac{\partial T_o}{\partial t} - \lambda \frac{\partial^2 T_o}{\partial z^2} - \rho c u \frac{\partial T_o}{\partial z} - b_{og}(T_o - T_g) = 0 \tag{12.93}$$

$$\rho c_g \frac{\partial T_g}{\partial t} - \lambda_g \frac{\partial^2 T_g}{\partial z^2} - b_{ig}(T_g - T_i) - b_{og}(T_g - T_o) = 0 \tag{12.94}$$

12.3.4.1   *Finite element discretization*

We use the weighted residual method to solve the coupled partial differential equations, Eqs. (12.92)–(12.94), subjected to the initial and boundary conditions, Eqs. (12.46)–(12.48). This gives

$$\int_t \int_\Omega WRd\Omega dt = 0 \tag{12.95}$$

where the residuals are

$$R_1 = \rho c \frac{\partial T_i}{\partial t} - \lambda \frac{\partial^2 T_i}{\partial z^2} + \rho c u \frac{\partial T_i}{\partial z} - b_{ig}(T_i - T_g)$$

$$R_2 = \rho c \frac{\partial T_o}{\partial t} - \lambda \frac{\partial^2 T_o}{\partial z^2} - \rho c u \frac{\partial T_o}{\partial z} - b_{og}(T_o - T_g) \tag{12.96}$$

$$R_3 = \rho c_g \frac{\partial T_g}{\partial t} - \lambda_g \frac{\partial^2 T_g}{\partial z^2} - b_{ig}(T_g - T_i) - b_{og}(T_g - T_o)$$

Substituting Eq. (12.96) into Eq. (12.95) gives
Pipe-in:

$$\int_t \int_V \left(W_i^T \rho c \frac{\partial T_i}{\partial t} - W_i^T \lambda \frac{\partial^2 T_i}{\partial z^2} + W_i^T \rho c u \frac{\partial T_i}{\partial z} \right) dV dt$$

$$- \int_t \int_S W_i^T b_{ig}(T_i - T_g) dS dt = 0 \tag{12.97}$$

Pipe-out:

$$\int_t \int_V \left(W_o^T \rho c \frac{\partial T_o}{\partial t} - W_o^T \lambda \frac{\partial^2 T_o}{\partial z^2} - W_o^T \rho c u \frac{\partial T_o}{\partial z} \right) dV dt$$

$$- \int_t \int_S W_o^T b_{og}(T_o - T_g) dS dt = 0 \tag{12.98}$$

Grout:

$$\int_t \int_V \left(W_g^T \rho c_g \frac{\partial T_g}{\partial t} - W_g^T \lambda_g \frac{\partial^2 T_g}{\partial z^2} \right) dV dt$$

$$- \int_t \int_S \left[b_{ig} W_g^T (T_g - T_i) + b_{og} W_g^T (T_g - T_o) \right] dS dt = 0 \tag{12.99}$$

Using integration by parts and the Green's theorem, we obtain
Pipe-in:

$$\int_t \int_V W_i^T \rho c \frac{\partial T_i}{\partial t} dV dt + \int_t \int_V \lambda \frac{dW_i^T}{dz} \frac{\partial T_i}{\partial z} dV dt$$

$$+ \int_t \int_V W_i^T \rho c u \frac{\partial T_i}{\partial z} dV dt - \int_t \int_S W_i^T b_{ig}(T_i - T_g) dS dt = 0 \tag{12.100}$$

Pipe-out:

$$\int_t \int_V W_o^T \rho c \frac{\partial T_o}{\partial t} dV dt + \int_t \int_V \lambda \frac{dW_o^T}{dz} \frac{\partial T_o}{\partial z} dV dt$$
$$- \int_t \int_V W_o^T \rho c u \frac{\partial T_o}{\partial z} dV dt - \int_t \int_S W_o^T b_{og}(T_o - T_g) dS dt = 0 \qquad (12.101)$$

Grout:

$$\int_t \int_V W_g^T \rho c \frac{\partial T_g}{\partial t} dV dt + \int_t \int_V \lambda_g \frac{dW_g^T}{dz} \frac{\partial T_g}{\partial z} dV dt - \int_t \int_S \lambda_g W_g^T \frac{\partial T_g}{\partial z} n_z dS dt$$
$$- \int_t \int_S [W_g^T b_{ig}(T_g - T_i) + W_g^T b_{og}(T_g - T_o)] dS dt = 0 \qquad (12.102)$$

There are many ways to approximate the solution of these equations. In what follows we use the Galerkin method, the upwind formulation, and the Petrov-Galerkin method to solve this problem.

12.3.4.2 *Galerkin formulation*
Using the standard Galerkin method, we have

$$\mathbf{W}(z) = \mathbf{N}(z) \qquad (12.103)$$

and

$$T(z,t) \equiv \mathbf{N}(z)\overline{\mathbf{T}}(t) \qquad (12.104)$$

Substituting Eq. (12.103) and Eq. (12.104) into Eqs. (12.100)–(12.102), the discretization of the spatial domain can be expressed as
Pipe-in:

$$\int_V \mathbf{N}^T \rho c \mathbf{N} dV \frac{d\overline{\mathbf{T}}_i}{dt} + \int_V \left(\frac{d\mathbf{N}^T}{dz} \lambda \frac{d\mathbf{N}}{dz} + \mathbf{N}^T \rho c u \frac{d\mathbf{N}}{dz} \right) dV \overline{\mathbf{T}}_i$$
$$- \int_S \mathbf{N}^T b_{ig} \mathbf{N} dS \overline{\mathbf{T}}_i + \int_S \mathbf{N}^T b_{ig} \mathbf{N} dS \overline{\mathbf{T}}_g = 0 \qquad (12.105)$$

Pipe-out:

$$\int_V \mathbf{N}^T \rho c \mathbf{N} dV \frac{d\overline{\mathbf{T}}_o}{dt} + \int_V \left(\frac{d\mathbf{N}^T}{dz} \lambda \frac{d\mathbf{N}}{dz} - \mathbf{N}^T \rho c u \frac{d\mathbf{N}}{dz} \right) dV \overline{\mathbf{T}}_o$$
$$- \int_S \mathbf{N}^T b_{og} \mathbf{N} dS \overline{\mathbf{T}}_o + \int_S \mathbf{N}^T b_{og} \mathbf{N} dS \overline{\mathbf{T}}_g = 0 \qquad (12.106)$$

Grout:

$$\int_V \mathbf{N}^T \rho c_g \mathbf{N} V \frac{d\overline{\mathbf{T}}_g}{dt} + \int_V \frac{d\mathbf{N}^T}{dz} \lambda_g \frac{d\mathbf{N}}{dz} dV \overline{\mathbf{T}}_g - \int_S \mathbf{N}^T b_{ig} \mathbf{N} dS \overline{\mathbf{T}}_g$$
$$+ \int_S \mathbf{N}^T b_{ig} \mathbf{N} dS \overline{\mathbf{T}}_i - \int_S \mathbf{N}^T b_{og} \mathbf{N} dS \overline{\mathbf{T}}_g + \int_S \mathbf{N}^T b_{og} \mathbf{N} dS \overline{\mathbf{T}}_o \qquad (12.107)$$
$$- \int_S \mathbf{N}^T b_{gs} \mathbf{N} dS \overline{\mathbf{T}}_g = - \int_S \mathbf{N}^T b_{gs} dS \overline{\mathbf{T}}_s$$

Putting these equations in a matrix format, gives a first order semi-discrete differential equation of the form

$$
\begin{pmatrix} \mathbf{M}_{ii} & 0 & 0 \\ 0 & \mathbf{M}_{gg} & 0 \\ 0 & 0 & \mathbf{M}_{oo} \end{pmatrix} \begin{pmatrix} \dot{\overline{\mathbf{T}}}_i \\ \dot{\overline{\mathbf{T}}}_g \\ \dot{\overline{\mathbf{T}}}_o \end{pmatrix} + \begin{pmatrix} \mathbf{K}_{ii} & \mathbf{K}_{ig} & 0 \\ \mathbf{K}_{gi} & \mathbf{K}_{gg} & \mathbf{K}_{go} \\ 0 & \mathbf{K}_{og} & \mathbf{K}_{oo} \end{pmatrix} \begin{pmatrix} \dot{\overline{\mathbf{T}}}_i \\ \dot{\overline{\mathbf{T}}}_g \\ \dot{\overline{\mathbf{T}}}_o \end{pmatrix} = \begin{pmatrix} q_{in} \\ F_{gs} \\ 0 \end{pmatrix} \qquad (12.108)
$$

where

$$
\mathbf{M}_{ii} = \int_V \mathbf{N}^T \rho c \mathbf{N} dV
$$

$$
\mathbf{M}_{oo} = \int_V \mathbf{N}^T \rho c \mathbf{N} dV \qquad (12.109)
$$

$$
\mathbf{M}_{gg} = \int_V \mathbf{N}^T \rho c_g \mathbf{N} dV
$$

$$
\mathbf{K}_{ii} = \int_V (\mathbf{B}^T \lambda \mathbf{B} + \mathbf{N}^T \rho c u \mathbf{B}) dV - \int_S \mathbf{N}^T b_{ig} \mathbf{N} dS
$$

$$
\mathbf{K}_{ig} = \int_S \mathbf{N}^T b_{ig} \mathbf{N} dS
$$

$$
\mathbf{K}_{oo} = \int_V (\mathbf{B}^T \lambda \mathbf{B} - \mathbf{N}^T \rho c u \mathbf{B}) dV - \int_S \mathbf{N}^T b_{og} \mathbf{N} dS
$$

$$
\mathbf{K}_{og} = \int_S \mathbf{N}^T b_{og} \mathbf{N} dS
$$

$$
\mathbf{K}_{gg} = \int_V \mathbf{B}^T \lambda_g \mathbf{B} dV - \int_S \mathbf{N}^T b_{ig} \mathbf{N} dS - \int_S \mathbf{N}^T b_{og} \mathbf{N} dS - \int_S \mathbf{N}^T b_{gs} \mathbf{N} dS
$$

$$
\mathbf{K}_{gi} = \int_S \mathbf{N}^T b_{ig} \mathbf{N} dS
$$

$$
\mathbf{K}_{go} = \int_S \mathbf{N}^T b_{og} \mathbf{N} dS
$$

$$
\mathbf{F}_{gs} = -\int_S \mathbf{N}^T b_{gs} dS \overline{\mathbf{T}}_s \qquad (12.110)
$$

with $\dot{\overline{\mathbf{T}}} = d\overline{\mathbf{T}}/dt$ and $\mathbf{B} = d\mathbf{N}/dz$.

The semi-discrete equation, Eq. (12.108), can be written in a compact form as

$$
\mathbf{M}\dot{\overline{\mathbf{T}}} + \mathbf{K}\overline{\mathbf{T}} = \mathbf{F} \qquad (12.111)
$$

This equation can be solved in time using any of the time integration schemes presented in Chapter 11. Here we utilize the finite element time integration scheme. In this case, the temperature in the time domain can be approximated as

$$
\mathbf{T}(t) = \sum_j N_j(t) T_j \qquad (12.112)
$$

in which $N_j(t)$ is the temporal shape function. Using a two-node linear element in time $N(t) \in (t_n, t_{n+1}]$ gives

$$
\mathbf{T}(t) = N_n \mathbf{T}_n + N_{n+1} \mathbf{T}_{n+1} \qquad (12.113)
$$

where

$$N_n = 1 - \frac{\tau}{\Delta t}, \quad N_{n+1} = \frac{\tau}{\Delta t}$$
$$\tau = t - t_n \quad , \quad \Delta t = t_{n+1} - t_n$$

(12.114)

Following the derivation presented in Chapter 11, we obtain

$$\left(\frac{\mathbf{M}}{\Delta t} + \theta \mathbf{K}\right) \mathbf{T}_{n+1} = \left(\frac{\mathbf{M}}{\Delta t} - (1-\theta)\mathbf{K}\right) \mathbf{T}_n + (1-\theta)\mathbf{F}_n + \theta \mathbf{F}_{n+1}$$

(12.115)

where the algorithmic parameter θ is expressed as

$$\theta = \frac{1}{\Delta t} \frac{\int_{t_n}^{t_{n+1}} \mathbf{W}(t)\tau dt}{\int_{t_n}^{t_{n+1}} \mathbf{W}(t)dt}$$

(12.116)

in which $\mathbf{W}(t)$ is the temporal weighting function. Following the principle of the weighted residual method, the weighting function in Eq. (12.116) can be any arbitrary function that satisfies the *admissibility* conditions for completeness and continuity. As the integral in Eq. (12.116) does not involve derivatives of the weighting function, the continuity condition is satisfied even by using any arbitrary scalar constant. This implies that we are able to choose any algorithmic parameter, which is capable of producing a stable algorithm. Preferably, it is chosen in a form that provides the means to control the dissipation in the high frequency range. If we choose $\mathbf{W}(t) = N_n(t)$, and solve Eq. (12.116), we obtain $\theta = 1/3$, which is unstable. If however, we choose $\mathbf{W}(t) = N_{n+1}(t)$, we obtain $\theta = 2/3$, which is unconditionally stable. It is also possible to adapt the value of the algorithmic parameter for each time step in order to obtain a stable solution and in the meanwhile less dissipative. For this, see Al-Khoury *et al.* (2011).

12.3.4.3 *Upwind formulation*

To eliminate the possible occurrence of oscillation due to a high Peclet number of an element, the energy equation is modified by including a balancing diffusion term. In Chapter 11 we have seen that by utilizing the truncation error of the difference equation as a diffusion term, it is possible that we obtain a stable and smooth solution. Proceeding along this line, Heinrich (1980) has shown that, for the transient case, the balancing diffusion can be of the form

$$\beta d \frac{\partial^3 T}{\partial z^2 \partial t}$$

(12.117)

in which the coefficient d is proportional to $uh\Delta t$ (to have a dimensional consistency). Similar to Eq. (12.64), and adding (12.117) to the transient BHE heat equations, we obtain

$$\rho c \frac{\partial T_i}{\partial t} - \left(\lambda + \rho c \frac{\alpha u h}{2}\right) \frac{\partial^2 T_i}{\partial z^2} + \rho c u \frac{\partial T_i}{\partial z} + \beta d \frac{\partial^3 T_i}{\partial z^2 \partial t} - b_{ig}(T_i - T_g) = 0$$

$$\rho c \frac{\partial T_o}{\partial t} - \left(\lambda - \rho c \frac{\alpha u h}{2}\right) \frac{\partial^2 T_o}{\partial z^2} - \rho c u \frac{\partial T_o}{\partial z} + \beta d \frac{\partial^3 T_o}{\partial z^2 \partial t} - b_{og}(T_o - T_g) = 0$$

(12.118)

$$\rho c_g \frac{\partial T_g}{\partial t} - \lambda_g \frac{\partial^2 T_g}{\partial z^2} - b_{ig}(T_g - T_i) - b_{og}(T_g - T_o) = 0$$

where you notice that we do not add the balancing diffusion to the grout heat equation because heat flow in the grout is only conductive, and standard Galerkin method is sufficient for this purpose. Applying the Galerkin finite element procedure, gives

Pipe-in:

$$\int_V \mathbf{N}^T \rho c \mathbf{N} dV \frac{d\overline{\mathbf{T}}_i}{dt}$$

$$+ \int_V \left(\frac{\partial \mathbf{N}^T}{\partial z} \lambda \frac{\partial \mathbf{N}}{\partial z} + \frac{\partial \mathbf{N}^T}{\partial z} \rho c \frac{\alpha u h}{2} \frac{\partial \mathbf{N}}{\partial z} + \mathbf{N}^T \rho c u \frac{\partial \mathbf{N}}{\partial z} + \beta d \frac{\partial}{\partial t} \left(\frac{\partial \mathbf{N}}{\partial z} \right) \right) dV \overline{\mathbf{T}}_i \qquad (12.119)$$

$$- \int_S \mathbf{N}^T b_{ig} \mathbf{N} dS \overline{\mathbf{T}}_i + \int_S \mathbf{N}^T b_{ig} \mathbf{N} dS \overline{\mathbf{T}}_g = 0$$

Pipe-out:

$$\int_V \mathbf{N}^T \rho c \mathbf{N} dV \frac{d\overline{\mathbf{T}}_o}{dt}$$

$$+ \int_V \left(\frac{\partial \mathbf{N}^T}{\partial z} \lambda \frac{\partial \mathbf{N}}{\partial z} - \frac{\partial \mathbf{N}^T}{\partial z} \rho c \frac{\alpha u h}{2} \frac{\partial \mathbf{N}}{\partial z} - \mathbf{N}^T \rho c u \frac{\partial \mathbf{N}}{\partial z} + \beta d \frac{\partial}{\partial t} \left(\frac{\partial \mathbf{N}}{\partial z} \right) \right) dV \overline{\mathbf{T}}_o \qquad (12.120)$$

$$- \int_S \mathbf{N}^T b_{og} \mathbf{N} dS \overline{\mathbf{T}}_o + \int_S \mathbf{N}^T b_{og} \mathbf{N} dS \overline{\mathbf{T}}_g = 0$$

Note that if the shape function, $\mathbf{N}$, is a function of space only, the last term of the second integral in both equations vanishes. Coupling the above two equations to the grout heat equation, the finite element formulation can then be written as

$$\begin{pmatrix} \mathbf{M}_{ii} & 0 & 0 \\ 0 & \mathbf{M}_{gg} & 0 \\ 0 & 0 & \mathbf{M}_{oo} \end{pmatrix} \begin{pmatrix} \dot{\overline{\mathbf{T}}}_i \\ \dot{\overline{\mathbf{T}}}_g \\ \dot{\overline{\mathbf{T}}}_o \end{pmatrix} + \begin{pmatrix} \mathbf{K}_{ii} & \mathbf{K}_{ig} & 0 \\ \mathbf{K}_{gi} & \mathbf{K}_{gg} & \mathbf{K}_{go} \\ 0 & \mathbf{K}_{og} & \mathbf{K}_{oo} \end{pmatrix} \begin{pmatrix} \overline{\mathbf{T}}_i \\ \overline{\mathbf{T}}_g \\ \overline{\mathbf{T}}_o \end{pmatrix} = \begin{pmatrix} q_{\text{in}} \\ \mathbf{F}_{gs} \\ 0 \end{pmatrix} \qquad (12.121)$$

where

$$\mathbf{M}_{ii} = \int_V \mathbf{N}^T \rho c \mathbf{N} dV$$

$$\mathbf{M}_{oo} = \int_V \mathbf{N}^T \rho c \mathbf{N} dV$$

$$\mathbf{M}_{gg} = \int_V \mathbf{N}^T \rho c_g \mathbf{N} dV$$

$$\mathbf{K}_{ii} = \int_V \left(\mathbf{B}^T \lambda \mathbf{B} + \mathbf{B}^T \rho c \frac{\alpha u h}{2} \mathbf{B} + \mathbf{N}^T \rho c u \mathbf{B} + \beta d \frac{d\mathbf{B}}{dt} \right) dV$$

$$\qquad - \int_S \mathbf{N}^T b_{ig} \mathbf{N} dS$$

$$\mathbf{K}_{ig} = \int_S \mathbf{N}^T b_{ig} \mathbf{N} dS$$

$$\mathbf{K}_{oo} = \int_V \left(\mathbf{B}^T \lambda \mathbf{B} - \mathbf{B}^T \rho c \frac{\alpha u h}{2} \mathbf{B} - \mathbf{N}^T \rho c u \mathbf{B} + \beta d \frac{d\mathbf{B}}{dt} \right) dV$$

$$\qquad - \int_S \mathbf{N}^T b_{og} \mathbf{N} dS$$

$$\mathbf{K}_{og} = \int_S \mathbf{N}^T b_{og} \mathbf{N} dS \qquad (12.122)$$

$$\mathbf{K}_{gg} = \int_V \mathbf{B}^T \lambda_g \mathbf{B} dV - \int_S \mathbf{N}^T b_{ig} \mathbf{N} dS - \int_S \mathbf{N}^T b_{og} \mathbf{N} dS$$

$$- \int_S \mathbf{N}^T b_{gs} \mathbf{N} dS$$

$$\mathbf{K}_{gi} = \int_S \mathbf{N}^T b_{ig} \mathbf{N} dS$$

$$\mathbf{K}_{go} = \int_S \mathbf{N}^T b_{og} \mathbf{N} dS$$

$$\mathbf{F}_{gs} = -\int_S \mathbf{N}^T b_{gs} dS \overline{\mathbf{T}}_s \qquad (12.123)$$

12.3.4.4 *Petrov-Galerkin formulation*

As discussed in Chapter 11, one important approach to obtain an improved algorithm for solving highly transient convective problems is the use of the Petrov-Galerkin finite element method. A space-time Petrov-Galerkin element is one of the effective products of this method. The weighting function of this element is formulated by perturbing the involved shape functions with their derivatives. We begin with writing Eqs. (12.119) and (12.120) in the form

Pipe-in:

$$\int_V \mathbf{N}^T \rho c \frac{\partial \mathbf{T}_i}{\partial t} dV$$

$$+ \int_V \frac{\partial \mathbf{N}^T}{\partial z} \lambda \frac{\partial \mathbf{T}_i}{\partial z} dV + \int_V \rho c u \left(\frac{\alpha h}{2} \frac{\partial \mathbf{N}^T}{\partial z} + \mathbf{N}^T + \beta \frac{d}{\rho c u} \frac{\partial}{\partial t} \left(\frac{\partial}{\partial z} \left(\frac{\partial \mathbf{N}}{\partial z} \right) \right) \right) \frac{\partial \mathbf{T}_i}{\partial z}$$

$$- \int_S \mathbf{N}^T b_{ig} \mathbf{T}_i dS + \int_S \mathbf{N}^T b_{ig} \mathbf{T}_g dS = 0 \qquad (12.124)$$

Pipe-out:

$$\int_V \mathbf{N}^T \rho c \frac{\partial \mathbf{T}_o}{\partial t} dV$$

$$+ \int_V \frac{\partial \mathbf{N}^T}{\partial z} \lambda \frac{\partial \mathbf{T}_o}{\partial z} dV - \int_V \rho c u \left(\frac{\alpha h}{2} \frac{\partial \mathbf{N}^T}{\partial z} - \mathbf{N}^T - \beta \frac{d}{\rho c u} \frac{\partial}{\partial t} \left(\frac{\partial}{\partial z} \left(\frac{\partial \mathbf{N}}{\partial z} \right) \right) \right) \frac{\partial \mathbf{T}_o}{\partial z}$$

$$- \int_S \mathbf{N}^T b_{og} \mathbf{T}_o dS + \int_S \mathbf{N}^T b_{og} \mathbf{T}_g dS = 0 \qquad (12.125)$$

where the shape function is a function of time and space, $\mathbf{N}(z,t)$. According to Heinrich (1980), the third term between brackets, $d/\rho c u$ in both equations, can be replaced by $h \Delta t/4$, and the Petrov-Galerkin weighting function can be chosen to be the terms between brackets, i.e.

Pipe-in:

$$\mathbf{W}_i(z,t) = \mathbf{N}(z,t) + \frac{\alpha h}{2} \frac{d\mathbf{N}(z,t)}{dz} + \frac{\beta h \Delta t}{4} \frac{d^2 \mathbf{N}(z,t)}{dz dt} \qquad (12.126)$$

Pipe-out:

$$\mathbf{W}_o(z,t) = \mathbf{N}(z,t) - \frac{\alpha h}{2} \frac{d\mathbf{N}(z,t)}{dz} + \frac{\beta h \Delta t}{4} \frac{d^2 \mathbf{N}(z,t)}{dz dt} \qquad (12.127)$$

For the grout, the standard Galerkin finite element weighting function can be utilized, as

$$\mathbf{W}_g(z,t) = \mathbf{N}(z,t) \qquad (12.128)$$

Finite element discretization leads then to
Pipe-in:

$$\int_t \int_V \mathbf{W}_i \rho c \frac{\partial \mathbf{T}_i}{\partial t} dVdt + \int_t \int_V \left(\frac{\partial \mathbf{W}_i^T}{\partial z} \lambda \frac{\partial \mathbf{N}}{\partial z} + \mathbf{W}_i^T \rho cu \frac{\partial \mathbf{N}}{\partial z} \right) dVdt \overline{\mathbf{T}}_i$$

$$- \int_t \int_S \mathbf{W}_i^T b_{ig} \mathbf{N} dSdt \overline{\mathbf{T}}_i + \int_t \int_S \mathbf{W}_i^T b_{ig} \mathbf{N} dSdt \overline{\mathbf{T}}_g = 0 \qquad (12.129)$$

Pipe-out:

$$\int_t \int_V \mathbf{W}_o \rho c \frac{\partial \mathbf{T}_o}{\partial t} dVdt + \int_t \int_V \left(\frac{\partial \mathbf{W}_o^T}{\partial z} \lambda \frac{\partial \mathbf{N}}{\partial z} - \mathbf{W}_o^T \rho cu \frac{\partial \mathbf{N}}{\partial z} \right) dVdt \overline{\mathbf{T}}_o$$

$$- \int_t \int_S \mathbf{W}_o^T b_{og} \mathbf{N} dSdt \overline{\mathbf{T}}_o + \int_t \int_S \mathbf{W}_o^T b_{og} \mathbf{N} dSdt \overline{\mathbf{T}}_g = 0 \qquad (12.130)$$

Grout:

$$\int_t \int_V \mathbf{N} \rho c_g \frac{\partial \mathbf{T}_g}{\partial t} dVdt + \int_t \int_V \frac{\partial \mathbf{N}^T}{\partial z} \lambda_g \frac{\partial \mathbf{N}}{\partial z} dVdt \overline{\mathbf{T}}_g$$

$$- \int_t \int_S (\mathbf{N}^T b_{ig} \mathbf{N} + \mathbf{N}^T b_{og} \mathbf{N}) dSdt \overline{\mathbf{T}}_g + \int_t \int_S \mathbf{N}^T b_{ig} \mathbf{N} dSdt \overline{\mathbf{T}}_i \qquad (12.131)$$

$$+ \int_t \int_S \mathbf{N}^T b_{og} \mathbf{N} dSdt \overline{\mathbf{T}}_o - \int_t \int_S \mathbf{N}^T b_{gs} \mathbf{N} dS \overline{\mathbf{T}}_g = - \int_t \int_S \mathbf{N}^T b_{gs} \overline{\mathbf{T}}_s dS$$

Note that the temperature derivative with space is carried out such that

$$\frac{\partial \mathbf{T}(z,t)}{\partial z} = \frac{\partial}{\partial z} [\mathbf{N}(z,t) \overline{\mathbf{T}}(t)]$$

$$= \frac{\partial \mathbf{N}(z,t)}{\partial z} \overline{\mathbf{T}}(t) \qquad (12.132)$$

The temperature derivative with time is

$$\frac{\partial \mathbf{T}(z,t)}{\partial t} = \frac{\partial}{\partial t} [\mathbf{N}(z,t) \overline{\mathbf{T}}(t)]$$

$$= \frac{\partial \mathbf{N}(z,t)}{\partial t} \overline{\mathbf{T}}(t) + \mathbf{N}(z,t) \frac{\partial \overline{\mathbf{T}}(t)}{\partial t} \qquad (12.133)$$

Substituting Eq. (12.133) into the corresponding terms in Eqs. (12.129)–(12.131), yields
Pipe-in:

$$\int_t \int_V \mathbf{W}_i \rho c \mathbf{N} \frac{\partial \overline{\mathbf{T}}_i}{\partial t} dVdt$$

$$+ \int_t \int_V \left(\mathbf{P}_i^T \lambda \mathbf{B} + \mathbf{W}_i^T \rho cu \mathbf{B} + \mathbf{W}_i^T \rho c \frac{\partial \mathbf{N}}{\partial t} \right) dVdt \overline{\mathbf{T}}_i$$

$$- \int_t \int_S \mathbf{W}_i^T b_{ig} \mathbf{N} dSdt \overline{\mathbf{T}}_i + \int_t \int_S \mathbf{W}_i^T b_{ig} \mathbf{N} dSdt \overline{\mathbf{T}}_g = 0 \qquad (12.134)$$

Pipe-out:

$$\int_t \int_V \mathbf{W}_o^T \rho c \mathbf{N} \frac{\partial \overline{\mathbf{T}}_o}{\partial t} dV dt$$

$$+ \int_t \int_V \left(\mathbf{P}_o^T \lambda \mathbf{B} - \mathbf{W}_o^T \rho c u \mathbf{B} + \mathbf{W}_o^T \rho c \frac{\partial \mathbf{N}}{\partial t} \right) dV dt \overline{\mathbf{T}}_o$$

$$- \int_t \int_S \mathbf{W}_o^T b_{og} \mathbf{N} dS dt \overline{\mathbf{T}}_o + \int_t \int_S \mathbf{W}_o^T b_{og} \mathbf{N} dS dt \overline{\mathbf{T}}_g = 0 \qquad (12.135)$$

Grout:

$$\int_t \int_V \mathbf{N} \rho c_g \mathbf{N} \frac{\partial \overline{\mathbf{T}}_g}{\partial t} dV dt + \int_t \int_V \left(\mathbf{B}^T \lambda_g \mathbf{B} + \mathbf{N}^T \rho c_g \frac{\partial \mathbf{N}}{\partial t} \right) dV dt \overline{\mathbf{T}}_g$$

$$- \int_t \int_S (\mathbf{N}^T b_{ig} \mathbf{N} + \mathbf{N}^T b_{og} \mathbf{N}) dS dt \overline{\mathbf{T}}_g$$

$$+ \int_t \int_S \mathbf{N}^T b_{ig} \mathbf{N} dS dt \overline{\mathbf{T}}_i + \int_t \int_S \mathbf{N}^T b_{og} \mathbf{N} dS dt \overline{\mathbf{T}}_o$$

$$- \int_t \int_S \mathbf{N}^T b_{gs} \mathbf{N} dS \overline{\mathbf{T}}_g = - \int_t \int_S \mathbf{N}^T b_{gs} \mathbf{T}_s dS \qquad (12.136)$$

in which $\mathbf{P}_i = \partial \mathbf{W}_i / \partial z$, etc.

Collecting terms, the finite element equation can then be expressed as

$$\begin{bmatrix} \mathbf{M}_{ii} & 0 & 0 \\ 0 & \mathbf{M}_{gg} & 0 \\ 0 & 0 & \mathbf{M}_{oo} \end{bmatrix} \begin{bmatrix} \dot{\overline{\mathbf{T}}}_i \\ \dot{\overline{\mathbf{T}}}_g \\ \dot{\overline{\mathbf{T}}}_o \end{bmatrix} + \begin{bmatrix} \mathbf{K}_{ii} & \mathbf{K}_{ig} & 0 \\ \mathbf{K}_{gi} & \mathbf{K}_{gg} & \mathbf{K}_{go} \\ 0 & \mathbf{K}_{og} & \mathbf{K}_{oo} \end{bmatrix} \begin{bmatrix} \overline{\mathbf{T}}_i \\ \overline{\mathbf{T}}_g \\ \overline{\mathbf{T}}_o \end{bmatrix} = \begin{bmatrix} q_{in} \\ F_{gs} \\ 0 \end{bmatrix} \qquad (12.137)$$

in which

$$\mathbf{M}_{ii} = \int_t \int_V \mathbf{W}_i^T \rho c \mathbf{N} dV dt$$

$$\mathbf{M}_{gg} = \int_t \int_V \mathbf{N}^T \rho c_g \mathbf{N} dV dt$$

$$\mathbf{M}_{oo} = \int_t \int_V \mathbf{W}_o^T \rho c \mathbf{N} dV dt$$

$$\mathbf{K}_{ii} = \int_t \int_V \left(\mathbf{P}_i^T \lambda \mathbf{B} + \mathbf{W}_i^T \rho c u \mathbf{B} + \mathbf{W}_i^T \rho c \frac{\partial \mathbf{N}}{\partial t} \right) dV dt$$

$$- \int_t \int_S \mathbf{W}_i^T b_{ig} \mathbf{N} dS dt$$

$$\mathbf{K}_{oo} = \int_t \int_V \left(\mathbf{P}_o^T \lambda \mathbf{B} - \mathbf{W}_o^T \rho c u \mathbf{B} + \mathbf{W}_o^T \rho c \frac{\partial \mathbf{N}}{\partial t} \right) dV dt$$

$$- \int_t \int_S \mathbf{W}_o^T b_{og} \mathbf{N} dS dt$$

$$\mathbf{K}_{gg} = \int_t \int_V \left(\mathbf{B}^T \lambda_g \mathbf{B} + \mathbf{N}^T \rho c_g \frac{\partial \mathbf{N}}{\partial t} \right) dV dt$$

$$- \int_t \int_S (\mathbf{N}^T b_{ig} \mathbf{N} + \mathbf{N}^T b_{og} \mathbf{N} + \mathbf{N}^T b_{sg} \mathbf{N}) dS dt \qquad (12.138)$$

$$\mathbf{K}_{ig} = \int_t \int_S \mathbf{W}_i^T b_{ig} \mathbf{N} dS dt$$

$$\mathbf{K}_{gi} = \int_t \int_S \mathbf{N}^T b_{ig} \mathbf{N} dS dt$$

$$\mathbf{K}_{go} = \int_t \int_S \mathbf{N}^T b_{og} \mathbf{N} dS dt$$

$$\mathbf{K}_{og} = \int_t \int_S \mathbf{W}_o^T b_{og} \mathbf{N} dS dt$$

$$\mathbf{F}_{gs} = -\int_t \int_S \mathbf{N}^T b_{gs} \overline{\mathbf{T}}_s dS dt \tag{12.139}$$

Using the theta-method, the semi-discrete equation, Eq. (12.137), can be written as

$$(\mathbf{M} + \theta \Delta t \mathbf{K})\overline{\mathbf{T}}_{n+1} = [\mathbf{M} - (1 - \theta)\Delta t \mathbf{K}]\overline{\mathbf{T}}_n + \Delta t[\theta \mathbf{F}_{n+1} + (1 - \theta)\mathbf{F}_n] \tag{12.140}$$

where $0 \le \theta \le 1$, and the subscript n represents a calculation time step.

Using the time-space Petrov-Galerkin method, it is possible to perform only implicit algorithms. This means that we should utilize $1/2 \le \theta \le 1$. This might appear as a disadvantage of this method. However, for heat flow in a shallow geothermal system, which requires analyses for short and long terms, using unconditionally stable algorithms might be the only way to conduct calculations that are computationally affordable.

Using a fully implicit scheme, i.e. $\theta = 1$, the final finite element equation might be expressed as

$$\begin{bmatrix} \frac{1}{\Delta t}\mathbf{M}_{ii} + \mathbf{K}_{ii} & \mathbf{K}_{ig} & 0 \\ \mathbf{K}_{gi} & \frac{1}{\Delta t}\mathbf{M}_{gg} + \mathbf{K}_{gg} & \mathbf{K}_{go} \\ 0 & \mathbf{K}_{og} & \frac{1}{\Delta t}\mathbf{M}_{oo} + \mathbf{K}_{oo} \end{bmatrix} \begin{bmatrix} \overline{\mathbf{T}}_i^{n+1} \\ \overline{\mathbf{T}}_g^{n+1} \\ \overline{\mathbf{T}}_o^{n+1} \end{bmatrix}$$

$$= \frac{1}{\Delta t} \begin{bmatrix} \mathbf{M}_{ii} & 0 & 0 \\ 0 & \mathbf{M}_{gg} & 0 \\ 0 & 0 & \mathbf{M}_{oo} \end{bmatrix} \begin{bmatrix} \overline{\mathbf{T}}_i^n \\ \overline{\mathbf{T}}_g^n \\ \overline{\mathbf{T}}_o^n \end{bmatrix} + \begin{bmatrix} q_{in}^{n+1} \\ \mathbf{F}_g^{n+1} + \mathbf{F}_{gs}^{n+1} \\ 0 \end{bmatrix} \tag{12.141}$$

12.3.4.5 *Double U-tube borehole heat exchanger*

The double U-tube borehole heat exchanger consists of five pipe components: two pipes-in, $i1$ and $i2$; two pipes-out, $o1$ and $o2$; and grout, g. The control volume of such a system is shown in Figure 12.3.

The coupled heat equations for this type of borehole heat exchangers can be expressed as
Pipes-in:

$$\rho c \frac{\partial T_{i1}}{\partial t} - \lambda \frac{\partial^2 T_{i1}}{\partial z^2} + \rho c u \frac{\partial T_{i1}}{\partial z} = b_{ig1}(T_{i1} - T_g)$$

$$\rho c \frac{\partial T_{i2}}{\partial t} - \lambda \frac{\partial^2 T_{i2}}{\partial z^2} + \rho c u \frac{\partial T_{i2}}{\partial z} = b_{ig2}(T_{i2} - T_g) \tag{12.142}$$

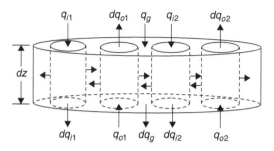

Figure 12.3   Control volume of a double U-tube BHE.

Pipes-out:

$$\rho c \frac{\partial T_{o1}}{\partial t} - \lambda \frac{\partial^2 T_{o1}}{\partial z^2} - \rho c u \frac{\partial T_{o1}}{\partial z} = b_{og1}(T_{o1} - T_g)$$

$$\rho c \frac{\partial T_{o2}}{\partial t} - \lambda \frac{\partial^2 T_{o2}}{\partial z^2} - \rho c u \frac{\partial T_{o2}}{\partial z} = b_{og2}(T_{o2} - T_g)$$

(12.143)

Grout:

$$\rho c_g \frac{\partial T_g}{\partial t} - \lambda_g \frac{\partial^2 T_g}{\partial z^2} = b_{ig1}(T_g - T_{i1}) + b_{ig2}(T_g - T_{i2})$$

$$+ b_{og1}(T_g - T_{o1}) + b_{og2}(T_g - T_{o2})$$

(12.144)

in which b_{ig1} is the reciprocal of the thermal resistance between pipe-in(1) and grout, etc. and b_{og1} is the reciprocal of the thermal resistance between pipe-out(1) and grout, etc. Applying the space-time finite element discretization procedure outlined above, we obtain

$$
\begin{bmatrix}
\mathbf{K}_{11} & 0 & \mathbf{K}_{ig1} & 0 & 0 \\
0 & \mathbf{K}_{22} & \mathbf{K}_{ig2} & 0 & 0 \\
\mathbf{K}_{gi1} & \mathbf{K}_{gi2} & \mathbf{K}_{33} & \mathbf{K}_{go1} & \mathbf{K}_{go2} \\
0 & 0 & \mathbf{K}_{og1} & \mathbf{K}_{44} & 0 \\
0 & 0 & \mathbf{K}_{og2} & 0 & \mathbf{K}_{55}
\end{bmatrix}
\begin{bmatrix}
\mathbf{T}_{i1}^{n+1} \\
\mathbf{T}_{i2}^{n+1} \\
\mathbf{T}_{g}^{n+1} \\
\mathbf{T}_{o1}^{n+1} \\
\mathbf{T}_{o2}^{n+1}
\end{bmatrix}
$$

(12.145)

$$
= \frac{1}{\Delta t}
\begin{bmatrix}
\mathbf{M}_{ii1} & 0 & 0 & 0 & 0 \\
0 & \mathbf{M}_{ii2} & 0 & 0 & 0 \\
0 & 0 & \mathbf{M}_{gg} & 0 & 0 \\
0 & 0 & 0 & \mathbf{M}_{oo1} & 0 \\
0 & 0 & 0 & 0 & \mathbf{M}_{oo2}
\end{bmatrix}
\begin{bmatrix}
\mathbf{T}_{i1}^{n} \\
\mathbf{T}_{i2}^{n} \\
\mathbf{T}_{g}^{n} \\
\mathbf{T}_{o1}^{n} \\
\mathbf{T}_{o2}^{n}
\end{bmatrix}
+
\begin{bmatrix}
\mathbf{F}_{i1}^{n+1} \\
\mathbf{F}_{i2}^{n+1} \\
\mathbf{F}_{g}^{n+1} + \mathbf{F}_{gs}^{n+1} \\
\mathbf{F}_{o1}^{n+1} \\
\mathbf{F}_{o2}^{n+1}
\end{bmatrix}
$$

in which

$$\mathbf{K}_{11} = \frac{1}{\Delta t}\mathbf{M}_{ii1} + \mathbf{K}_{ii1}$$

$$\mathbf{K}_{22} = \frac{1}{\Delta t}\mathbf{M}_{ii2} + \mathbf{K}_{ii2}$$

$$\mathbf{K}_{33} = \frac{1}{\Delta t}\mathbf{M}_{gg} + \mathbf{K}_{gg} \qquad (12.146)$$

$$\mathbf{K}_{44} = \frac{1}{\Delta t}\mathbf{M}_{oo1} + \mathbf{K}_{oo1}$$

$$\mathbf{K}_{55} = \frac{1}{\Delta t}\mathbf{M}_{oo2} + \mathbf{K}_{oo2}$$

$$\mathbf{M}_{ii\pi} = \int_t \int_V \mathbf{W}_i^T \rho c \mathbf{N} dV dt$$

$$\mathbf{M}_{oo\pi} = \int_t \int_V \mathbf{W}_o^T \rho c \mathbf{N} dV dt \qquad (12.147)$$

$$\mathbf{M}_{gg} = \int_t \int_V \mathbf{N}^T \rho c_g \mathbf{N} dV dt$$

$$\mathbf{K}_{ii\pi} = \int_t \int_V \left(\lambda \mathbf{P}_i^T \mathbf{B} + \rho c u \mathbf{W}_i^T \mathbf{B} + \rho c \mathbf{W}_i^T \frac{\partial \mathbf{N}}{\partial t} \right) dV dt$$
$$- \int_t \int_S b_{ig} \mathbf{W}_i^T \mathbf{N} dS dt$$

$$\mathbf{K}_{oo\pi} = \int_t \int_V \left(\lambda \mathbf{P}_o^T \mathbf{B} - \rho c u \mathbf{W}_o^T \mathbf{B} + \rho c \mathbf{W}_o^T \frac{\partial \mathbf{N}}{\partial t} \right) dV dt$$
$$- \int_t \int_S b_{og} \mathbf{W}_o^T \mathbf{N} dS dt \qquad (12.148)$$

$$\mathbf{K}_{gg} = \int_t \int_V \left(\lambda_g \mathbf{B}^T \mathbf{B} + \rho c_g \mathbf{N}^T \frac{\partial \mathbf{N}}{\partial t} \right) dV dt$$
$$- \int_t \int_S \left(2b_{ig} \mathbf{N}^T \mathbf{N} + 2b_{og} \mathbf{N}^T \mathbf{N} + b_{sg} \mathbf{N}^T \mathbf{N} \right) dS dt$$

$$\mathbf{K}_{ig\pi} = \int_t \int_S b_{ig\pi} \mathbf{W}_i^T \mathbf{N} dS dt$$

$$\mathbf{K}_{gi\pi} = \int_t \int_S b_{ig\pi} \mathbf{N}^T \mathbf{N} dS dt$$

$$\mathbf{K}_{og\pi} = \int_t \int_S b_{og\pi} \mathbf{W}_o^T \mathbf{N} dS dt \qquad (12.149)$$

$$\mathbf{K}_{go\pi} = \int_t \int_S b_{og\pi} \mathbf{N}^T \mathbf{N} dS dt$$

$$\mathbf{F}_{gs} = - \int_t \int_S b_{sg} \overline{\mathbf{T}}_s \mathbf{N}^T dS dt$$

where π represents pipe 1 or 2, $\mathbf{P}_i = \partial \mathbf{W}_i / \partial z$, $\mathbf{B}_i = \partial \mathbf{N}_i / \partial z$ etc.

The above formulation is general and describes different inner pipe geometries and materials. However, in practice, the double U-tube is made of two single U-tubes made of the same material.

Applying this condition to the above equations results to

$$\mathbf{K}_{ii1} = \mathbf{K}_{ii2} = \mathbf{K}_{ii}$$
$$\mathbf{K}_{ig1} = \mathbf{K}_{ig2} = \mathbf{K}_{ig} \qquad (12.150)$$
$$\mathbf{K}_{gi1} = \mathbf{K}_{gi2} = \mathbf{K}_{gi}$$

and hence $\overline{T}_{i1} = \overline{T}_{i2} = \overline{T}_i$. The same is valid for pipe-out. Applying this condition to the grout, the third row in the left hand side of Eq. (12.145), yields

$$2\mathbf{K}_{gi}\overline{\mathbf{T}}_i + \mathbf{K}_{gg}\overline{\mathbf{T}}_g + 2\mathbf{K}_{go}\overline{\mathbf{T}}_o = \mathbf{F}_g + \mathbf{F}_{gs} \qquad (12.151)$$

Introducing this equation to pipe-in and pipe-out heat equations, it can readily be shown that the number of degrees-of-freedom can be reduced from 5 to 3 and the finite element equation becomes

$$
\begin{bmatrix}
\dfrac{1}{\Delta t}\mathbf{M}_{ii} + \mathbf{K}_{ii} & \mathbf{K}_{ig} & 0 \\[2ex]
2\mathbf{K}_{gi} & \dfrac{1}{\Delta t}\mathbf{M}_{gg} + \mathbf{K}_{gg} & 2\mathbf{K}_{go} \\[2ex]
0 & \mathbf{K}_{og} & \dfrac{1}{\Delta t}\mathbf{M}_{oo} + \mathbf{K}_{oo}
\end{bmatrix}
\begin{bmatrix}
\overline{\mathbf{T}}_i^{n+1} \\[2ex]
\overline{\mathbf{T}}_g^{n+1} \\[2ex]
\overline{\mathbf{T}}_o^{n+1}
\end{bmatrix}
$$

$$
= \dfrac{1}{\Delta t}
\begin{bmatrix}
\mathbf{M}_{ii} & 0 & 0 \\
0 & \mathbf{M}_{gg} & 0 \\
0 & 0 & \mathbf{M}_{oo}
\end{bmatrix}
\begin{bmatrix}
\overline{\mathbf{T}}_i^n \\
\overline{\mathbf{T}}_g^n \\
\overline{\mathbf{T}}_o^n
\end{bmatrix}
+
\begin{bmatrix}
\mathbf{F}_i^{n+1} \\
\mathbf{F}_g^{n+1} + \mathbf{F}_{gs}^{n+1} \\
\mathbf{F}_o^{n+1}
\end{bmatrix}
\qquad (12.152)
$$

in which

$$\mathbf{K}_{gg} = \int_t \int_V \left(\lambda_g \mathbf{B}^T \mathbf{B} + \rho c_g \mathbf{N}^T \frac{\partial \mathbf{N}}{\partial t} \right) dV dt$$

$$- \int_t \int_S (2b_{ig} \mathbf{N}^T \mathbf{N} + 2b_{og} \mathbf{N}^T \mathbf{N} + b_{sg} \mathbf{N}^T \mathbf{N}) dS dt \qquad (12.153)$$

12.3.4.6   *1D Transient line element*

As for the steady-state, the borehole heat exchanger is described using a line element. We use a line element which is quadratic in time and linear in space. The pipe element has a length of h with pipe-in cross sectional area A_i and surface area P_i, pipe-out cross sectional area A_o and surface area P_o, and grout cross sectional area A_g and surface area P_g. The space-time shape functions for $0 \le z \le h$, and $0 \le t \le \Delta t$ are (Heinrich and Pepper 1999)

$$N_1(z,t) = 4 \left(1 - \frac{z}{h} \right) \frac{t}{\Delta t} \left(1 - \frac{t}{\Delta t} \right)$$

$$N_2(z,t) = 4 \frac{z}{h} \frac{t}{\Delta t} \left(1 - \frac{t}{\Delta t} \right) \qquad (12.154)$$

$$N_3(z,t) = N_2(z,t)$$

$$N_4(z,t) = N_1(z,t)$$

Note that only two shape functions are necessary to describe the element. The corresponding Petrov-Galerkin weighting functions, using Eqs. (12.126)–(12.128), are

Pipe-in:

$$W_{i1} = 4\left(1 - \frac{z}{h}\right)\frac{t}{\Delta t}\left(1 - \frac{t}{\Delta t}\right) - 2\alpha\frac{t}{\Delta t}\left(1 - \frac{t}{\Delta t}\right) + \frac{\beta}{\Delta t}(2t - \Delta t)$$

$$W_{i2} = 4\frac{z}{h}\frac{t}{\Delta t}\left(1 - \frac{t}{\Delta t}\right) + 2\alpha\frac{t}{\Delta t}\left(1 - \frac{t}{\Delta t}\right) - \frac{\beta}{\Delta t}(2t - \Delta t) \tag{12.155}$$

Pipe-out:

$$W_{o1} = 4\left(1 - \frac{z}{h}\right)\frac{t}{\Delta t}\left(1 - \frac{t}{\Delta t}\right) + 2\alpha\frac{t}{\Delta t}\left(1 - \frac{t}{\Delta t}\right) - \frac{\beta}{\Delta t}(2t - \Delta t)$$

$$W_{o2} = 4\frac{z}{h}\frac{t}{\Delta t}\left(1 - \frac{t}{\Delta t}\right) - 2\alpha\frac{t}{\Delta t}\left(1 - \frac{t}{\Delta t}\right) + \frac{\beta}{\Delta t}(2t - \Delta t) \tag{12.156}$$

Grout:

$$W_{g1} = N_{g1} = 4\left(1 - \frac{z}{h}\right)\frac{t}{\Delta t}\left(1 - \frac{t}{\Delta t}\right)$$

$$W_{g2} = N_{g2} = 4\frac{z}{h}\frac{t}{\Delta t}\left(1 - \frac{t}{\Delta t}\right) \tag{12.157}$$

Applying these definitions to Eq. (12.152), and integrating over the corresponding volumes, surface areas, and time analytically, we obtain

$$\int_t \int_V \mathbf{P}_i^T \mathbf{B} dV dt = \frac{8\Delta t}{15}\frac{A_i}{L}\begin{pmatrix} 1 & -1 \\ -1 & 1 \end{pmatrix} \tag{12.158}$$

$$\mathbf{P}_o^T \mathbf{B} = \mathbf{P}_i^T \mathbf{B}$$
$$\mathbf{B}^T \mathbf{B} = \mathbf{P}_i^T \mathbf{B} \tag{12.159}$$

$$\int_t \int_V \mathbf{W}_i^T \mathbf{B} dV dt = \frac{8\Delta t}{15}A_i\begin{pmatrix} -\frac{1}{2} + \frac{1}{2}\alpha & \frac{1}{2} - \frac{1}{2}\alpha \\ -\frac{1}{2} - \frac{1}{2}\alpha & \frac{1}{2} + \frac{1}{2}\alpha \end{pmatrix} \tag{12.160}$$

$$\int_t \int_V \mathbf{W}_o^T \mathbf{B} dV dt = \frac{8\Delta t}{15}A_o\begin{pmatrix} -\frac{1}{2} - \frac{1}{2}\alpha & \frac{1}{2} + \frac{1}{2}\alpha \\ -\frac{1}{2} + \frac{1}{2}\alpha & \frac{1}{2} - \frac{1}{2}\alpha \end{pmatrix} \tag{12.161}$$

$$\int_t \int_V \mathbf{W}_i^T \mathbf{N} dV dt = \frac{8\Delta t}{15}A_i h\begin{pmatrix} -\frac{1}{12}(-4 + 3\alpha) & -\frac{1}{12}(-2 + 3\alpha) \\ \frac{1}{12}(2 + 3\alpha) & \frac{1}{12}(4 + 3\alpha) \end{pmatrix} \tag{12.162}$$

$$\int_t \int_V \mathbf{W}_o^T \mathbf{N} dV dt = \frac{8\Delta t}{15}A_o h\begin{pmatrix} \frac{1}{12}(4 + 3\alpha) & \frac{1}{12}(2 + 3\alpha) \\ -\frac{1}{12}(-2 + 3\alpha) & -\frac{1}{12}(-4 + 3\alpha) \end{pmatrix} \tag{12.163}$$

$$\int_t \int_V \mathbf{W}_i^T \dot{\mathbf{N}}_i dV dt = \frac{2}{3}A_i h\begin{pmatrix} -\beta & -\beta \\ \beta & \beta \end{pmatrix} \tag{12.164}$$

$$\int_t \int_V \mathbf{W}_o^T \dot{\mathbf{N}}_o dV dt = \frac{2}{3}A_o h\begin{pmatrix} \beta & \beta \\ -\beta & -\beta \end{pmatrix} \tag{12.165}$$

$$\mathbf{N}^T \dot{\mathbf{N}}_g = 0 \tag{12.166}$$

$$\int_t \int_S \mathbf{W}_i^T \mathbf{N} dS dt = \frac{8\Delta t}{15} P_i h \begin{pmatrix} -\frac{1}{12}(-4+3\alpha) & -\frac{1}{12}(-2+3\alpha) \\ \frac{1}{12}(2+3\alpha) & \frac{1}{12}(4+3\alpha) \end{pmatrix} \tag{12.167}$$

$$\int_t \int_S \mathbf{W}_o^T \mathbf{N} dS dt = \frac{8\Delta t}{15} P_o h \begin{pmatrix} \frac{1}{12}(4+3\alpha) & \frac{1}{12}(2+3\alpha) \\ -\frac{1}{12}(-2+3\alpha) & -\frac{1}{12}(-4+3\alpha) \end{pmatrix} \tag{12.168}$$

$$\int_t \int_S \mathbf{N}^T \mathbf{N} dS dt = \frac{8\Delta t}{15} P h \begin{pmatrix} \frac{1}{3} & \frac{1}{6} \\ \frac{1}{6} & \frac{1}{3} \end{pmatrix} \tag{12.169}$$

12.4   NUMERICAL IMPLEMENTATION

Ground source heat pumps constitute a coupling between one or more borehole heat exchangers and a surrounding soil mass. In standard finite element procedure, each individual BHE component and the soil mass are described separately. The spatial coupling between them is conducted via the assembling technique of the finite element method. This technique, and due to the slenderness of the BHE and the dominance of convection, results to a large system of equations, which needs to be stored and solved numerically. To obtain a feasible computational requirements and CPU time it is important to reduce the number of elements. In our modeling of the borehole heat exchanger we explicitly included the spatial thermal interaction between the BHE components in the governing partial differential equations. We took advantage of the uniformity and the cylindrical shape of the pipe to discretize these equations using a simple line element. This departure from the standard finite element procedure circumvents the need to discretize the individual BHE components separately, and hence saving enormous amount of computational power.

Two finite elements have been developed: a BHE element and a soil element. The BHE element is a 1D line element with the number of degrees of freedom (DOF) equal to the number of its components. That is, for a single U-tube BHE consisting of pipe-in, pipe-out and grout, there are three DOF at each node. The soil element is a 3D element with one DOF at each node. In describing a geothermal system, these elements are assembled and the resulting system of equations is solved numerically. There are different ways to solve these equations. Here, we introduce two numerical schemes: sequential and static condensation.

12.4.1   *Sequential scheme*

Coupling between the borehole heat exchanger elements and the soil mass elements is made along the contact surface area between the two domains. However, the physical processes in the two domains are different and can be treated separately. This kind of coupling can be modeled elegantly using a sequential solution scheme, also known as *staggered* scheme.

In the sequential scheme, the temperature distribution in the BHE is first calculated, independently from the soil, using the initial temperature distribution of the soil. Once the solution of the BHE system is obtained, its values are substituted to the soil system of equations, and the soil system is solved independently. The new soil temperature will now be substituted to the BHE equations and the system is re-solved again. The process is repeated several times between Eq. (12.42) and for example Eq. (12.152), until convergence takes place. This scheme is simple, but requires an iterative solver. Using such a scheme many possibilities are open, including (Zienkiewicz *et al.* 2005):

- Completely different formulation, numerical procedure and methodology could be used for each subsystem.
- Independent coding of the subsystems can be performed using independent subroutines.
- For each subsystem, different solvers can be used.

The coupling of the soil-BHE system of equations can be described as

$$
\begin{bmatrix}
\mathbf{K}_{\text{pipe}} & \mathbf{K}_{\text{pipe-soil}} \\
\mathbf{K}_{\text{soil-pipe}} & \mathbf{K}_{\text{soil}}
\end{bmatrix}
\left\{
\begin{matrix}
\mathbf{T}_{\text{pipe}} \\
\mathbf{T}_{\text{soil}}
\end{matrix}
\right\}
=
\left\{
\begin{matrix}
\mathbf{q}_{\text{pipe}} \\
\mathbf{q}_{\text{soil}}
\end{matrix}
\right\}
\tag{12.170}
$$

in which the subscript "pipe" indicates the BHE. In the conventional staggered scheme, this equation is decomposed into two equations of the form:

$$
\begin{aligned}
\mathbf{K}_{\text{pipe}}\mathbf{T}_{\text{pipe}} &= \mathbf{q}_{\text{pipe}} - \mathbf{K}_{\text{pipe-soil}}\mathbf{T}_{\text{soil}} \\
\mathbf{K}_{\text{soil}}\mathbf{T}_{\text{soil}} &= \mathbf{q}_{\text{soil}} - \mathbf{K}_{\text{soil-pipe}}\mathbf{T}_{\text{pipe}}
\end{aligned}
\tag{12.171}
$$

This system requires explicit formulation of the coupling terms, $\mathbf{K}_{\text{pipe-soil}}$ and $\mathbf{K}_{\text{soil-pipe}}$. To avoid this formulation, the two subsystems are calculated independently but the calculated heat flux from one subsystem becomes part of the force vector for the other, such that

$$
\begin{aligned}
\mathbf{K}_{\text{pipe}}\mathbf{T}_{\text{pipe}} &= \mathbf{q}_{\text{pipe}} + \mathbf{q}_{\text{soil}} \\
\mathbf{K}_{\text{soil}}\mathbf{T}_{\text{soil}} &= \mathbf{q}_{\text{soil}} + \mathbf{q}_{\text{pipe}}
\end{aligned}
\tag{12.172}
$$

Physically, this implies that the two subsystems meet each other at their boundaries and each one of them acts as a heat source to the other. Numerically, this entails that two global matrix systems are assembled and solved independent on each other.

Coding the sequential algorithm requires two nesting loops: an outer loop, for solving the soil set of equations; and an inner loop, for solving the pipe set of equations. In the outer loop, temperature distribution in the soil mass due to initial conditions is first calculated. Calculated temperatures along the pipe are singled out. In the inner loop, these temperatures are included in the force vector of the pipe, term $\mathbf{F}_{gs}$ in Eq. (12.149), and comes into the second term on the right-hand side of the first equation of Eq. (12.172), $\mathbf{q}_{\text{soil}}$. Using this equation, temperatures in pipe-in, pipe-out and grout are calculated, from which the grout temperatures are singled out and moved to the outer loop. They come into the force vector of the soil, the second term on the right-hand side of the second equation of Eq. (12.172), $\mathbf{q}_{\text{pipe}}$. Then the temperature in the soil is re-calculated. This process continues until a state of equilibrium has been achieved by fulfilling a certain convergence criterion. For this, any non-linear numerical solver can be used. The simple *Picard* iterative scheme is sufficient.

12.4.2 *Static condensation scheme*

The sequential scheme, described above, is very simple to implement, but requires an iterative solver. In an alternative way, Diersch (2011a and 2011b) proposed solving Eq. (12.170), directly, using the static condensation algorithm, also known as the *substructuring* algorithm. This algorithm implies eliminating the variables of one of the subsystems, solving for the other subsystem independently, and then substituting the results into the first subsystem.

The first equation of Eq. (12.170) can be written as

$$
\mathbf{K}_{\text{pipe}}\mathbf{T}_{\text{pipe}} + \mathbf{K}_{\text{pipe - soil}}\mathbf{T}_{\text{soil}} = \mathbf{q}_{\text{pipe}}
\tag{12.173}
$$

This implies that

$$\mathbf{T}_{\text{pipe}} = \mathbf{K}_{\text{pipe}}^{-1} \left\{ \mathbf{q}_{\text{pipe}} - \mathbf{K}_{\text{pipe-soil}} \mathbf{T}_{\text{soil}} \right\} \tag{12.174}$$

Substituting Eq. (12.174) into the second equation of Eq. (12.170), gives

$$\tilde{\mathbf{K}} \mathbf{T}_{\text{soil}} = \tilde{\mathbf{F}} \tag{12.175}$$

in which

$$\tilde{\mathbf{K}} = \mathbf{K}_{\text{soil}} - \mathbf{K}_{\text{soil-pipe}} \mathbf{K}_{\text{pipe}}^{-1} \mathbf{K}_{\text{pipe-soil}}$$

$$\tilde{\mathbf{F}} = \mathbf{q}_{\text{soil}} - \mathbf{K}_{\text{soil-pipe}} \mathbf{K}_{\text{pipe}}^{-1} \mathbf{q}_{\text{pipe}} \tag{12.176}$$

This system of equations can be solved independently from the BHE subsystem. After solving for $\mathbf{T}_{\text{soil}}$, the temperature in the BHE can be determined from Eq. (12.174). This scheme is more complicated than the sequential scheme, but requires no iteration to solve.

12.5   VERIFICATIONS AND NUMERICAL EXAMPLES

Numerical and experimental verifications of the finite element model together with numerical examples representing some engineering applications can be found in Al-Khoury et al. (2005, 2006 and 2010). In these examples, finite element simulations of quite large geometries consisting of several borehole heat exchangers embedded in multilayer systems are conducted using considerably coarse meshes.

References

Abate, J. & Valko, P.P.: Multi-precision Laplace transform inversion. *International Journal for Numerical Methods in Engineering* 60 (2004), pp.979–993.

Abramowitz, M. & Stegun, I.A.: *Handbook of Mathematical Functions*. Dover Publication, New York, USA, 1972.

Aifantis, E.C.: A new interpretation of diffusion in high-diffusivity paths: a continuum approach. *Acta Metallurgica* 27 (1979), pp.683–691.

Al-Khoury, R.: Spectral framework for geothermal borehole heat exchangers. *International Journal of Numerical Methods for Heat and Fluid Flow* 20:7 (2010), pp.773–793.

Al-Khoury, R. & Bonnier, P.G.: Efficient finite element formulation for geothermal heating systems. Part II: Transient. *International Journal for Numerical Methods in Engineering* 67 (2006), pp.725–745.

Al-Khoury, R., Bonnier, P.G. & Brinkgreve, B.J.: Efficient finite element formulation for geothermal heating systems. Part I: Steady state. *International Journal for Numerical Methods in Engineering* 63 (2005), pp.988–1013.

Al-Khoury, R., Kolbel, T. & Schramedei, R.: Efficient numerical modeling of borehole heat exchangers. *Computer and Geosciences* 36:10 (2010), pp.1301–1315.

Al-Khoury, R., Weerheijm, J., Dingerdis, K. & Sluys L.J.: An adaptive time integration scheme for blast loading on a saturated soil mass. *Computers and Geotechnics* 38:4 (2011), pp.448–464.

Arfken, G.B. & Weber, H.S.: *Mathematical Methods for Physicists*. 4th edn. Academic Press, San Diego, USA, 1995.

Bandyopadhyay, G., Gosnold, W. & Mann, M.: Analytical and semi-analytical solutions for short-time transient response of ground heat exchangers. *Energy and Buildings* 40 (2008), pp.1816–1824.

Biot, M.A.: General theory of three-dimensional consolidation. *Journal of Applied Physics* 12 (1941), pp.155–164.

Bowen, R.M.: Theory of mixtures. In: A.C. Eringen (ed): *Continuum Physics, Vol. 3*. Academic Press, New York, USA 1976, pp.1–127.

Bozzoli, F., Pagliarini, G., Rainieri, S. & Schiavi, L.: Estimation of soil and grout thermal properties through a TSPEP (two-step parameter estimation procedure) applied to TRT (thermal response test) data. *Energy* 36:2 (2011), pp.839–846.

Brigham, E.O.: *The fast Fourier transform and its applications*. Prentice Hall, New Jersey, USA, 1988.

Bromberg, M., & Shirtliffe, C.J.: (1978). Influence of moisture gradients on heat transfer through porous building materials. In: R.P. Tye (ed.): *Thermal Transition Measurements of Insulation*. ASTM, Materials Park, OH, USA. ASTM STP 660, pp.211–233.

Bronson, R. & Gabriel, C.: *Differential Equations* (3rd edn.). Schaum's Outlines, McGraw-Hill, USA, 2006.

Brooks, A.N. & Hughes, T.J.R.: Streamline upwind/Petrov-Galerkin formulations for convective dominated flows with particular emphasis on the incompressible Navier-Stokes equations. *Computer Methods in Applied Mechanics and Engineering* 32 (1982), pp.199–259.

Carslaw, H.S. & Jaeger, J.C.: *Conduction of heat in solids"*. (2nd ed). Oxford University Press, London, UK, 1959.

Chaudhry, M.A. & Zubair, S.M.: Generalized incomplete Gamma functions with applications. *Journal of Computational and Applied Mathematics* 68 (1994), pp.849–866.

Christie, I., Griffiths, D.R., Mitchell, A.R. & Zienkiewicz, O.C.: Finite element methods for second order differential equation with significant first derivatives. *International Journal for Numerical Methods in Engineering* 10 (1976), pp.1389–1396.

Christie, I. & Mitchell, A.R.: Upwinding of high order Galerkin methods in conduction-convection problems. *International Journal for Numerical Methods in Engineering* 12 (1978), pp.1764–1771.

Claesson, J. & Eskilson, P.: *Conductive heat extraction by a deep borehole, analytical study*. Department of Mathematical Physics and Building Technology. University of Lund, Sweden, 1987.

Clauser, C. (ed.): *Numerical simulation of reactive flow in hot aquifers, SHEMAT and processing SHEMAT*. Springer, Berlin, Germany, 2003.

Cook, R.D., Malkus, D.S. & Plesha, M.E.: *Concepts and applications of finite element analysis* (3rd ed.). Wiely, New York, USA, 1989.

Cooley, J.W. & Tukey, J.W.: An algorithm for the machine calculation of complex Fourier series. *Mathematics of Computation* 19:90 (1965), pp.297–301.

Coussy, O., Dormieux, L. & Detournay, E.: From mixture theories to Biot's theory. *International Journal of Solids and Strucutres* 35 (1998), pp.4619–4635.

Diersch, H.-J.G., Bauer, D., Heidemann, W., Ruhaak, W. & Schatzl, P.: Finite element modeling of borehole heat exchanger systems: Part 1. Fundamentals. *Computers & Geosciences*, 37:8 (2011a), pp. 1122–1135.

Diersch, H.-J.G., Bauer, D., Heidemann, W., Ruhaak, W. & Schatzl, P.: Finite element modeling of borehole heat exchanger systems: Part 2. Numerical simulation. *Computers & Geosciences*, 37:8 (2011b), pp. 1136–1147.

Dincer, I, & Rosen, M.A.: *Thermal energy storage, systems and applications*. Wiley, West Sussex, UK, 2011.

Doyle, J.F.: A spectrally formulated finite elements for longitudinal wave propagation. *International Journal of Analytical and Experimental Modal Analysis* 3 (1988a), pp.1–5.

Doyle, J.F.: *Wave propagation in structures: Spectral analysis using fast discrete Fourier transforms* (2nd edn). Springer-Verlag, New York, USA, 1997.

Doyle, J.F.: Spectral analysis of coupled thermoelastic waves. *Journal of Thermal Stresses*, 11 (1988b), pp.175–185.

Dubner, H. & Abate, J.: Numerical inversion of Laplace transforms by relating them to the finite Fourier Cosine transform. *Journal of the ACM* 15:1 (1968), pp.115–123.

Eskilson, P.: Thermal analysis of heat extraction boreholes. Ph.D. Thesis, University of Lund, Sweden, 1987.

Eskilson, P. & Claesson, J.: Simulation model for thermally interacting heat extraction boreholes. *Numerical Heat Transfer* 13 (1988), pp.149–165.

FEFLOW, DHI-WASY: Available at www.feflow.info (accessed January 2011).

Hart, P. & Couvillion, R.: Earth coupled heat transfer. *National Water Well Association* (1986) pp.372–379.

Hassanizadeh, S.M. & Gray, W.G.: General conservation equations for multiphase systems: I, averaging procedure. *Advances in Water Resources* 2 (1979), pp.131–144.

Hassanizadeh, S.M. & Gray, W.G.: Mechanics and thermodynamics of multiphase flow in porous media including interface boundaries. *Advances in Water Resources* 13:4 (1990), pp.169–186.

Heinrich, J.C.: On quadratic elements in finite element solutions of steady-state convection diffusion equation. *International Journal for Numerical Methods in Engineering* 15 (1980), pp.1041–1052.

Heinrich, J.C. & Pepper, D.W.: *Intermediate finite element method, fluid flow and heat transfer applications*. Taylor & Francis, NY, USA, 1999.

Hughes, T.J.R.: *The finite element method, linear static and dynamic finite element analysis*. Dover Publications, Inc. NY, USA, 2000.

Hughes, T.J.R.: A simple scheme for developing upwind finite elements. *International Journal for Numerical Methods in Engineering* 12 (1978), pp.1359–1365.

Idelsohn, S.R., Heinrich, J.C. & Onate, E.: Petrov-Galerkin methods for the transient advective-diffusive equation with sharp gradients. *International Journal for Numerical Methods in Engineering* 39 (1996a), pp.1455–1473.

Idelsohn, S.R., Nigro, N., Storti, M. & Buscaglia, G.: A Petrove-Galerkin formulation for advection-diffusion problems. *Computer Methods in Applied Mechanics and Engineering* 136 (1996b), pp.27–46.

IMSL, Fortran Numerical Library, MATH/LIBRARY, 6.0: Available at www.intel.com (accessed June 2010).

Ingersoll, L.R. & Plass, H.J.: Theory of the ground pipe heat sources for the heat pump. *Heating, Piping and Air conditioning* 20 (1948), pp.119–122.

Ingersoll, L.R., Zobel O.J. & Ingersoll A.C.: *Heat conduction with engineering, geological, and other applications*. McGraw-Hill, NY, USA, 1954.

Kaviany, M.: *Principles of heat transfer in porous media* (2nd ed). Springer-Verlag, NY, USA, 1995.

Kreyszig, E.: *Advanced engineering mathematics* (7th ed). John Wiley & Sons inc. NY, USA, 1993.

Krupiczka, R.: Analysis of thermal conductivity in granular materials. *International Chemical Engineering* 7 (1967), pp.122–144.

Lamarche, L. & Beauchamp, B.: A new contribution to the finite line-source model for geothermal boreholes. *Energy and Buildings* 39 (2007a), pp.188–198.

Lamarche, L. & Beauchamp, B.: New solutions for the short-time analysis of geothermal vertical boreholes. *International Journal of Heat and Mass Transfer* 50 (2007b), pp.1408–1419.

Lamarche, L., Kajl, S., & Beauchamp, B.: A review of methods to evaluate borehole thermal resistances in geothermal heat-pump systems. *Geothermics* 39 (2010), pp.187–200.

Lee, C.K. & Lam, H.N.: Computer simulation of borehole ground heat exchangers for geothermal heat pump systems. *Renewable Energy* 33:6 (2008), pp.1286–1296.

Lee, U.: *Spectral element method in structural dynamics*. Wiley, Singapore, 2009.

Lewis, R.W. & Schrefler, B.A.: *The finite element method in the static and dynamic deformation and consolidation of porous media*. (2nd ed), John Wiley & Sons, 2000.

Lewis, R.W., Nithiarasu, P. & Seetharamu, K.N.: *Fundamentals of the finite element method for heat and fluid flow*. John Wiley & Sons, West Sussex, UK, 2004.

Mands, E. & Sanner, B.: *Shallow geothermal energy*, 2005. Available at www.ubeg.de/Downloads/Shallow GeothEngl.pdf (accessed January 2011).

MAPLE 13.0, Maplesoft, Waterloo Maple Inc. Available at http://www.maplesoft.com (accessed April 2010).

Marcotte, D., & Pasquier, P.: Fast fluid and ground temperature computation for geothermal ground-loop heat exchanger systems. *Geothermics* 37 (2008), pp.651–665.

Marcotte, D. & Pasquier, P.: The effect of borehole inclination on fluid and ground temperature. *Geothermics* 38:4 (2009), pp.392–398.

Morgensen, P.: Fluid to duct wall heat transfer in duct system heat storage. Proceedings of the *International Conference on Surface Heat Transfer in Theory and Practice*, Stockholm, 1983, pp.652–657.

Muraya, N.K.: Numerical modelling of the transient thermal interface of vertical U-tube heat exchangers. Ph.D thesis, Texas A&M University, TX, USA, 1994.

Murli, A. & Rizzardi, M.: Algorithm 682: Talbot's method of the Laplace inversion problems. *ACM Transitions on Mathematical Software* 16:2 (1990), pp.158–168.

Narayanan, G.V. & Beskos, D.E.: Use of dynamic influence coefficients in forced vibration problems with the aid of fast Fourier transform. *Computers & Structures* 9:2 (1978), pp.145–150.

Nield, D.A. & Bejan, A.: *Convection in porous media* (2nd edn.). Springer-Verlag, NY, USA, 1999.

Ozisik, M.N.: *Boundary value problems of heat conduction*. Dover Phoenix Editions, NY, USA, 1968.

Patera, A.: A spectral element method for fluid dynamics: Laminar flow in a channel expansion. *Journal of Computational Physics* 54:3 (1984), pp.468–488.

Philippe, M., Bernier, M. & Marchio, D.: Validity ranges of three analytical solutions to the transfer in the vicinity of single boreholes. *Geothermics* 38 (2009), pp.407–413.

Pitts, D. & Sisson L.: *Heat transfer* (2nd edn.). Schaum's Oulines. McGraw-Hill, NY, USA, 1998.

Quintard, M., Kaviany, M. & Whitaker, S.: Two-medium treatment of heat transfer in porous media: numerical results for effective properties. *Advances in Water Resources* 20:2–3 (1997), pp.77–94.

Rizzi, S.A. & Doyle, J.F.: A spectral element approach to wave motion in layered solids. *Journal of Vibration and Acoustics* 114:4 (1992), pp.569–577.

Roache, P.J.: *Computational Fluid Dynamics*. Albuquerque Hermosa Publishers, 1976.

Rohsenow, W.M., Hartnett, J.P. & Cho, Y.I.: *Handbook of Heat Transfer* (3rd edn.). McGraw-Hill, NY, USA, 1998.

Schiavi, L.: 3D simulation of the thermal response test in a U-tube borehole heat exchanger. Proceedings of *COMSOL conference*, Milan, Italy 2009. Avilable at: http://www.comsol.com/papers/6825/download/ Schiavi.pdf, (accessed at May 2011).

Serway, R.A.: *Physics for scientists and engineers* (4th ed.). Saunders College Publishing, Orlando, FL, USA, 1996.

Sharqawy, M.H., Mokheimer, E.M. & Badr H.M.: Effective pipe-to-borehole thermal resistance for vertical ground heat exchangers. *Geothermics* 38 (2009), pp.271–277.

Slattery, J.C.: Single-phase flow through porous media. *AIChE Journal* 15 (1969), pp.866–872.

Sliwa, T. & Gonet, A.: Theoretical model of borehole heat exchanger. *Journal of Energy Resources Technology, ASME* 27 (2005), pp.142–148.

Sneddon, I.N.: *Fourier transforms*. McGraw-Hill Book Company Inc. NY, USA 1951. (Reprinted by Dover Books on Mathematics, NY, USA, 1995.)

Spiegel, M.R.: *Laplace transforms*. Schume's Outlines Series, McGraw-Hill, NY, USA, 1965.

Stehfest, H.: Algorithm 368: Numerical inversion of Laplace transforms. *Communications of the ACM* 13 (1970), pp.47–49 and pp.624.

Stroud, K.A.: *Further engineering mathematics* (2nd edn.). Macmillan Education, Hong Kong, 1990.

Sutton, M.G., Nutter, D.W. & Couvillion, R.J.: A ground resistance for vertical bore heat exchangers with groundwater flow. *Journal of Energy Resources Technology* 125 (2003), pp.183–189.

Talbot, A.: The accurate numerical inversion of Laplace transforms. *Journal of the Institute of Mathematics and Its Applications* 23 (1979), pp.97–120.

van der Meer, F.P., Al-Khoury, R. & Sluys L.J.: Time-dependent shape functions for modeling highly transient geothermal systems. *International Journal for Numerical Methods in Engineering* 77 (2009), pp.240–260.

van Genuchten, M.T. & Alves, W.J.: Analytical solutions of the one-dimensional convective-dispersive solute transport equation. *Technical Bulletin Number 1661,* US Dept. Agriculture, (1982).

Wagner, R. & Clauser, C.: Estimating both thermal conductivity and thermal capacity from thermal response tests. *Journal of Geophysics and Engineering* 2 (2005), pp.349–356.

Westernik, J.J. & Sheta, D.: Consistent high degree Petrov-Galerkin methods for the solution of the transient convection-diffusion equation. *International Journal for Numerical Methods in Engineering* 28 (1989), pp.1077–1102.

Wikipedia (a): *Fundamental theorem of calculus.* Available at: http://en.wikipedia.org/wiki/Fundamental_theorem_of_calculus (accessed January 2011).

Wikipedia (b): *Geothermal electricity.* Available at: http://en.wikipedia.org/wiki/Geothermal_electricity (accessed January 2011).

Wikipedia (c): *Nusselt number.* Available at: http://en.wikipedia.org/wiki/Nusselt_number (accessed January 2011).

Wood, W.L.: *Practical time-stepping schemes.* Clarendon Press, Oxford, UK, 1990.

Yavuzturk, C.: Modeling of vertical ground loop heat exchangers for ground source heat pump systems. Ph.D. Thesis, Oklahoma State University, Oklahoma, USA, 1999.

Yavuzturk, C. & Spitler, J.: A short time step response factor model for vertical ground loop heat exchangers. *ASHRAE Transactions* 105:2 (1999), pp.475–485.

Yener, Y. & Kakac S.: *Heat conduction* (4th edn.). Taylor and Francis Group. NY, USA, 2008.

Yu, C.C. & Heinrich J.C.: Petrov-Galerkin methods for the time-dependent convective transport equation. *International Journal for Numerical Methods in Engineering* 23 (1986), pp.883–901.

Yu, M.Z., Peng, X.F., Li, X.D. & Fang, Z.H.: A simplified model for measuring thermal properties of deep ground soil. *Experimental Heat Transfer* 17 (2004), pp.119–130.

Zeng, H., Diao, N. & Fang, Z.: A Finite line-source model for boreholes in geothermal heat exchangers. *Heat Transfer Asian Research* 31:7 (2002), pp.558–567.

Zeng, H., Diao, N. & Fang, Z.: Heat transfer analysis of boreholes in vertical ground heat exchangers. *International Journal of Heat and Mass Transfer* 46:23 (2003), pp.4467–4481.

Zienkiewicz, O.C., Taylor, R.L. & Zhu, J.Z.: *The finite element method: Its bassis and fundamentals* (6th edn.). Elsevier, USA, 2005.

Zill, D.G. & Cullen, M.R.: *Advanced engineering mathematics* (2nd edn.). Jones and Bartlett Publishers, Sudbury, Massachusetts, USA, 2000.

Author index

Subject index

Multiphysics Modeling

Series Editors: Jochen Bundschuh & Mario César Suárez Arriaga

ISSN:1877-0274

Publisher: CRC/Balkema, Taylor & Francis

1. Numerical Modeling of Coupled Phenomena in Science and Engineering:
    Practical Use and Examples
    Editors: M.C. Suárez Arriaga, J. Bundschuh & F.J. Domínguez-Mota
    2009
    ISBN: 978-0-415-47628-72.

2. Introduction to the Numerical Modeling of Groundwater and Geothermal Systems:
    Fundamentals of Mass, Energy and Solute Transport in Poroelastic Rocks
    J. Bundschuh & M.C. Suárez Arriaga
    2010
    ISBN: 978-0-415-40167-83.

3. Drilling and Completion in Petroleum Engineering: Theory and Numerical Applications
    Editors: Xinpu Shen, Mao Bai & William Standifird
    2011
    ISBN: 978-0-415-66527-8

4. Computational Modeling of Shallow Geothermal Systems
    Rafid Al-Khoury
    2011
    ISBN: 978-0-415-59627-5